Agricultural Markets in a Transitioning Economy: An Albanian Case Study

FSC
www.fsc.org
MIX
Paper from responsible sources
FSC® C013604

Agricultural Markets in a Transitioning Economy: An Albanian Case Study

Edited by

Catherine Chan-Halbrendt

University of Hawai'i at Mānoa

and

Jean Fantle-Lepczyk

University of Hawai'i at Mānoa

www.cabi.org

CABI is a trading name of CAB International

CABI
Nosworthy Way
Wallingford
Oxfordshire OX10 8DE
UK
Tel: +44 (0)1491 832111
Fax: +44 (0)1491 833508
E-mail: info@cabi.org
Website: www.cabi.org

CABI
Chauncey Street
Suite 1002
Boston, MA 02111
USA
Tel: +1 617 395 4056
Tel: +1 617 354 6875
E-mail: cabi-nao@cabi.org

A catalogue record for this book is available from the British Library, London, UK.

Library of Congress Cataloging-in-Publication Data

Agricultural markets in a transitioning economy : an Albanian case study / [edited by Catherine Chan-Halbrendt].
p. cm.
Includes bibliographical references.
ISBN 978-1-78064-100-3
1. Agriculture--Economic aspects--Albania--Case studies. 2. Post-communism--Albania--Case studies. 3. Albania--Economic policy--Case studies.
I. Chan-Halbrendt, Catherine, 1949-

HD2040.5.A35 2013
381'.41094965--dc23

2012026675

ISBN-13: 978 1 78064 100 3

Commissioning editor: Rachel Cutts
Editorial assistant: Emma McCann
Production editor: Simon Hill

Typeset by SPi, Pondicherry, India.
Printed and bound in the UK by CPI Group (UK) Ltd, Croydon, CR0 4YY.

Contents

Contributors vii

Foreword ix
Jim Barnhart

Preface xi

1. Twenty Years of Transition in Central and Eastern Europe—an Overview 1
Nancy Cochrane and Kristaq Jorgji

2. Agriculture and Land Reform in Albania 18
Edvin Zhllima and Fatmir Guri

3. Agricultural Extension in Transition 35
Halina M. Zaleski

4. Albania's Agriculture Industry 52
Brinton Foy Reed and Engjell Skreli

5. Agricultural Production and the Food Processing Industry in Albania 75
Ilir Kapaj, Remzi Keco and Ana Kapaj

6. Distributions and Markets in Albania 93
Ana Kapaj, Drini Imami and Bahri Musabelliu

7. Quality Assurance and Food Safety in Albania 109
Luciano Leonetti, Ilir Kapaj and Florian Xhafa

8. Consumer Preferences for Selected Agricultural Products in Albania 127
Drini Imami, Edvin Zhllima and Catherine Chan-Halbrendt

9. Trade in Albania 148
Engjell Skreli and Alex McCalla

10. Albania's Agriculture Sector: Trends and Projections 173
Luciano Leonetti

11. The Effects of EU Integration on a Transitioning Economy 220
Birgit Schaefer, Roland Cela and Shkelzen Marku

12. Albanian Research and Extension Services in Transition 241
Edvin Zhllima, Drini Imami, Tatjana Dishnica, Carl Evensen and Bahri Musabelliu

Index 257

Contributors

Roland Cela, German Agency for International Cooperation (Deutsche Gesellschaft für Internationale Zusammenarbeit, GIZ), c/o Ministry of Agriculture, Food and Consumer Protection, Sheshi Skenderbej 2, Tirana, Albania. E-mail: Roland.Cela@giz.de

Catherine Chan-Halbrendt, AHEED Project, Department of Natural Resources and Environmental Management, College of Tropical Agriculture and Human Resources, University of Hawai'i at Mānoa, 1910 East-West Rd, Sherman 101, Honolulu, HI 96822, USA. E-mail: chanhalb@hawaii.edu

Nancy Cochrane, USDA Economic Research Service, 1400 Independence Ave., SW, Mail Stop 1800, Washington, DC 20250, USA. E-mail: Cochrane@ers.usda.gov

Tatjana Dishnica, Ministry of Agriculture, Food and Consumer Protection, Sheshi 'Skenderbej' Nr 2, Tirana, Albania. E-mail: tana@icc-al.org

Carl Evensen, Cooperative Extension Service, College of Tropical Agriculture and Human Resources, University of Hawai'i at Mānoa, 3050 Maile Way, Gilmore Hall 203, Honolulu, HI 96822, USA. E-mail: evensen@hawaii.edu

Jean Fantle-Lepczyk, AHEED Project, Department of Natural Resources and Environmental Management, College of Tropical Agriculture and Human Resources, University of Hawai'i at Mānoa, 1910 East-West Rd, Sherman 101, Honolulu, HI 96822, USA. E-mail: fantle@hawaii.edu

Fatmir Guri, European Commission – Joint Research Centre, Institute for Prospective Technological Studies, Seville, Spain and Agricultural University of Tirana. Correspondence address: Faculty of Economy and Agribusiness, Agricultural University of Tirana, Koder Kamez, Tirana, Albania. E-mail: fatmirguri@yahoo.fr

Drini Imami, Faculty of Economy and Agribusiness, Agricultural University of Tirana, Koder Kamez, Tirana, Albania. driniimami@yahoo.com

Kristaq Jorgi, USAID/Albania, American Embassy, Rr. Elbasanit, Nr 103, Tirana, Albania. E-mail: kjorgji@usaid.gov

Ana Kapaj, Faculty of Economy and Agribusiness, Agricultural University of Tirana, Koder Kamez, Tirana, Albania. amane@ubt.edu.al

Ilir Kapaj, Faculty of Economy and Agribusiness, Agricultural University of Tirana, Koder Kamez, Tirana, Albania. E-mail: ikapaj@ubt.edu.al

Remzi Keco, Faculty of Economy and Agribusiness, Agricultural University of Tirana, Koder Kamez, Tirana, Albania. E-mail: rkeco@ubt.edu.al

Luciano Leonetti, Development Solution Associates (DSA), Rr. Gjon Pali, Nr 11/1, Kati 4, A. 2 Tirana, Albania. E-mail: lkns@hotmail.com

Shkelzen Marku, Partnership for Development Foundation, Tirana, Albania. Correspondence address: Mountain Area Development Agency (MADA), Rr. 'Shinasi Dishnica' Nr 100, Tirana, Albania, E-mail: smarku@redeval.org

Alex McCalla, Department of Agricultural and Resource Economics, 3121 Social Sciences and Humanities Bldg, University of California, Davis (UC Davis), 1 Shields Ave, Davis, CA 95616, USA. E-mail: alex@primal.ucdavis.edu

Bahri Musabelliu, Faculty of Economy and Agribusiness, Agricultural University of Tirana, Koder Kamez, Tirana, Albania. E-mail: abahri_musabelliu@yahoo.com

Brinton Foy Reed, Department of Natural Resources and Environmental Management, College of Tropical Agriculture and Human Resources, University of Hawai'i at Mānoa, 1910 East-West Rd, Sherman 101, Honolulu, HI 96822, USA. E-mail: beefreed@yahoo.com

Birgit Schaeffer, Naccon GbR, Wilonstr. 140, 72072 Tuebingen, Germany. E-mail: schaefer@naccon.de

Engell Skreli, Faculty of Economy and Agribusiness, Agricultural University of Tirana, Koder Kamez, Tirana, Albania. E-mail: eskreli@ubt.edu.al

Florian Xhafa, Department of Business Law, Tirana Business University, Rruga e Kavajës, 1023 Tirana, Albania. Correspondence address: Rr. Brigada VIII, Nd. 19, H. 3, Ap. 6, 1019 Tirana, Albania. E-mail: florian.xhafa@gmail.com

Halina Zaleski, Department of Human Nutrition, Food and Animal Sciences, College of Tropical Agriculture and Human Resources, University of Hawai'i at Mānoa, 1955 East-West Rd, Ag Sci 216B, Honolulu, HI 96822, USA. E-mail: halina@hawaii.edu

Edvin Zhllima, Faculty of Economy and Agribusiness, Agricultural University of Tirana, Koder Kamez, Tirana, Albania. E-mail: edvinzh@yahoo.com

Foreword

The United States Agency for International Development is proud to have supported the fruitful partnership of the University of Hawaii at Manoa and the Agricultural University of Tirana.

Agriculture is fundamental to Albania's economic future. Agriculture provides jobs for almost half of Albania's workforce and serves as a buffer of stability during times of economic uncertainty. As agricultural practices modernize, trade practices improve, and the quality and quantity of output increases, rural incomes rise and the Albanian economy prospers.

Albania has made impressive progress in improving its quality and diversity in agriculture over the past twenty years. The remaining challenges revolve around the need for Albanian farmers and producers to have access to the knowledge, best practices and know-how to compete regionally and globally.

This volume articulates clearly the accomplishments thus far, and argues persuasively for continued investment in practical academic research and agriculture extension services in Albania.

Jim Barnhart, Ph. D.
Mission Director, USAID/Albania

Preface

When the Albania/Hawaii Higher Education and Economic Development (AHEED) Project began back in 2008, we had very optimistic plans and a real desire to make a difference for faculty, students, and ultimately, all participants in Albania's agricultural industry. We at the University of Hawai'i at Mānoa (UHM), with the generous support of a Higher Education for Development (HED)/United States Agency for International Development (USAID) grant, were eager to partner with faculty and staff at the Agricultural University of Tirana (AUT) and to mentor them in order to strengthen the capacity of AUT's Faculty of Economy and Agribusiness (FEA) to provide quality graduate education and to build capacity for effective training, consulting and advising to the extension services of the Albanian Ministry of Agriculture.

As the AHEED project enters its fourth year, we are proud of our many accomplishments. During this period, we have helped to bring about great change for the FEA at AUT in terms of new and improved graduate program curricula, novel and cutting-edge research ideas and techniques, and greater exposure to and investment in the international scientific community. Of particular impact on the FEA has been its participation in international conferences, the organization of the Annual Agribusiness Conference at AUT, publication in international journals and the development of a small grants program in order to encourage research activities.

Of course, along the way we've faced many challenges, and have not always completely managed to meet all of our goals and objectives. This book, however, is decidedly not one of those moments. Rather, it is a culmination of four years of effort, growth and learning, and a very exciting next step for the project. This endeavor is a natural outcome of all the work the AHEED project has done in the last four years and we are anxious to share all that we have learned with a much wider audience. Seeing this book published is of great importance to us for many reasons. One really meaningful aspect for the project is that our Albanian cooperators are highly involved in the writing of each chapter, representing their great personal and professional growth as a result of this project. We feel the effort and time every one has put into this book shows the type of capacity building that is sustainable, and we are incredibly proud to see this evidence of the sort of self-motivation to pursue academic excellence that we had hoped to foster. Furthermore, this is the first book of its type to use Albania as a case study to analyze the current state of agriculture in similar economies, while simultaneously highlighting the latest research activities and discussing the roles of education and extension in advancing the agricultural industry.

This book focuses on the major challenges and opportunities facing agriculture sectors in the wake of the transition from a planned to a market economy and potential entry to the European

Union (EU). Using Albania as a case study, it examines the shift from a centrally planned economy to free markets, and the lasting effects of such change on the agricultural sector, extension and universities. It also looks at the future of agriculture in transitioning economies and presents findings from recent academic research activities funded by the project. A consistent theme throughout the text is the changing roles of universities and extension, and their importance to the future growth of economies such as that of Albania. The book will highlight the primary research supported by the AHEED project and give readers an accurate portrayal of the path that lies ahead for many developing countries, while making a case for the importance of ongoing education and extension.

At the beginning of this project, we could not have foreseen its success or its impacts. Of course, we did not get to this point, the production of this book, without the support and participation of many colleagues along the way. None of this would have been possible without the support of HED and USAID. The faculty and administration of the FEA at AUT has also heavily invested in this project, and their enthusiasm and willingness to learn and grow has been an inspiration. Finally, we have had much support state side as well. We are very thankful to all the professionals both at UHM and at other US institutions who have generously donated their time both to the AHEED project and the book itself. Overall, this has been a wonderful and enlightening process for us and we hope you learn as much from reading this book as we did from editing it.

Catherine Chan-Halbrendt
Jean Fantle-Lepczyk
University of Hawai'i at Mānoa

1 Twenty Years of Transition in Central and Eastern Europe—an Overview

Nancy Cochrane[1] and Kristaq Jorgji[2]
[1]*USDA Economic Research Service, Washington DC, USA;*
[2]*USAID, Tirana, Albania*

It has been just over 20 years since the countries of Central and Eastern Europe (CEE) began their transition to democratic and free market economies. The early years of the transition were marked by severe economic recession, declining agricultural output, particularly in the livestock sectors and, in some countries, by a rise in subsistence farming. By the end of the 1990s, however, many of the countries were on the road to recovery—output began to recover, incomes were rising, and foreign investment was flowing in.

Most important, perhaps, all the Eastern European countries, with the exception of the former Yugoslavia, had begun preparing for accession to the European Union (EU). The need to meet the stringent requirements of the EU for acceding members forced an acceleration of reform and restructuring of the agricultural and food sectors. But the process also drew several million euros of pre-accession assistance. As of this writing, ten of the formerly communist Eastern European countries are EU members. Croatia will become the 28th member in 2013. Serbia, Montenegro, Macedonia, Bosnia and Herzegovina, and Albania are either candidates or striving for candidate status.

Membership in the EU has brought substantial increases in farm income, but there have been negative impacts as well. A number of New Member States (NMS) have had difficulties competing in a single market and have suffered increasing imbalances in their agricultural trade with other Member States. Bulgaria and Romania have had a particularly difficult time since accession. Moreover, the accession treaties of 2004 and 2007 resulted in lower payments to NMS farmers than are received by their counterparts in the former EU-15 (the number of member countries in the EU before the accession of ten candidate countries on 1 May 2004), which is a cause for continuing resentment.

Looking ahead to the next decade, there are many reasons for optimism. But there also are a number of lingering issues across Eastern Europe. Because of weaknesses in market institutions, small farmers throughout the region continue to be isolated from markets. Aspiring new members in the Balkans have seen their bids for EU candidacy held up by uneven progress towards market reform, lingering corruption, and political instability. The recession that began in 2008 and the more recent euro crisis have reduced access to credit on the part of farmers, raised poverty and unemployment, and slowed the flow of foreign investment. Still, with the exception of Hungary, most of the Eastern European countries have so far avoided the debt problems afflicting Greece and Spain.

1.1 The Early Years: Supply and Demand Side Shocks

The transition from centrally planned to market-based economies created severe shocks on both the demand and the supply side of the transitioning economies. Under central planning, most prices were set by the state, state-owned enterprises accounted for most production, and levels of output were set by government planners rather than market forces. In addition, both production and consumption tended to be heavily subsidized, with retail food prices and prices for farm inputs set well below market levels. In a number of the former communist countries, prices for agricultural output were set above market levels, providing a subsidy to producers.

With the liberalization of prices that began in the early 1990s, prices at all levels adjusted to world market levels. The result was generally a worsening of terms of trade for farmers as input prices rose to world levels and output prices dropped. The main supply-side shocks have been changes in relative prices faced by producers—output prices compared with input prices, as well as relative prices between inputs. The two policies most responsible for these relative price changes were price liberalization and integration into the world economy.

The main shock on the demand side was the reduction in consumer income and purchasing power brought on by economic reform. First, reform increased both unemployment and underemployment. Consumer income dropped because price liberalization, the lead policy of economic reform, caused prices to rise more than wages and salaries, thereby decreasing consumers' real income and lowering their purchasing power. In the pre-reform period, consumption of most foodstuffs—livestock products in particular but many other consumer goods as well—was heavily subsidized, with consumer prices often far below the real cost of production. Price liberalization eliminated most of these consumer subsidies, causing a jump in consumer prices to reflect full production costs. Within 4–5 years of the beginning of economic reform, per capita real income had decreased significantly throughout the region: from 20% to 25% in Poland and Hungary to about 40% in Romania. Demand for meat and other livestock goods, being more sensitive to changes in income (income elastic) fell sharply, while demand for staple products, such as bread and potatoes, did not fall as much.

The impacts of these shocks were evident in the livestock sectors of the CEE countries (see Bjornlund *et al.*, 2002, for more detail). Livestock producers were squeezed by rising feed costs and falling consumer demand. Cattle numbers declined the most: Poland's cattle inventory in 1998 was 37% below the level of 1989; Romania's declined 51% during the same period. Pig and poultry numbers declined less in Poland, but Romania experienced substantial declines in pig and poultry numbers, as well as in cattle numbers. There were corresponding declines in meat output.

Romania suffered greater declines than Poland, in part because of the dramatic changes in farm structure that occurred after the fall of communism. Most cattle during the communist period were kept on cooperative farms, which were liquidated early in the transition. The cattle were redistributed to private farmers, who lacked the financial means to raise them. Most of Romania's hogs were kept on the state farms, which remained largely intact during the early transition. However, these farms experienced increasingly severe financial constraints with the withdrawal of subsidies and the need to pay market prices for feed, and hog inventories shrank in response.

Another casualty of the transition was fruit production. Bulgaria's grape production declined by 39% between 1989 and 2000; Bulgarian apple output fell 80% during the same period—output dropped by two-thirds just from 1990 to 1991 (information from FAOSTAT – the Food and Agriculture Organization statistics database). In Albania, the number of fruit trees declined from 20 million to 7 million between 1990 and 1992 (Agolli, 2000). Reasons included the older than optimal age of the trees before the transition, lack of investment in new trees, use of old technologies, and disruptions to surrounding land redistribution. Another reason, in the case of Bulgaria, was that its major export market for fruits and vegetables during the communist period was the former Soviet Union. That market dried up after the breakup of the Soviet

Union, and Bulgaria couldn't meet the quality standards demanded by other potential importers.

The adjustments to world markets were accompanied by rapid moves towards privatization and land redistribution. During the communist period, only Poland and the former Yugoslavia maintained widespread private ownership of land—in Poland, nearly 80% of the land was in private hands during that period. In the other countries, most land was confiscated from previous owners and organized into large state and cooperative farms. The countries took different routes towards land privatization. Hungary issued vouchers to former landowners, and the state and cooperative farms were transformed into various types of stock-holding companies. Romania restituted all land on cooperative farms to previous owners. State farms remained intact and were eventually sold to private investors. Bulgaria also carried out widespread restitution of land to former owners. Albania, in 1992, passed a law calling for the distribution of agricultural land on a per capita basis to families registered as village inhabitants before 31 July 1992. The restitution or compensation to former owners of agricultural land was subject to another law.

The result of all this was a bipolar farm structure, with large numbers of very small, semi-subsistence farms and a small number of large ones. Land fragmentation was a particular problem in Romania, Bulgaria, and Albania, where many private farms were 2 ha or less. Both Poland and the former Yugoslavia had imposed size restrictions on private farms, so these were also quite small—generally 10 ha or less. Most of these small farmers were not well integrated into markets. Their production often failed to meet quality standards, the farmers had difficulty transporting their goods to markets, and they lacked market information and extension advice. Those that wanted to expand their holdings and invest in value-added production were hampered by poorly developed land markets and lack of credit.

Since the early 1990s, there has been a trend towards farm consolidation and growing numbers of commercial farms. But the farm consolidation process has been slow, and land fragmentation continues to be a serious issue in many of the CEE countries. In Albania and the successor states of former Yugoslavia, not only are land parcels very small, but many farms are fragmented into several non-contiguous plots. The average farm size for Albania is even smaller than in the former Yugoslav Republics—just 1.21 ha in 2010. Furthermore, the average Albanian farm consists of four to five parcels, so that individual parcels actually average only 0.26 ha (Sallaku *et al.*, 2010; MAFCP, 2010).

One obstacle to land consolidation has been a lack of fully functioning land markets. These markets were often hampered by lack of clear ownership rights, difficult access to credit, and the poor profitability of agriculture. In a number of countries, land consolidation has largely been achieved through leasing. Holders of small pieces of land living in urban areas and uninterested in farming were happy to lease their land to others.

1.2 EU Accession Intensifies Pressures for Reform

At the very beginning of the transition in the early 1990s, discussions began about eventual accession of the Eastern European countries to the EU. This became a reality for eight of the countries in 2004 and for two more (Bulgaria and Romania) in 2007. Croatia is slated to join in 2013, and Serbia, Macedonia, Montenegro, Albania, Bosnia and Herzegovina, and possibly Kosovo, are expected to eventually follow. The promise of eventual accession was a major driving force for reform. In order to qualify for EU membership, the countries were required to be fully functioning democracies, complete the transition to market economies, and raise the quality of their agricultural output to meet stringent EU quality standards.

At the beginning of the transition, the impacts of EU membership were believed to be overwhelmingly positive. Farmers of CEE countries eagerly anticipated higher prices and the subsidies enjoyed by farmers in the former EU-15. At that time, EU commodity prices were substantially higher than CEE prices, and EU farmers received generous income support.

However, as the accession date grew closer, farmers began to understand that not all the anticipated benefits would become a reality, and that EU accession would bring costs as well as benefits (see Cochrane and Seeley, 2004, for more discussion of the period leading up to accession.) For one thing, by the early 2000s, the gap between EU and CEE prices had narrowed considerably, for a number of reasons. The currencies of the candidate countries had gradually strengthened against the euro. The EU itself had undergone significant agricultural policy reform. Reforms in 1992 reduced intervention prices (price supports) and introduced a system of direct payments to producers, which were decoupled from production, to compensate for the lost income. Agenda 2000, introduced in 1999, further reduced intervention prices. At the same time, the CEE governments, in an effort to align their policies with those of the EU, had begun to intervene strongly in some markets, resulting in higher CEE prices. Poland maintained an aggressive intervention program for wheat, rye, sugar, and dairy products. Hungary and the Czech Republic were providing support to their livestock sectors.

In addition, the price gaps of the early 1990s reflected quality differences more than policy differences. Pork and beef prices reported by the EU are for the top three grades—in terms of lean meat content—of the EU grading system (SEUROP). CEE statistical offices had been reporting average prices for all grades. Throughout the 1990s though, the average lean meat content had been increasing, and by 2004, more CEE pork and beef met the top three EU grades. CEE prices for pork and poultry were now, as a result, on par with EU prices.

A major bone of contention at the time of accession that has led to enduring bitterness among CEE farmers was the level of direct payments that CEE producers would receive. The EU Commission realized that it would be impossible to provide the full range of direct payments to CEE farmers without violating the budget limits agreed upon in Agenda 2000. For this reason, the final compromise provided for a 10-year phase in of payments. The EU provided only 25% of the payments from the Common Agricultural Policy (CAP) budget during the first year; this share was to increase by 5% each year until CEE farmers receive 100% of EU payments. However, national governments were allowed to top off these payments by a maximum of 30% each year, so that payments during the first year of accession could be as much as 55% of what farmers in the former EU-15 received.

Even after they are fully phased in, direct payments remain lower for CEE countries because of the way the payments are computed. Payments were tied to the yields associated with a reference period—1995–1999—and to a reference area. Because of the disruptions caused by the transition from central planning to free markets, CEE yields during 1995–1999 were substantially lower than those of the EU, which will keep CEE payment levels lower relative to EU payment levels (Fig. 1.1.)

The NMS also had to comply with the full body of EU regulations, known as the *acquis communitaire*. There are approximately 80,000 pages of EU laws and regulations relating to market regulation, veterinary and sanitary controls, animal welfare, and the administrative structures needed to implement EU price and income support programs. Grain producers, for example, had to meet minimum quality requirements to sell grain into EU intervention stocks. Livestock breeders now have to raise, transport, and track all animals according to the animal welfare regulations and record-keeping requirements of the EU. All these measures increased production costs, eroding the net returns of producers. In addition, acceding CEE governments had to incur large administrative costs, as agencies were now required to maintain detailed databases on production, animal numbers, and other pertinent information for each farm that will receive EU payments.

As the date of accession drew closer and CEE farmers became more aware of the costs associated with EU membership, many were bitterly opposed to joining. They were apprehensive about higher costs, wary about increased competition from the EU-15, and feared they would be driven out of business. Poland, in particular, witnessed several disruptive demonstrations, road blocks and other forms of farmer protest. In early 2004, just

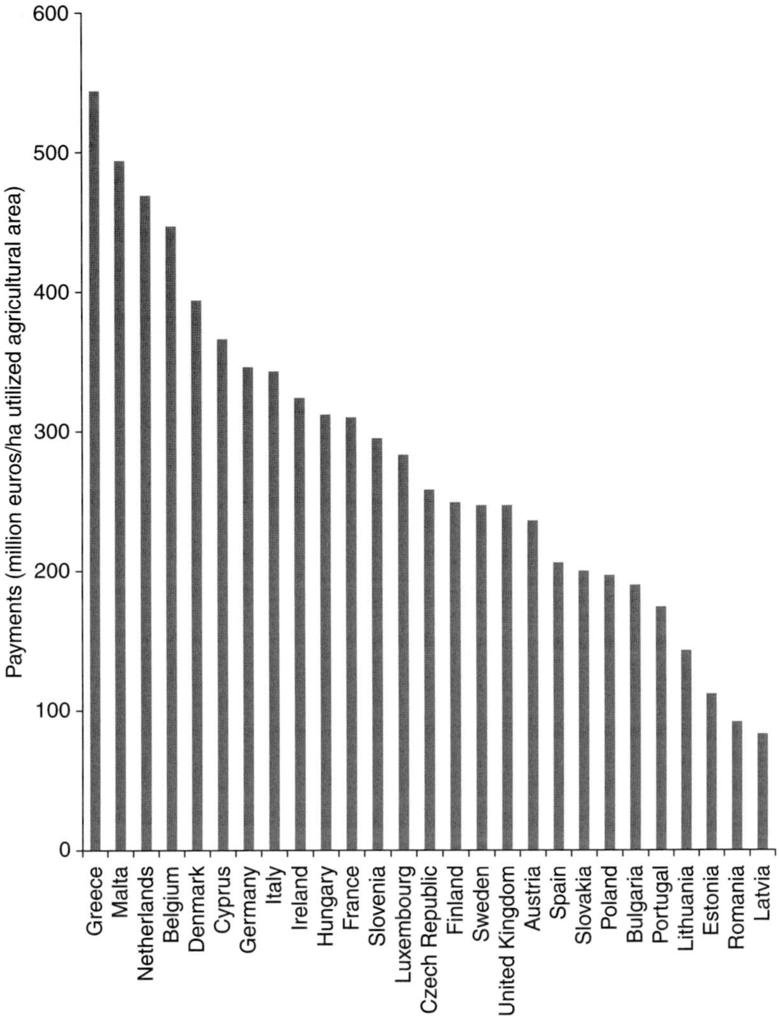

Fig. 1.1. Wide disparities in per hectare direct payments, Pillar 1 payments (payments to farmers) in 2013. Source: Farm Accountancy Data Network, EU Commission.

before accession, 49% of farmers were against EU membership, fearing that their farms would fail, unable to compete with the West, and food prices would soar (Kawonczyk and Figurska, 2009).

1.3 So What Happened?

The experience of the CEE countries following EU accession is mixed. Many farmers have done quite well. A year after joining the EU, average farm income throughout the NMS was up by 54%; Poland's farm income rose 74%. Between 2000 and 2008, Poland's farm income rose by 90%, while Czech farm income more than doubled (information from European Commission's Farm Accountancy Data Network, FADN). This jump in farm income came despite the lower direct payments.

Conversely, CEE farmers have seen increasingly stiff competition from the former EU-15. Poland has become a net agricultural exporter, but has also seen increasing imports from the EU-15. Hungary, a net exporter before accession, has become a smaller net exporter. Bulgaria and Romania have seen substantial rises in imports from other EU states and are increasingly net importers. Farmers throughout the region have complained about the expanding presence of foreign-owned supermarkets, resulting in strict quality standards and delayed payments.

Agricultural performance since accession has varied considerably across countries. In general, the grain sectors haven't changed much. Production rises and falls in response to weather and world prices: joining the EU seems to have had little impact. Poland remains a net importer, while Hungary and Bulgaria continue to export. Hungary has had some difficulties, however, as its status as a landlocked country raises the costs of moving the grain out of the country. In the winter of 2012, the problems were compounded by the harsh winter, which caused the Danube to freeze over and stopped all shipments for several weeks. Bulgaria, Serbia and Romania were also affected by that.

The principal change in the crop sector has been a dramatic increase in rapeseed output (Fig. 1.2). This was mainly the result of EU subsidies for biodiesel feedstock.

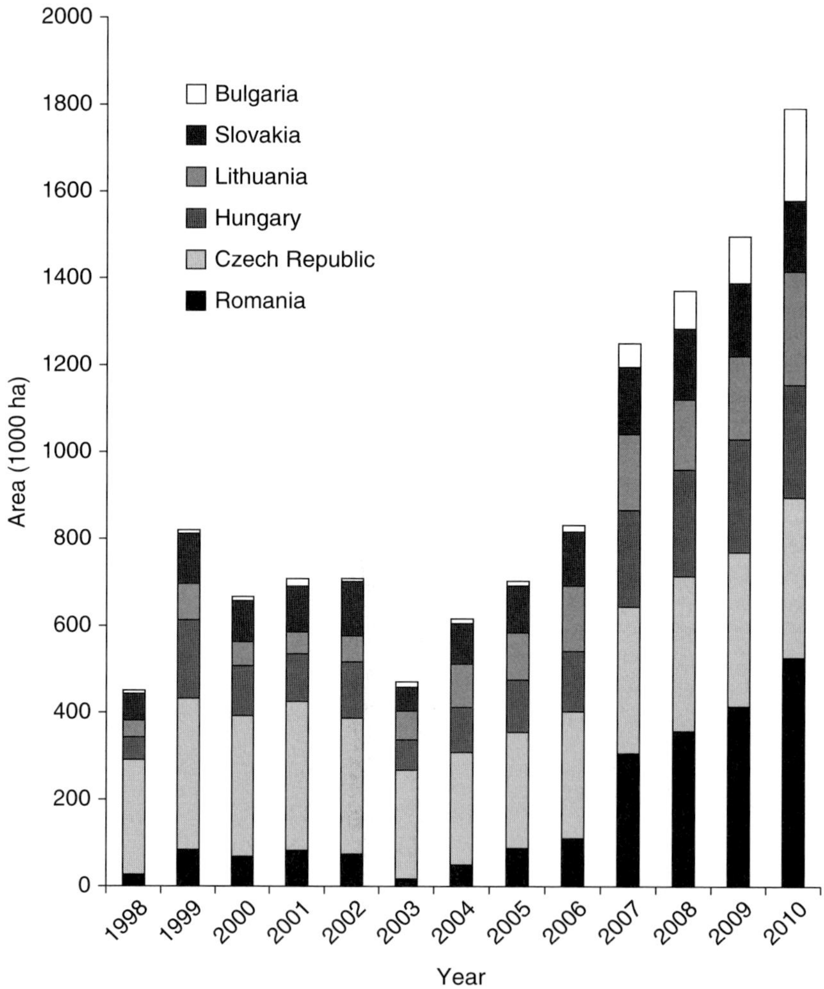

Fig. 1.2. New Member States rapeseed area rises sharply after EU accession. Source: Eurostat.

Impacts have been more visible in the livestock sectors. Pork output has declined throughout the NMS, mainly as a result of higher feed costs (Fig. 1.3). But Hungarian pig farmers were benefiting from export subsidies before accession, and net returns declined once those disappeared following accession. Moreover, all of the NMS found it difficult to compete with low-cost producers in Denmark and the Netherlands. Imports from those two countries surged, and domestic producers could not compete.

In most of the NMS, cattle and beef production continued the declining trend that began during the 1990s. Hungarian cattle breeders also found themselves disadvantaged by the formula used for EU direct payments. Before the 2003 CAP reform, the CAP included a complex array of headage payments and slaughter premiums for cattle and sheep, in addition to the area payments. The 2003 CAP reform converted most of these payments in a Single Farm Payment, decoupled from production (see Kelch and Normile, 2004, for an analysis of the 2003 CAP reform). Hungary, like most of the NMS, chose to implement a simplified area payment scheme (known as SAPS), similar but slightly different from the Single Farm Payment. Under this scheme, farms receive a single per hectare payment, not linked to current production or animal herds. Cattle breeders tended to operate fewer hectares than crop farmers and therefore received lower direct payments.

The notable exception is Poland, which saw increases both in cattle numbers and beef production (Fig. 1.4). Poland was exporting live cattle to the former EU-15 before accession, but was constrained by the quota imposed by its trade agreement. With the removal of that quota, exports surged, and with that cattle numbers (Fig. 1.5).

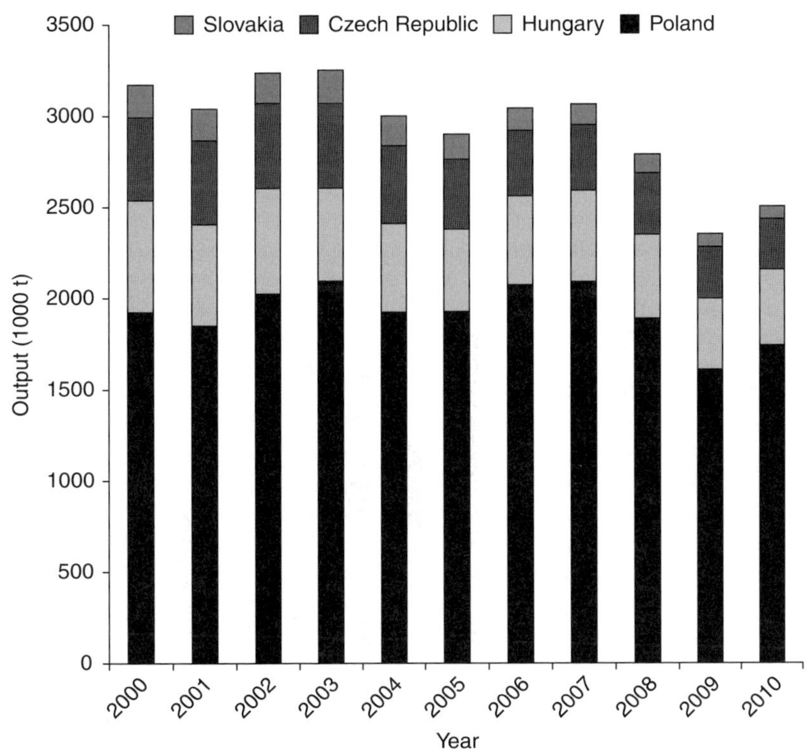

Fig. 1.3. New Member States pork output declines after accession. Source: Eurostat.

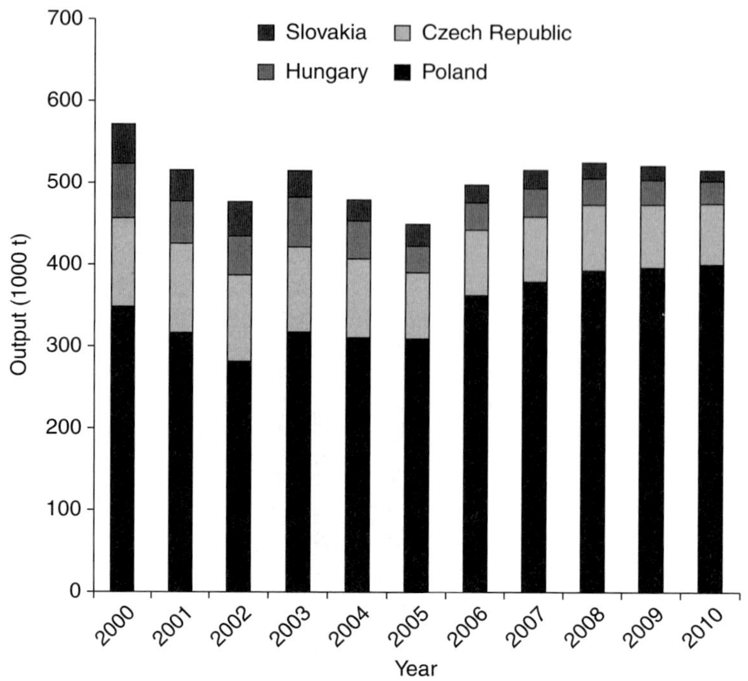

Fig. 1.4. Beef output rises in Poland; falls in other New Member States. Source: FAO.

In more recent years, however, Poland has been exporting lower numbers of live cattle and increasing volumes of beef as meat (Fig. 1.6). This is in part the result of investment in better meat breeds of cattle. Some Polish experts also say that this is the result of more stringent animal welfare regulations that raise the cost of transporting live cattle (author Cochrane's conversations with Polish experts).

All the NMS have found it difficult to compete in the single EU market. Poland has managed to maintain a strong surplus in agricultural trade, but has a deficit in its trade with the EU-15. That deficit has been offset by exports to other NMS. Hungary and Bulgaria have managed to maintain agricultural trade surpluses, but these surpluses are smaller than in the past. Farmers throughout the region, even in those countries maintaining a trade surplus, have been protesting the flood of what they perceive to be cheap, poor-quality imports. From the first year after accession through the present, Hungarian poultry producers and processors have been complaining about the rise in poultry imports. In fact, while poultry imports have risen, Hungary remains a substantial net exporter of poultry—the industry only seems to see the increase in foreign competition.

There is a growing concern among the NMS about the increasing dominance of large supermarket chains. According to Csaki and Jambor (2009), the share of the top ten supermarket chains in NMS retail sectors was 43% in Bulgaria, 88% in Hungary and 99% in Slovenia. Farmers throughout the region complain of exacting quality standards, downward pressure on prices, delayed payments, and other practices that make it difficult to market to these stores. One Serbian

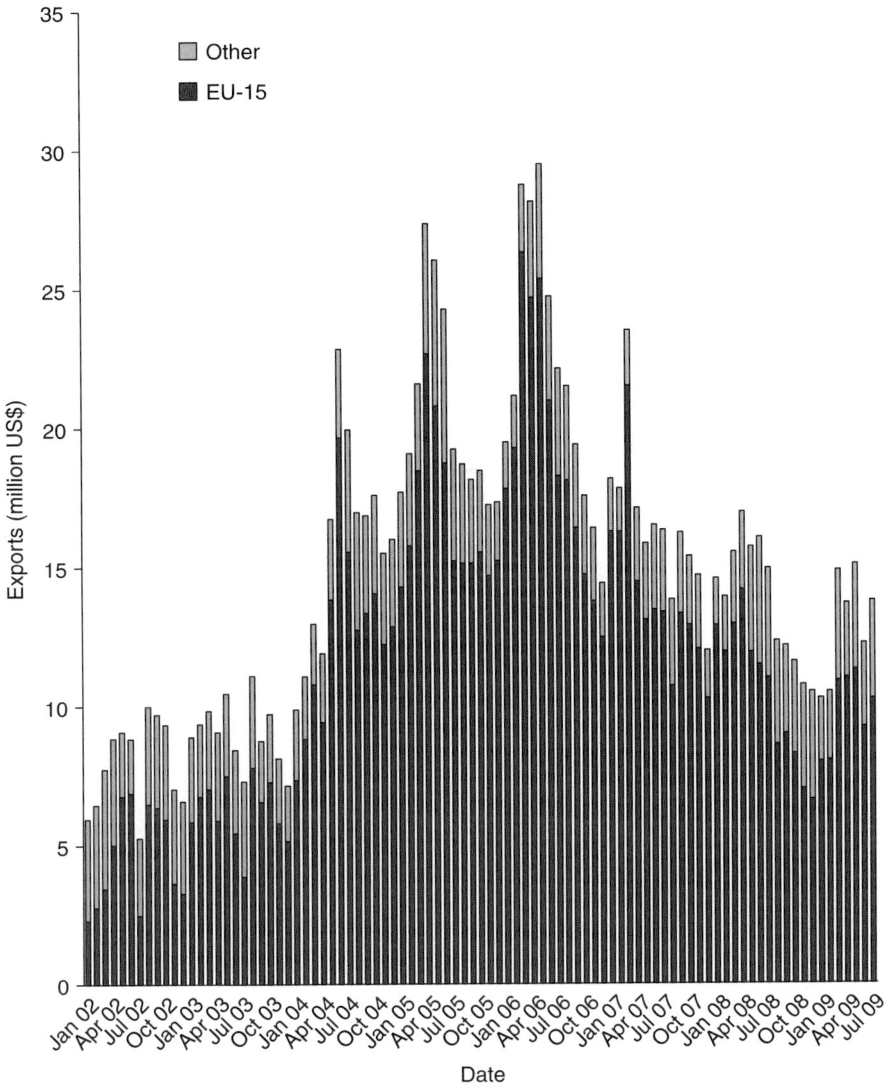

Fig. 1.5. Polish live cattle exports surge after accession. EU-15, exports to original EU member countries; Other, exports to other countries. Source: World Trade Atlas.

mushroom producer complained that she was forced to take back the produce that the supermarket she delivered to failed to sell (conversation with author Cochrane). Supermarkets are also blamed for the rising share of imported goods on NMS markets. Csaki and Jambor cite figures showing that the share of domestically produced goods on the Hungarian retail market declined from 92% in 2003 to 75% in 2010.

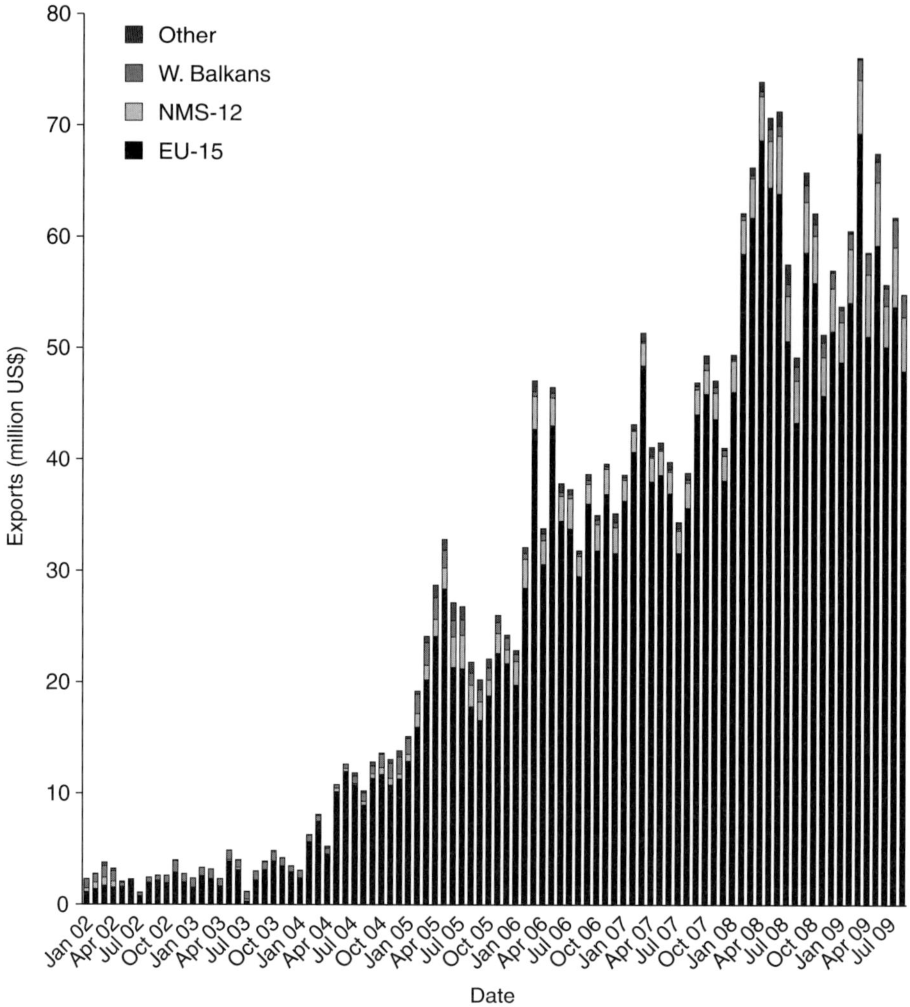

Fig. 1.6. Polish exports shift from cattle to beef. EU-15, exports to original EU member countries; NMS-12, exports to New Member States (joined 2004–2007); W. Balkans, exports to Western Balkan countries; Other, exports to other countries. Source: World Trade Atlas.

1.4 Bulgaria and Romania Experience Difficulties Adapting to the Single Market

Romania and Bulgaria have had more difficulty adjusting to EU membership than other NMS. Both have found it difficult to compete in the single market. Bulgaria has managed to maintain a small agricultural trade surplus, but has still seen rapidly rising imports from other EU countries. Romania has suffered a widening agricultural trade deficit (Fig. 1.7). Romanian production hasn't changed overall: grain output rises and falls in response to weather conditions, pork output has remained fairly constant, while poultry output has risen, and beef output has declined. But while Romania has remained a significant corn exporter, it continues to be a large net importer of livestock products, mainly from the EU-15 (Fig. 1.8).

Part of the problem is that a large share of Romania's pig population is still found on small semi-subsistence farms. Romania's pig sector is split between large, industrialized

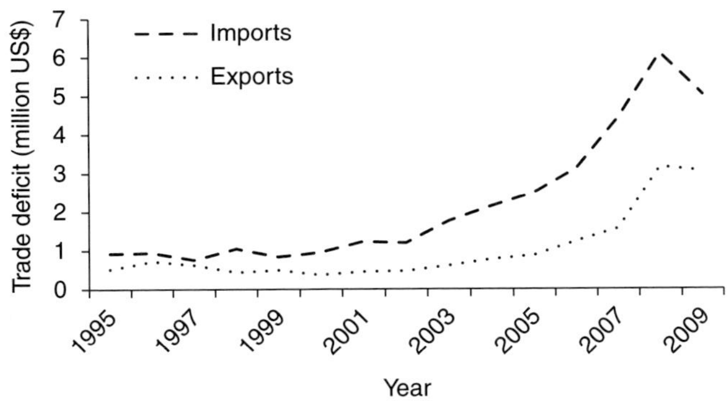

Fig. 1.7. Romania's agricultural trade deficit widens immediately after accession. Source: FAO.

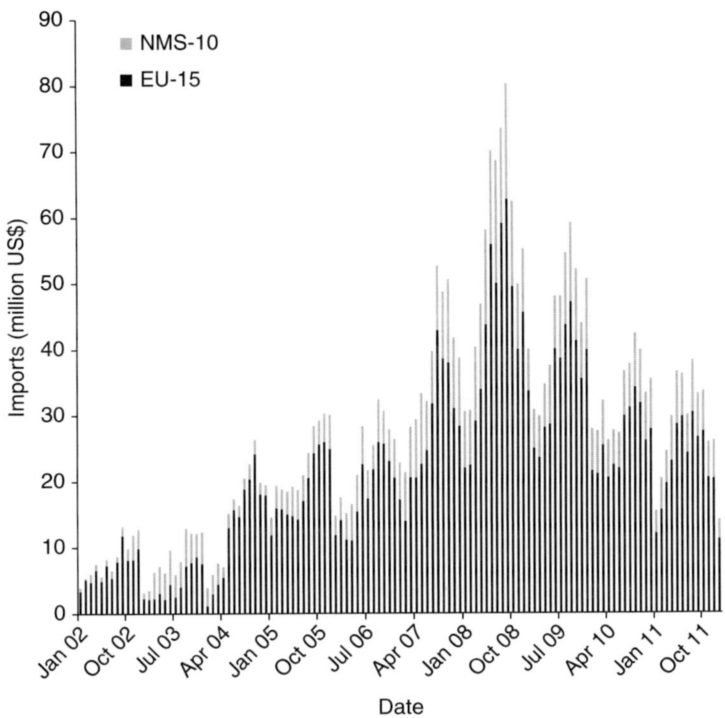

Fig. 1.8. Romania's pork imports surge following EU accession. EU-15, imports from original EU member countries; NMS-10, imports from New Member States (joined 2004). Source: World Trade Atlas.

farms and small household plots. Only 40% of the pigs slaughtered in Romania are estimated to be slaughtered in industrial plants (Nistor *et al.*, 2010). The pigs from the household sector, for the most part, can't meet the quality standards of the large plants. Another problem has been an outbreak of swine fever that led to a ban on imports by Romania's trading partners.

Both Romania and Bulgaria have also endured numerous instances of corruption and fraud. The EU Commission was concerned about these issues before those countries acceded, and the terms of accession called for strict surveillance and the possibility that the EU Commission could withhold funds if it saw evidence of spending irregularities. The Bulgarian press, in particular, has published numerous stories of fraudulent applications for direct payments. According to one account (*Banker Weekly English*, 5 March 2010):

> at the beginning of this year, Shoumen Regional Court sentenced a farmer who has tried to siphon money from Eurofunds. The farmer … pleaded guilty for submitting at the regional paying agency a wrong application for subsidies for nearly 170 decares (one tenth of a hectare). He has asked for subsidies for a total 24 land plots in Veliki Preslav, as well as in the villages of Kyulevcha, Nova Byala Reka and Konevo, which he declared as cultivated lands. However, the investigation revealed that the data on 14 of these land plots were false and [the farmer] was thus sentenced to probation and fined 600 Bulgarian levs (300 euros).
>
> In the spring of 2010, another four people that have allegedly tried to gain from subsidies were brought to trial…. Using false documents, they managed to get an Agriculture Fund subsidy for the purchase of 582 tonnes of ammonium nitrate that they resold afterwards.
>
> Exactly because of such kind of attempts, the European Court of Auditors will fine Bulgaria EUR20.78 million. At the end of 2009, Court of Auditors' inspectors checked a total 34 land plots that subsidies have been received for. Of them, 12 plots turned out to be barren lands. Money has been provided even for "pastures" that in the end appeared to be abandoned village stadiums, warehouses or even cemeteries …. And quite often exactly the fund's officials are the ones that conceal producers who provide for their own families with money earmarked for production or investments. This was also revealed by an internal audit that was carried out at the paying agency. According to the results, some officials at the fund have gained tens of kilogrammes of cheese and mutton for doing quite dubious favors.

In 2009, evidence of misuse of EU rural development funds led the EU Commission to withdraw €220 million (US$315 million) after finding irregularities in 34 EU-funded projects (*Bosnia Daily*, 5 January 2009).

Another major issue facing Bulgaria and Romania is the continuing isolation of many of its small producers from the market, and this is an issue that will confront the Western Balkan nations as they prepare to enter the EU. Agriculture in both countries is polarized between a small number of large enterprises and large numbers of very small plots—some as small as half a hectare. Products produced on the small farms often don't meet the quality standards demanded by the EU and thus cannot be marketed through official channels.

The rural development funds available from the EU were partially intended to help small producers become more competitive. Before accession, the NMS benefited from the SAPARD (Special Accession Program for Agriculture and Rural Development) Program, which was designed to help farmers and agribusiness make the investments needed to compete in the single market. Now they have access to Pillar 2 funding from the CAP for rural development. So-called Axis 1 of Pillar 2, aimed at 'improving the competitiveness of agriculture and forestry', provides funding for investments in product quality and new technologies, support for advisory services, support for farmers attempting to meet EU standards, as well as assistance for semi-subsistence farmers in the NMS.

Many CEE farmers did benefit and continue to benefit from this assistance. But accessing these funds is more difficult for the smallest farmers. Application procedures are complex, there are co-financing requirements, and beneficiaries do not receive payment until the completion of the project. Also, the funds can be reduced if the beneficiaries do not adhere strictly to the terms of the project. Applicants for EU rural development funds in current member states often hire private consultants to help prepare the paperwork. For Polish extension advisors, it is almost a full-time job to help farmers prepare applications. As a result of such difficulties, the rate of absorption of EU Pillar 2 funds has been low in both Romania and Bulgaria. In Bulgaria, for example, only 39% of the available funds were contracted (Forty-First National Assembly of the Republic of Bulgaria, 2011).

1.5 What Next—EU Accession for the Western Balkans

All the countries of the Western Balkans (Serbia, Croatia, Bosnia and Herzegovina, Macedonia, Montenegro, Kosovo and Albania) are at various stages of preparation for eventual EU accession. Croatia is set to join in 2013; Serbia, Montenegro, and Macedonia are candidates, while Albania, Kosovo and Bosnia are 'potential candidates'. Serbia and Montenegro are perhaps the best prepared of the candidate countries, but neither is expected to join until the end of the current decade.

There are a number of outstanding issues that need to be resolved before the EU will accept any of the candidate or potential candidates for membership. Many of these issues involve reforms of the judicial systems and tighter controls on corruption. Albania's application for membership was rejected in 2011, and in 2012, the EU imposed conditions that Albania must meet before it can become a candidate. Most involved demands for greater political stability, but the EU is also demanding resolution of property issues. Bosnia is under pressure to strengthen its national institutions. Serbia is under pressure to improve its relations with Kosovo.

But there are also issues connected with agricultural competitiveness. Serbia, the only net agricultural exporter in the region, is in a stronger position than its neighbors. The other countries are significant net agricultural importers, particularly of processed food products. Albania has the largest trade imbalance. The value of agricultural imports in 2012 exceeded exports by a ratio of 10 to 1. However, even Serbia has considerable work to do before it can fully compete in a single EU market.

The Balkan countries are all characterized by a substantial number of small farmers who are not well integrated into markets. Serbia has a number of large, vertically integrated enterprises, mostly in the northern province of Vojvodina. These are former state-owned enterprises, known as *agro-kombinats*, which have been sold to private investors. These typically take advantage of modern technology and produce high quality goods that can be sold on European markets. A number of fruit growers—especially raspberry producers in central Serbia—are also well integrated into foreign markets. In contrast, other parts of Serbia are dominated by small semi-subsistence producers who are isolated from markets. This phenomenon is even wider spread in the other Balkan countries.

The large number of producers that are isolated from markets also hampers the growth of processing enterprises. There are a number of strong food processing firms throughout the Western Balkans. Each year, more plants are licensed to export to the EU; implementation of HACCP (hazard analysis and critical control points) and other food safety regimes is widespread. A common complaint among plant managers though is that they cannot procure enough raw products that meet their standards on the local markets. At the same time, producers complain of low prices and difficulties in selling to local processors. The strength of the food processing sector is a positive legacy from the former Yugoslavia, which had a long tradition of supplying high quality goods to Western markets. But there are also negative legacies, which manifest themselves in the isolation of many producers from the markets.

For example, the managers of a Macedonian pork plant interviewed by one of the authors (Cochrane) in 2008 had invested in their own pig farm rather than try to buy hogs from local producers. The managers insisted this was the only way they can ensure consistent supplies of the requisite quality. The small producers either slaughter on farm or sell to smaller slaughterhouses. For the same reasons, the major Montenegrin slaughterhouses, interviewed by the same author, prefer to import live cattle from Serbia rather than buy from local producers. Cattle production is widespread in Montenegro, but production is pasture based, most farmers own no more than two to three cows, and buyers for one plant maintained that the domestic market cannot even supply enough animals for 1 day of production.

In Albania, there has been considerable investment in the meat processing sector, and there is now a good core of meat processing facilities that can supply the local market and part of neighbor countries using the ethnic and religion preferences. However, these too have difficulty in ensuring reliable supplies of local raw materials, and their costs of production are much higher than those of imports. As a result, the processing industry imports about 90% of its raw products, and domestic animals mainly provide meat for home use.

Some of the enduring weaknesses in the Western Balkans include those described in Sections 1.5.1–1.5.9 below.

1.5.1 Fragmented farm structure

The average farm size throughout the former Yugoslavia is barely more than 5 ha, a legacy from the communist period when there was a 10 ha maximum for private holdings. Most animal breeders own only a few animals and mainly for their own consumption; many cattle farmers have only two to three cows. A large farm is considered to be one of 20 cows. This situation greatly increases the transaction costs of purchasing raw materials.

1.5.2 Weak land markets

The ownership status of the former socialized farms remains unresolved in many cases, and private owners often lack proper documentation of ownership. The result is that farmers who wish to grow larger are often blocked by the inability to buy additional land. The issue is more serious in Albania than in the former Yugoslavia. Albania continues to have unresolved issues about claims of former owners and how best to compensate them. One World Bank study blames poorly functioning land markets for the high rate of land abandonment—according to this study, as much as 10% of Albania's productive land is left uncultivated for reasons other than crop rotation (Deininger *et al.*, 2012).

1.5.3 Poor infrastructure

Many roads, particularly in mountainous areas, are still in very poor condition. The main road from Serbia into Montenegro is a two lane mountain road that is subject to numerous shutdowns for repairs or accidents. Irrigation is inadequate throughout the region. Attempts to organize small producers into water associations have met with limited success. Wholesale markets are undeveloped; they tend to be large parking lots and lack any cold storage facilities.

1.5.4 Lack of market information

Accurate and timely price information is critical to smoothly functioning markets. Such information helps producers understand what is demanded on the market, and allows buyers to source their raw materials more efficiently. The lack of such information makes it more difficult for buyers and sellers to find each other. The agricultural ministries of Croatia, Serbia, Montenegro, Macedonia and Albania publish regular reports of prices from agricultural markets, Bosnia, however, still does not have a market information system. The agricultural ministries of the two entities of Bosnia and Herzegovina established by the 1995 Dayton Peace Accords—the Federation of Bosnia and Herzegovina and Republika Srpska—still have not agreed on a way to work together to create a national system. But even in those countries which have a market information system, dissemination remains inadequate. The information is available free of charge on the agricultural ministries' websites, but many producers lack access to the Internet. Radio and newspaper dissemination of market information is mostly haphazard.

1.5.5 Lack of quality and food safety standards

There is no uniform system of quality standards in the region. Slaughterhouses must implement the SEUROP grading system for meat (carcass) classification (Superior, Excellent, Very good, Good, Fair, Poor) in order to be licensed for export to the EU, and a number of plants throughout the region have implemented this system, although producers remain largely unaware of it. In most developed countries, slaughterhouses pay producers on a carcass weight basis after grading the carcasses. But throughout the Western Balkans, producers don't trust the slaughterhouses to pay on this basis, because they cannot directly observe the grading process that takes place inside the plant. Instead, most slaughterhouse buyers pay producers cash on a live-weight basis, basing their payment on a visual inspection of quality of the live animal.

1.5.6 Weak tradition of farm associations and cooperatives

This is a well-known problem among experts working in the region—many have observed that farmers have little trust in the idea of cooperatives because of their communist past. There have been numerous international projects seeking to support the creation of true cooperatives or associations, based on the premise that this is the only way for small producers to gain any market power. Most of these projects have not been successful. As soon as international donors withdraw their financial support, the cooperatives tend to collapse.

1.5.7 Need for farmer training

There is a critical role for agricultural extension in explaining the requirements of the markets to the farmers and helping them meet those requirements. While each of the countries in question has an extension service, these tend to be underfunded and unable to provide the needed services. Many of the extension staff continue to focus primarily on technical advice and do not offer adequate training in farm management or marketing. In some of the countries, such as Bosnia and Montenegro, the extension staff are also responsible for distributing government subsidies and do not have time to devote to farmer training.

1.5.8 Inadequate collection and reporting of basic agricultural statistics

There is an urgent need throughout the region to upgrade the central statistical offices. Bosnian officials complain that they have no clear idea just how much agricultural land is being cultivated. Serbia has no reliable data on poultry numbers (however, Serbian officials hope this will be resolved after the completion of the ongoing agricultural census), as there has been no attempt to count the chickens in people's backyards. The Montenegrin statistical office provides no information on raspberry production, a sector the Ministry of Agriculture and Rural Development is actively trying to promote.

1.5.9 Haphazard, ad hoc policymaking

Ministries are seeking to boost production through a complex array of subsidies, price supports, and border measures. These are introduced in response to pressure from producer groups and are often based on inadequate or wrong information. For example, the Montenegrin Ministry is trying to subsidize beef cattle production in order to reduce imports from Serbia, despite the fact that it is not clear whether beef cattle can be profitable in the country, because feed must be imported. The Ministry is under pressure to subsidize feed production, even though all raw materials must be imported, and it seems unlikely that domestic feed production will ever be profitable. One major Montenegrin feed producer interviewed by one of the authors (Cochrane) in 2009 was seriously thinking about giving up production and becoming a distributor of imported feed. No analysis has been done to evaluate the effectiveness of the Ministry's support.

To counteract the above weaknesses, EU membership will bring needed investment to the region and will do much to boost farm income, although it will also bring more foreign competition and stricter quality standards. Many of the small slaughterhouses in Macedonia will likely close, and the small hog farmers will have to meet the requirements of the large buyers or cease producing for the market. A number of unique dairy products could be eliminated from the market if they don't meet the strict hygienic standards of the EU.

The EU has made substantial pre-accession funds available to farmers and agribusinesses in the candidate countries that can be used to overcome these problems. The Western Balkan countries, once they achieve candidate status, are eligible for the Instrument for Pre-accession Assistance for Rural Development (IPARD Program). But conditions for accessing those funds are largely the same as those faced by current EU farmers trying to access EU rural development funds, and small farmers will find access difficult without assistance.

1.6 Conclusions: Future Prospects are Mixed

In the 20 years since the fall of communism, the agricultural sectors of the CEE countries have done quite well overall. Market oriented reform is complete nearly everywhere, and EU accession has brought significant benefits to the farmers of the NMS. Pressures to conform to EU standards have led to considerable consolidation within the processing sector and a growing share of production is being produced on commercialized farms. In general, the CEE farmers have fared considerably better than their counterparts in the former Soviet Union, where governments have been far slower to embrace market reforms.

There are some uncertainties going forward. The global recession that began in 2008, compounded by the recent euro crisis, has created difficulties for all the CEE countries, both current and hopeful EU member countries. The CEE countries have been spared the severe debt problems afflicting Greece, Spain or Portugal, and by 2011 they were beginning to recover from the 2008 downturn (information from Eurostat). This year, however, they are suffering fallout from the euro crisis. In the first quarter of 2012, most CEE countries once again experienced negative growth—Hungary's GDP (gross domestic product) fell by 1.3% in the first quarter (Wagstyl, 2012). Agriculture has been less affected than other sectors, but all the governments have had to sharply reduce expenditures, including funds to support agriculture. Hungary has been threatened with a cutoff of EU funds because of its failure to control its budget deficit. Credit is more difficult to access as banks, most of them branches of Western European banks, have grown more cautious in their lending. Unemployment is rising throughout the region, and real wages have fallen, which reduces both domestic and foreign demand for agricultural output.

In addition, the current EU budget cycle is scheduled to end in 2013, and the EU Member States are debating the shape of the new budget, slated to begin in 2014. Stakeholders in the NMS are loudly demanding greater equity in the formula for determining direct payments—NMS farmers want to receive payments equal to what their counterparts in the former EU-15 are receiving. In response, the CAP reform proposals put forward in 2011 specify that the formula for computing direct payments will be adjusted so that 'all member states with direct payments below 90% of the EU-27 (the member countries of the EU from 1 January 2007) average will, over the period, close one-third of the gap between their current level and 90% of the EU average direct payments' (EC, 2011). To compensate, payments for countries whose farmers receive more than the average will be reduced by 1–7%. It is by no means clear that all the EU Member States will agree to this formula. Moreover, the proposal includes provisions linking the payments to stricter environmental requirements, which could raise production costs.

References

Agolli, S. (2000) *Bujqesia Shqiptare ne Vite: Shifra, Fakte, Komente [Review of Albanian Agriculture: Figures, Facts, Comments]*. Shtepia Botuse and Shtypshkronja "Pegi", Tirana, Albania.

Bjornlund, B., Cochrane, N., Haley, M., Hoskin, R., Liefert, O. and Paarlberg, P. (2002) *Livestock Sectors in the Economies of Eastern Europe and the former Soviet Union: Transition from Plan to Market and the Road Ahead*. Agricultural Economic Report No. 798, US Department of Agriculture Economic Research Service, Washington, DC.

Cochrane, N. and Seeley, R. (2004) *EU Enlargement: Implications for New Member Countries, the United States, and World Trade*. Outlook Report No. WRS040501, US Department of Agriculture Economic Research Service, Washington, DC.

Csaki, C. and Jambor, A. (2009) *The Diversity of Effects of EU Membership on Agriculture in New Member States*. Policy Studies on Rural Transition No. 2009–4, Food and Agriculture Organization Regional Office for Europe and Central Asia, Budapest, Hungary.

Deininger, K., Savastano, S. and Carletto, C. (2012) *Land Fragmentation, Cropland Abandonment, and Land Market Operation in Albania*. World Bank Policy Research Working Paper 6032, World Bank, Washington, DC.

EC (2011) *Communication from the Commission to the European Parliament and the Council: Enlargement Strategy and Main Challenges 2011–2012*. Brussels, 12.10.2011, COM(2011) 666 final. European Commission, Brussels. Available at: http://ec.europa.eu/enlargement/pdf/key_documents/2011/package/strategy_paper_2011_en.pdf (accessed 29 August 2012).

Forty-First National Assembly of the Republic of Bulgaria (2011) *Interim Report on the Absorption of European Funds in the Republic of Bulgaria in 2011, Summary, September, 2011*. Committee on European Affairs and Oversight of the European Funds, Forty-First National Assembly of the Republic of Bulgaria, Sofia, Bulgaria. Available at: http://www.ipex.eu/IPEXL-WEB/dossier/files/download/082dbcc532f7e552013301dfd6cc0678.do (accessed 28 August 2012).

Kawonczyk, A. and Figurska, J. (2009) *Polish Farmers Count Benefits and Losses After 5 Years in the EU*. GAIN Report Number: PL 9010. Global Agricultural Information Network, US Department of Agriculture, Foreign Agricultural Service, Washington, DC. Available at: http://gain.fas.usda.gov/Recent%20GAIN%20Publications/Polish%20farmers%20count%20benefits%20and%20losses%20after%205%20years%20in%20the%20EU._Warsaw_Poland_5-12-2009.pdf (accessed 28 August 2012).

Kelch, D. and Normile, M.A. (2004) *CAP Reform of 2003–04*. Outlook Report No. WRS-04-07, US Department of Agriculture Economic Research Service, Washington, DC.

MAFCP (2011) *Statistical Yearbook 2010*. Ministry of Agriculture, Food and Consumer Protection, Tirana, Albania.

Nistor, E., Bampidis, V., Pet, L. and Ciolac, V. (2010) Impact of EU enlargement on the Romanian meat industry. *Scientific Papers: Animal Science and Biotechnologies* (Romania) 43, 364–368.

Sallaku, F., Jojiç, E., Tota, O., Huqi, B. and Fortuzi, S. (2010) The role of land consolidation activities in the sustainable rural development in Albania. *Research Journal of Agricultural Science* (Romania) 42, 825–832.

Wagstyl, S. (2012) CEE: GDP numbers shock investors. *Financial Times* (London), 15 May 2012. Available at: http://blogs.ft.com/beyond-brics/2012/05/15/cee-gdp-numbers-shock-investors/#axzz24vlqr2xS (accessed 29 August 2012).

2 Agriculture and Land Reform in Albania

Edvin Zhllima[1] and Fatmir Guri[2*]
[1]*Agricultural University of Tirana, Albania;* [2]*Agricultural University of Tirana, Albania and European Commission Joint Research Centre, Seville, Spain*

2.1 Introduction

The land reforms implemented in Albania during the last century were decisive in stimulating the distribution of land rights among Albanian farmers and other landholders (state, religious institutions and other large landholders), shaping the structure of farms and influencing the socioeconomics of agricultural society. Albania has experienced industrial development for only a short period of time and the population has always made its living through agriculture. In spite of Albania's poor agricultural performance (see Lorenzoni, 1930; Roucek, 1933; Vlora, 1973; Fishta and Toçi, 1984), the country's agricultural industry supported the livelihoods of more than three quarters of its population before World War II and more than 50% of the population in the remaining part of the century, until recently.

There has always been a need for better land reforms in Albania. However, the reforms have always been partially implemented, and attempted under short time constraints, and under strong political regime changes. The first land reform drafted after the independence of the country from the Ottoman Empire in 1912 was interrupted due to the end of a short-lived government. Then the period 1912–1920 was followed by ethnic strife, political instability, and war, which created an unfavorable environment for land reforms, and for state supported agricultural development. The strengthening of the role of state at the beginning of 1930 called again for a land reform, which was not extended to all the regions of the country. Reform and the better functioning of the land market between the two World Wars reduced the share of land possessed by large landowners. Later on, the entrance into the communist regime in 1945 was accompanied by a very intense collectivization, and the exit from the regime in 1991 again transformed the agriculture sector from state-owned large landholdings to very fragmented, family-based farming. Again, the incentives to nationalize the land (during the communism era in 1945) or privatize it (for moving toward a market economy in 1991) have always been made as a result of the short-term needs of the government and donors (achieving food security, reduction of poverty and inequality, etc.) rather than reflecting the real needs of the agriculture sector and respecting the peasant population's land rights and their customary ties with the land.

* The views expressed are purely those of the authors and may not in any circumstances be regarded as stating an official position of the European Commission.

Post-socialist countries in Central and Eastern Europe during the 1990s, have seen a complete redistribution of land from state and collective units to individual smallholders. Generally, land reform programs were aimed at institutionalizing land rights to the private sector, to increase investments, and create land markets, in order to allow individuals to have access to land ownership at its best use (Sikor and Müller, 2009). Thus, it is supposed that the stronger the security of land rights, the larger the investment in improving the productivity of the property (Furubotn and Richter, 1997). Such a program was used throughout the Central and Eastern Europe Countries (CEEC) and was the basic idea for the distribution of land to individuals, and the distribution of ownership titles for securing these new land rights, such as occurred in the 1980s and 1990s (Bouquet and Colin, 2009). In these reforms, the state took a primary role in promoting land redistribution and registration (de Janvry *et al.*, 2001).

The land distribution reforms had different impacts on the restructuring of agriculture in different CEEC. The breakup of large state and cooperative holdings as a result of the land reform program into individually privately owned farms was followed by the establishment of two types of land tenure structures. A domination of large agricultural holdings was established in countries such as Slovakia, Hungary, the Czech Republic, and in most states of the former Soviet Union, whereas in Western Balkan countries there emerged a greater domination of small-scale farming (Giovarelli and Bledsoe, 2001; Swinnen *et al.*, 2006). In most CEEC, land was given back and, in cases where this was not possible, compensated for in kind or distributed in shares to the rural dwellers (the case of Russia and Ukraine). The rest of the land was leased to those interested in farming (such as the cases of East Germany, the Czech and Slovak Republics, Bulgaria, etc.). There were also special cases, such as in Albania, where agricultural land, excepting pastures, was totally distributed to those residing in villages and working in agriculture (Swinnen *et al.*, 2006).

All of these CEEC have implemented land reforms through various legal mechanisms. In the majority of the countries, land reform was led by the state itself based on a set of established legal norms. In the case of Albania, land reform was led by the state in the plain areas, but it was more community driven in the mountainous areas. In the areas where government land reform was implemented by introducing the land reform Law No. 7501, a high degree of land fragmentation was created; in some of the villages long standing landownership insecurity resulted owing to the uncompleted compensation of pre-1945 owners, especially if pre-1945 owners of land were still living in the village. In contrast, in the mountainous areas, mostly in northeast Albania, the fragmentation was a result of the harsh and remote land situation, but the land rights insecurity is less evident as the land was distributed according to the will of the villages, based on the pre-1945 rights. However, the extent and history of fragmentation and landownership insecurity vary even among villages of the same region, from both the southwest to the northeast area of the country (Kodderitzsch, 1999). This is related to how communities accepted and implemented the land reform program, and to the criteria used to merge the land reform rules with the established rules inherited from the pre-communist times.

This chapter reviews some of the most important milestones in Albanian land tenure history. It starts with a description of the three main periods of Albania's land tenure history. From then on, the focus is on the period from the last state reform implemented in 1991—its policy objectives, implementation in practice, and influence on agricultural growth, and on the reduction of food vulnerability. The chapter emphasizes the unexpected effects of the land reform process, such as land rights insecurity; its effect on the land market and land investments; and structural effects such as fragmentation, and effects on land use.

2.2 Historical Background

2.2.1 The Ottoman legacy and the reform of King Zog I—the emergence of a landed middle class (1912–1944)

At the beginning of the 20th century, half of the agricultural land in Albania was divided into large land holdings called *çiftlik* and given temporarily to military and civil servants based on merits; the rest was given temporarily to villages and religious institutions. The structure of landownership was bimodal, dominated by either large estates or small farms. The *çiftlik* covered more than 55% of the land, the majority of which was owned by the Ottoman state and religious institutions. The remaining, approximately 45%, of land was owned by small landholders with less than 2 ha each. This allocation of land existed until the beginning of the World War II. During this period, several researchers travelling in Albania, such as Sedlmayer (1916; cited in Gambeta, 1999), Lorenzoni (1930), Roucek (1933), and Vlora (1973) witnessed a dismal situation in agriculture, with poor land operations, limited access to irrigation and drainage, and minimal mechanization.

During the period 1912–1932, two land redistribution reforms were attempted. The first attempt was in 1912, when the land of the Ottoman Empire was returned to the new Albanian state, as a result of Albanian independence. At that time, a fragile government, functioning in a period of chaos and ethnic conflicts, declared a land redistribution policy, which was short-lived. Again, in 1924, the Parliament of Albania approved a land reform law with the objective of distributing land among the northern Albanian (Kosovar) migrants who moved to the south after World War I, and other landless people. However, the lack of a clear mechanism for implementing the land reform law, and the pressure of the country's elite, who would have lost their land as a result of this reform, blocked any land reform attempt. In 1930, a land reform was undertaken by King Zog I (Lorenzoni, 1930), who, through the Agrarian Reform Act, called for a free distribution of land to landless peasants. The land, according to the Agrarian Reform Act, would allow purchases from large landowners with the provision of keeping at least 40 ha for each owner, plus 5 ha each for spouses and children. At the same time, a financial reform bill was created and was financed by the National Agricultural Bank for the financial compensation of the land lost by the landholders (Roucek, 1933). According to Civici (2003), this reform was not fully implemented because most land records, on which were based the criteria for land redistribution, were falsified. At the end of the reform, only 8109 ha of land (less than 2.2% of the available agricultural land) had been redistributed to the landless families, of which 3411 ha (or 42%) were formerly state owned and 4698 ha (or 58%) belonged to private landowners. Although agricultural land redistribution may not have been successful, Kareco (2008) through a historical review of the very little available data, revealed that the reform attempt increased the agricultural output more than the rate of population growth, despite the sharp reduction of the prices in the aftermath of the 1930s economic depression.

Despite the failure to achieve effective and complete land reform in this period, the structure of the land was much fragmented and less concentrated in few landlords, as a result of the better functioning of the land market. During the period 1912–1938, the average farm size decreased to 2.5 ha. The number of small and medium owners doubled due to the shift of land rights from the large landholdings to the small farm holders (Fig. 2.1) through a more active agricultural land market and the fragmentation of the large *çiftlik* due to the intergenerational transfer of land to the next generation (Sabates-Wheeler and Waite, 2003). By 1945, around 83% of the agricultural land was owned by small and medium landowners, who owned on average 1.8 ha of agricultural land. Still, land inequality was high as a few landlords, the state, and religious institutions owned most of the fertile land. The big landowners possessed an average of 2000 ha each and the middle class of landowners held 1.9 ha of land each. Approximately 14% of the total population was landless, which prompted a radical land reform after the end of the World War II.

2.2.2 Reforms after World War II—the first step toward the collectivization and the adoption of the socialist model of agriculture in 1946

The end of World War II found Albania in a very different position from that before the war. The communist regime that came into power in Albania in 1944 initiated an agrarian reform in 1945 that was implemented in two phases. The first phases, in 1944–1946, included land expropriation and redistribution to the landless peasantry. The main objective was to give land to more than 145,000 families who had less than 5 ha of property. The reform of 1945 brought dramatic changes in land ownerships. More than 19,000 large agricultural units belonging to large landowners and religious institutions were expropriated. This amounted to a total land area of 172,659 ha, with the distribution of 155,000 ha of land (30% of state land), 474,227 olive trees and 5923 draft animals to 70,211 landless peasant families (Morone, 1997; Stanfield *et al.*, 2002; Sabates-Wheeler and Waite, 2003). As a result of this expropriation, during the period 1945–1946 the share of land belonging to the large and middle-class landowners shrank from 52% to 16% of the total agricultural land.

However, possession of this land by farmers did not last because redistribution was followed immediately by the collectivization and then the nationalization of all Albanian land. This policy was copied from the Soviet regime ideology, in which land should be transferred to landless peasantry as well as to the working class for equity and efficiency principles. This was the second phase of the agrarian reform under the communist regime which started in 1946 and ended in 1967 in the northeastern mountainous areas of Albania. Collectivization became very intensive in the 1950s and was completed in 1959 in southwestern Albania and in 1967 in the country's remaining mountainous areas (Fishta and Toçi, 1984). During the period 1956–1970, more than 300,000 ha (74% of the country's total area) was consolidated (see Table 2.1). Thus, all of the country's agricultural production was organized into state farms and cooperatives. Cooperatives were created by the accumulation of land from pre-1945 landowners, including large, medium and small owners, whereas state farms were

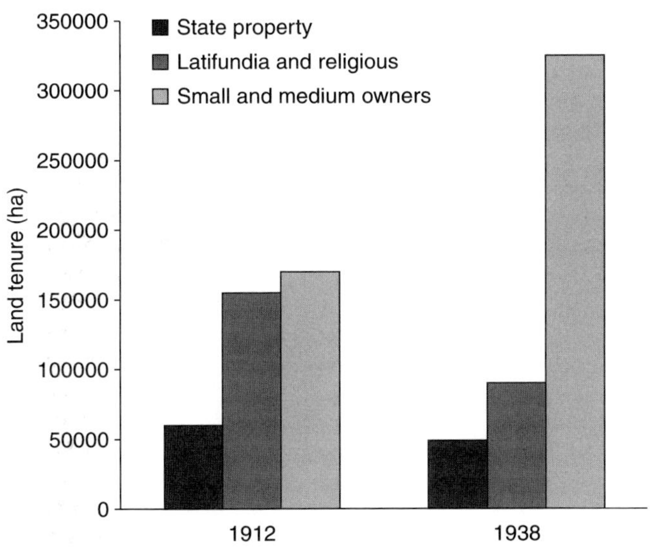

Fig. 2.1. Distributional changes in land tenure for the period 1912–1938. Sources: INSTAT (1991); Wheeler and Waite (2003).

Table 2.1. Albanian land ownership by area during the period 1950–1990 in thousand ha. Source: INSTAT, 1991.

Year	1950	1960	1970	1980	1990
State farms	13	65	124	151	170
Cooperatives	21	330	454	532	504
Peasant home plots	357	62	21	19	30

mainly created on land that had formerly belonged to religious institutions, nationalized foreign agricultural enterprises, or large landowners, or on land made productive after continuing improvements (irrigation and desalinization) under the regime.

Because the majority of the population was rural (70.5% in 1946) and so strongly linked to their land, the communists addressed the process of the agricultural land nationalization more slowly than it did other sectors of the economy, although this slow process was pursued with an aggressive propaganda. Sectors such as banking, light industry, trade and services were all nationalized in a period of just two years (Stanfield *et al.*, 2002).

Individual rights to use agricultural land were restricted to 2.7% of the total land in the form of small plots around dwellings. These were typically home gardens with a maximum surface area of 1100 m^2 per household (1500 m^2 in mountain areas due to the harsher conditions of production), and included a cow and/or ten small ruminants (Stanfield *et al.*, 2002; Civici, 2003; Sabates-Wheeler and Waite, 2003). The agricultural land, including the home gardens, was named 'a collective property of the people', preventing by law its sale and inheritance. The 1976 Constitution claimed every asset in rural areas a *de jure* 'state property' (Stanfield *et al.*, 2002), making Albania the only country in the CEEC to entirely eliminate private ownership of agricultural land and make the state its sole, exclusive owner (Cungu and Swinnen, 1999). Therefore, even the land inside the home plots was monitored by the agronomists and no sale or rental agreement was allowed.

During the period 1960–1990, the agricultural production cooperatives and state farms were further consolidated, thereby reducing farms in number and increasing them in size. Farm cooperatives became bigger, consequently reducing them in number from 1484 to 492. As a result of the state land expansion, the peasant home plots also reduced in size three times during the period 1960–1980. The development of irrigation made it possible to increase the proportion of arable land under irrigation from 10.5% in 1950 to 60% in 1989. The land consolidation, and improvements in input supply, extension, mechanization, and irrigation made possible a threefold increase in grain yield, fourfold increase in milk yield, fourfold increase in fruit and olive tree numbers, and an eightfold increase in vineyard surface area from 1950 to 1989 (INSTAT, 1991). However, land remained scarce, given that the population tripled during the same period, reducing the land surface per capita by 10% (to 0.2 ha per capita) (INSTAT, 1991). Thus, due to land scarcity at the end of the 1980s, the country experienced a diminishing return of scale from the investments in agriculture and a considerable underemployment in the agricultural sector. Researchers have evaluated that, despite the zero unemployment declared by the government in the second part of 1980, there was an underemployment which was mostly revealed in the agricultural sector (Sandstrom and Sjoberg, 1991).

The period from 1960 to 1989 was a period of self-reliance, a policy implemented in order to reduce the international market dependence and to offset the vulnerability from other countries or other previous market partners of CEEC. It was a strategy which was heavily promoted after Albania found herself isolated and without a large partner, such as was the case with the demise of the economic cooperation partnership with Russia in 1961 and with China in 1978. The deterioration of the agricultural sector in the second half of 1980s, coupled with droughts, pressured the government to further collectivize the people's agricultural assets by taking away their livestock (Lerman, 2000). This policy was also instituted to increase livestock production, reallocate the workload among cooperative members and reduce the hidden underemployment in cooperative production. The state failed as cattle production fell, due to slaughtering and demotivation in animal care, thereby reducing both meat and milk production by more than in the period before this intervention (respectively by 5.5% and 12%). Intensification of land use in the 1970s and 1980s and the reclamation of new land were also reducing the marginal rate of return from investments because newly opened land parcels during these decades were of low quality (Erebara, 2011). In the late 1980s, the agricultural sector went into a recessionary phase. While the rural population was growing at 3% per year, farm production was growing at only 0.03% as a result of low investments and poor quality of land. In 1988, the government tried to increase farmer incentives by transferring between 0.1 and 0.3 ha of land to peasant home plots to increase their total size from 19,000 ha in 1980 to 30,000 ha in 1990 (see Table 2.1). The government also distributed a cow and small ruminants to individual farmers and allowed a modest agricultural private market to function (Civici, 1997; de Waal, 2004), but this intervention did not gave a clear incentive to the farmers and therefore did not improve productivity or change agricultural production and increase farmers' motivation.

2.2.3 The land reform of 1991

The food crisis and the external political changes made the government accept the transition from a mono-party to the multi-party system in 1991, as well as the opening of the Albanian economy (Pashko, 1993), which was approved by parliament at the end of October 1991. The transition of the political and economic regime was accompanied by a package of laws, including laws influencing price and trade liberalization, investment deregulation, a convertible currency, and privatization of land and other immovable assets. As a result, the parliament adopted the law 'On Constitutional Principles' of 1991, which repealed the constitution of 1976 and served as a basic law for the privatization of land, and the establishment of economic relations based on market principles. The political will for a market-based economy required the transfer of land and other state ownership to private agents. A distribution of land for achieving social equity was endorsed by the government. This was based on a somewhat debatable law that defined eligible owners of land as those who lived on the (about) 610 agricultural cooperatives and 120 state farms (which were privatized and distributed 1 year later) before 31 July 1991. The land was to be distributed free of charge to families, on a per capita basis, based on the total area of each village, strictly respecting the quality of the land and other indicators of its value. The ultimate unit of the land divisions was the village, the surface area of which was divided up by the number of people in the civil register who inhabited the village before 31 July 1991. Before this decision had been made, a spontaneous de-collectivization had already been started by the rural population through the stealing of raw materials and products from cooperatives and state farms (de Waal, 2004).

A fundamental problem introduced by the agrarian reform of 1991 was the compensation to former owners. A national committee was created in 1993 for the redistribution of property to former owners and to compensate them with parcels of coastal land. The issue of compensation is central to the current controversy in Albania because it is one of the criteria placed by the European Commission for achieving future membership of Albania in the EU (EC, 2010, 2011). The state is giving more rights for restitution and compensation as a result of international pressure, thereby heightening the expectations of the former owners. The laws in different periods have acknowledged this by increasing the upper boundary of compensation for rural land. Thus, the government decided to give compensation of up to 100% of the land value for up to 15 ha in 1993, 60 ha in 2004, and 100 ha in 2006, after pressures from lobbying by pre-1945 owners. As of now, an acceptable solution to the problem has not yet been found and the state has not managed to compensate former owners, either in kind or in money. Over the years, governmental interventions have been unstable, and shown a lack of serious endeavor toward full solution of the pre-1945 claims.

The process of land restitution and compensation is still in discussion because of the absence of a long-term financial or land fund for compensation. The government has lately taken a step back by interrupting the independence of the institution created for the restitution and compensation process. The process is still unsolved, giving both the former and new owners a sense of low credibility vis-à-vis the deliverance of government policies. Recently, the government has been drafting another strategy for the enforcement of land rights and the accomplishment of compensation to the pre-1945 land owners in kind or in money.

The process of land distribution progressed quickly in the first 3 years, covering

88% of the land area that was expected to be distributed, including 91.5% (430,155 ha) of the cooperative land and 79% (114,560 ha) of the state farm land. The latest data (from 2010) show that 81.5% of all cadastral rural areas have been registered. Approximately 137,000 ha of land are still held by the state, of which 110,000 ha are administered by communes and municipalities, while 27,000 ha are defined as available to the state (EC, 2010, 2011). During this period, errors in land distribution, as well as land grabbing and squatting have occurred and generated land disputes and conflicts (World Bank, 2006; unpublished internal report on land reform progress, Ministry of Agriculture, Food and Consumer Protection (MAFCP), Tirana, Albania, 2007). These errors obliged parliament to officially end land distribution reform at the end of August 2008.

Although formally closed, land reform was not still fully adopted according to the principle of equal distribution by quality and equity principles, as the criteria for the rules of land distribution had not been correctly implemented in various areas of the country. Approximately 15–20% of agricultural land (mainly in the mountainous northeast areas) was distributed to the previous owners on grounds based on common will and agreement in respective communities (Morone, 1997; Bardhoshi, 2004), i.e. differently from those villages which distributed the land based on per capita distribution and quality categories without reference to the pre-1945 land family location. In some of these areas, a mixed approach was used, distributing the land per capita but respecting pre-collectivization boundaries (Kodderitzsch, 1999). Thus, since these communities did not implement the reform according to the Law No. 7501, which obliged the village to distribute the land based on per capita distribution of the land and quality categories without reference to the pre-1945 land family location, the state doesn't allow land holders the right to sell or use the land for collateral. This is especially true when the local land commissions have not provided the central government institutions with the documents reflecting an eligible practice of land distribution.

In order to regulate the process, the Parliament of Albania passed a law to verify the legitimacy of titles which were the subject of disputes during and after the land distribution. This legal instrument was created by policymakers to both guarantee ownership titles and stimulate development of a land market. However, the law does not regulate the land titles of those mountainous regions that have distributed the land based on non-legal practices. Therefore, these areas do not regulate their market mechanism through the law. An example of the bottlenecks in the northern eastern areas was the 'statalization' of land for the interests of the road infrastructure linking the capital of the country with the Kukes region. The progress in the compensation of farmers for the land lost was very slow owing to the lack of the documentation of ownership on the land plots.

2.3 Structural and Social Impacts of Land Privatization of Albanian Agriculture

2.3.1 Land reform influence on agriculture sector efficiency

Albania may be viewed as an interesting and unique example of transition. After the fall of communism in the early 1990s, it was one of the quickest countries to progress in terms of privatizing the land. By mid-1996, it was deriving the highest proportion (75%) of its agricultural output from private enterprises, equaled only to the Czech Republic (EBRD, 1996). Albanian reform had a particularly dramatic impact on the size and fragmentation of farm structures, given that Albania has one of the highest rural population densities (Lerman, 2000), and, as a consequence, has limited land per capita in rural areas (approximately 0.2 ha) (MAFCP, 2007a). The massive and rapid de-collectivization was justified by the extreme poverty and the social and political situation. According Cungu and Swinnen (1999), Albanian institutions elected to speed up agriculture privatization because of the high percentage of the rural (65%) to the total population and the high contribution (more than 35%) of agriculture to gross domestic product (GDP). According to Pashko (1993), the distribution of land was also supposed to be a tool for coping with the country's food shortage, as it would allow an increase, or at least stable growth, of the agriculture sector (Roselle and Swinnen, 2004). The primary figures on agricultural efficiency indicators show that the land reform had a positive influence on the agriculture sector. Among the countries of Eastern Europe, the former Soviet Union, and the Western Balkans, Albania achieved the highest gross agricultural output (GAO) level in the first 10 years of the post-land reform (see Fig. 2.2).

The GAO is not the only economic indicator used during the transition years to show the impact of land privatization. Another example is the level of agricultural labor productivity (ALP), which is the agricultural productivity by worker per unit of time (Fig. 2.3). This index increased at 1.6% a year during the first five years after transition, with a slight reduction of 0.8% in the period 1996–2001.

As clearly shown in Fig. 2.2 above, Albanian GAO and ALP grew faster than those of other Balkan countries during the same time span. Other countries did not totally reverse the collectivization heritage of the communist regime. Most regimes maintained large holdings, and distributed shares of land surface under possession in the form of vouchers to the members of cooperatives and farms, thereby maintaining their linkage with the same forms of agriculture production. According to researchers observing the trend during the beginning of the 1990s (Pashko, 1993), Albanian farmers did not enjoy being reminded of the former socialist infrastructure and therefore were more responsive than other countries in

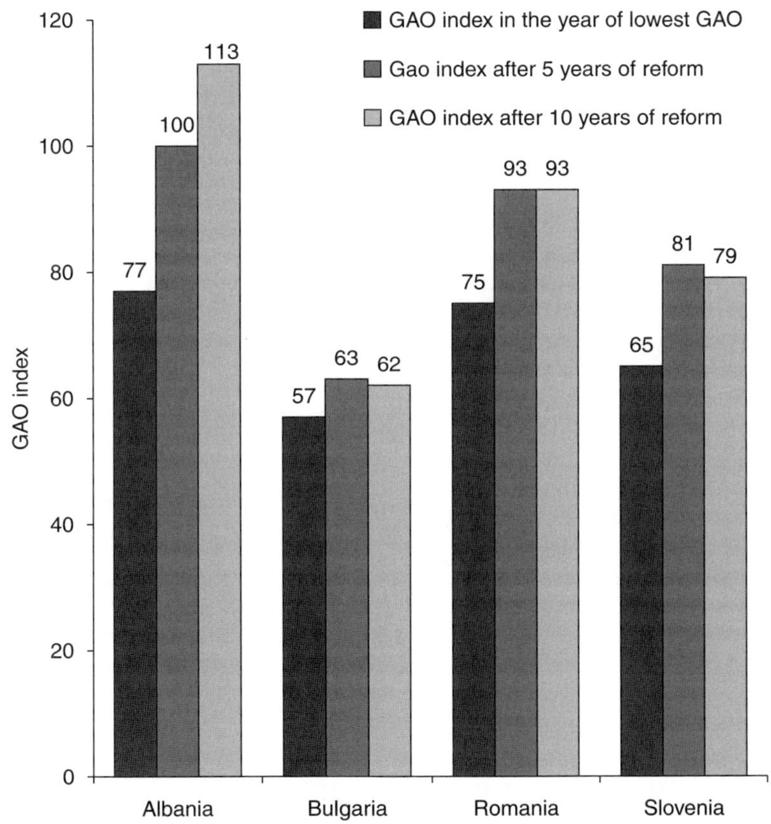

Fig. 2.2. Gross agriculture output (GAO) index in Albania compared with other countries of Central and Eastern Europe (CEE). Source: Roselle and Swinnen (2004).

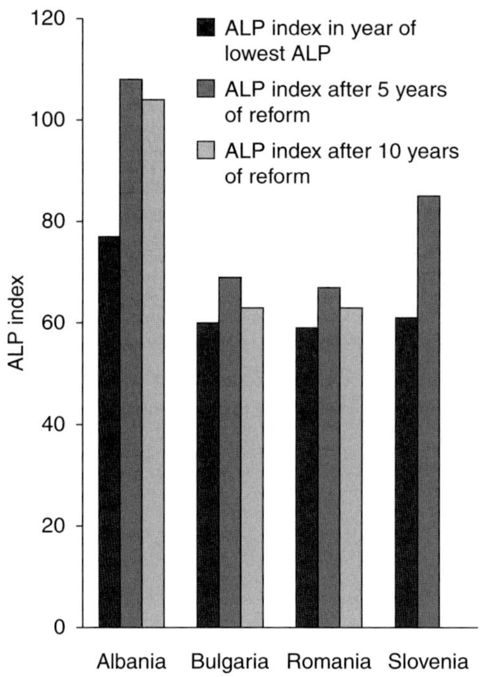

Fig. 2.3. Agricultural labor productivity (ALP) of Albania compared with other countries of Central and Eastern Europe (CEE). Source: Roselle and Swinnen (2004).

reacting to the newly opened orientation of the agricultural market. However, the relatively low efficiency of agriculture at the end of the period of collectivization in Albania created a misleading and lower benchmark than those in other CEEC, and therefore a higher rate of growth experienced by Albania in the beginning years of the reform. This could be a misrepresentation and might not show a better performance of the Albanian agricultural sector, but rather a catching up process. Therefore, when observing the trends in the first decade of the reform, the highly positive indicators in the first 5 years should be interpreted a bit more conservatively.

2.3.2 The consequences of the structural effects of land reform—land fragmentation and abandonment

Though the land reform movement did achieve its first objective of redistributing land, there were also several indirect and undesirable effects. One such negative consequence was the resulting fragmentation of farmland quality and size. The family farms that emerged after reform were, and continue to be, small (1.2 ha on average in 2011), and each farm is composed of many small parcels (an average of 3–4 plots/farm) which are often located far from each other and from farmhouses (from 1 to 10 km, Table 2.2). Much smaller farms are more common in the mountain and northern areas of the country, where the average farm size is 0.5–0.7 ha. Most studies carried out in Albania conclude that fragmentation is one of the most negative outcomes of the reform as it increases the input costs, performs weakly on economies of scale, increases the costs of transportation, reduces the effectiveness of irrigation, and limits access to mechanization (Lusho and Papa, 1998; Lemel, 2000; Civici, 2003; Guri and Jouve, 2003; MAFCP, 2007a).

One major negative outcome of land reform had been the loss of fertile agricultural land. The new Albanian farmers did not understand the legal obligations related to land reform Law No. 7501. Article 11 of Law No. 7501 maintains that those individuals that have land in ownership or in use are obligated to preserve and increase its productive capacity and to create systems of protection. Furthermore, the Law on the Protection of Fruit Trees prevents the conversion of orchards and vineyards into arable cropland without the permission from Land Protection Offices. Land redistribution was preceded and followed by a wave of cropland abandonment in marginal plots (MAFCP, 2010), the uprooting of vineyards, fruit trees and olives, the illegal construction of houses, and the destruction of irrigation and drainage infrastructures. The fruit sector

Table 2.2. The structural changes to agriculture due to land distribution in the earlier years of land reform. Source: Civici (2003).

Land distribution	1990	1994
Number of units	622	445,000
Unit surface	1,065 ha	1.2 ha
Plot surface	38 ha	0.2–0.3 ha
Land surface irrigated	50%	30%

was largely damaged, thereby reducing cash-cropping opportunities at the farm level.

Land abandonment continues to be a significant problem in the country. Surveys show that Albanian farmers left fallow 10–15% of the productive land because of low quality and long distances between plots (Deininger, 2003; World Bank, 2006).

2.3.3 Land rights security

Another negative aspect of land reform was its impact on land rights security. Land rights, along with land registration, are recognized as a means of creating secure land tenure and helping farmers to resolve any land disputes. Lack of authentic documentation, registration discrepancies, and inaccurate mapping has created insecurity in Albania (Swinnen and Mathijs, 1997; Stahl, 2007; MAFCP, 2007b). In the latest study on land security covering 630 farms in five regions of northwest, central and southeast Albania, Zhllima and Imami (2012) found that land titling contributed positively to land security perceptions of farmers.

Another source of insecurity stems from the inability of the Albanian government to quickly resolve the land claims of pre-collectivization (1945) owners from the very beginning of land reform in 1991. Platteau (1996) and de Soto (2001) show that in countries where distributional measures are implemented by the state, conflicts arise when property rights with little or no compensation are given to the disadvantaged parties—in the case of Albania, the pre-1945 owners. These groups will oppose the new arrangement, emphasizing the need for determined state intervention in the promotion and establishment of titles to land through compensation or restitution. Such factors heavily influenced Albania when the exclusion of the 'original owners' from the already implemented land distribution brought forth a restitution/compensation scheme. The scheme provided for land restitution when possible (excluding the agricultural land which in 1991 was subject to per capita distribution) and for compensation in kind or in cash. This policy has undergone several adjustments since its inception, but it still fails to address completely the rights of pre-collectivization owners.

Therefore, the landowners benefiting from the plots belonging to other families before 1945 found themselves subject to disputes and conflicts. In Albania, land and property disputes constitute a high proportion of the cases pending in the civil law courts, but local and customary mediators (such as local village elders, other municipal level officers, religious leaders and nongovernmental organizations) are also called upon to mediate land and property conflicts (OSCE, 2004; World Bank, 2006; Stahl *et al.*, 2009). In fact, more than half of all civil court cases are due to property disputes (World Bank, 2006), and escalations of conflicts have resulted in 8000 deaths since 1992 (Majko, 2012). The high number of conflicts has increased insecurity on owned land, which is termed (by Lemel, 2000) a 'subjective' insecurity. Zhllima *et al.* (2010) and Zhllima and Imami (2012) found that perceived tenure insecurity (i.e. the likelihood of losing ownership) is still common in rural Albania. In their survey, Zhllima and Imami (2012) found that insecurity is lower for plots acquired through customary rights (predominantly ancestral land), where there are no conflicting claims between post-collectivization and pre-1945 owners, than for plots acquired through state reform, which exhibit these property conflicts.

2.3.4 Land insecurity, land investments and the land market

Based on the literature, land security is an investment-enhancing factor (Demsetz, 1967; Feder and Feeny, 1991; Binswanger *et al.*, 1995; Feder and Nishio, 1996). Land rights insecurity is detrimental to land use and investment in land improvement and conservation, and especially on long-term impact decisions, such as the planting of fruit trees and vineyards (Lemel, 2000). In their study, Zhllima and Imami (2012) found that almost two-thirds of farm investments in perennial crops were made in plots that were perceived as very secure, as opposed to 5% that were invested in plots perceived as highly insecure. Other types of investments (including those in agriculture buildings, greenhouses, wells, etc.) are more likely to take place on plots perceived as secure or very secure (more than two-thirds of investments are made on such plots), while there are no investments made in plots perceived as highly insecure (Zhllima and Imami, 2012).

Having title to the land improves its transferability, either on a temporary basis through a lease or on a permanent basis through sale to other farmers who have the resources to make better use of it (Binswanger *et al.*, 1995; Deininger, 2003). In Albania, the low activity of the land market may be a negative result of land rights insecurity (Swinnen and Mathijs, 1997; Lemel, 2000; Giovarelli and Bledsoe, 2001; Civici, 2003; Stahl *et al.*, 2009; Zhllima *et al.*, 2010; Zhllima and Imami, 2012). Zhllima and Imami (2012) found that farmers prefer to purchase land plots previously held based on ancestral rights, which are perceived as more secure than simply officially distributed government land titles that are not combined with ancestral rights. The high costs of finding plots with such combination of rights reduce the ability of the sellers and buyers to make transactions on land. Data and surveys from the first decade after the beginning of the reform (Moor and Flyn, 1997; Kodderitzsch, 1999; Lemel, 2000; Sabates-Wheeler and Waite, 2003) to the recent past (World Bank, 2006, 2007; ARD, 2007) show a rural property market that has hardly developed in Albania. Since 1991, when the land was distributed, less than 15% of surveyed farms have made land transactions (World Bank, 2006; Zhllima and Imami, 2012). The low activity of the land market transactions reduces the possibility of land consolidation.

While the land sale market has not yet been able to develop in Albania, the rental market is more active. Surveys done in 1999 and 2002 show that 10% of the farmers enter into land leases (Swinnen *et al.*, 2006; World Bank, 2007; Zhllima and Imami, 2012). The rental market is important in Albania as it has been estimated that 90% of farms with more than 2 ha are based on lease holdings (MAFCP, 2003). Interviews with real estate companies reveal that the land rental rate varies from 10,000 ALL/ha per year in poor quality lands to 20,000–50,000 ALL/ha per year in the most fertile regions. There are cases of very highly fertile regions, such as Divjaka or Xara, where the land rental rate may even go into annual ranges of 70,000–130,000 ALL/ha, but these are exceptions.

In addition to perceived land (in)security, there are several others reasons behind the weakly developed land market. Farmers' limited access to information on land sale rights (Lusho and Papa, 1998), low willingness to sell land owing to strong emotional ties with the land (Guri and Jouve, 2003; Zhllima *et al.*, 2010), farmers' negative perception of the high costs of formal arrangements and complex administrative procedures (ILD, 2008; Stahl *et al.*, 2009; Zhllima *et al.*, 2010) have also been factors of negative influence on land market activity.

Last, but certainly not least, the high price of land due to weak land zoning (it is relatively easy to transform agricultural into development land) has increased land prices, and, therefore, has reduced farmer incentive to purchase land for purely agricultural purposes. This is especially true in the central and coastal areas of the country, where urbanization pressure has increased the average price of the agricultural land (Guri, 2007; Guri and Jouve, 2009; Zhllima *et al.*, 2010). Interviews with real estate companies support this evidence, as for the period of 2009–2011, land sale price for agriculture varied from 1 million ALL/ha in hilly land or lowland distant from the urban center, to 3-4 million ALL/ha for the lowland areas that are in the fertile southwest and southeast plains, which tend to border the urban areas. The price is higher if land is under irrigation as it increases the production potentials. The price for land compensation to the pre-1945 owners is 1.6 million ALL/ha, which is a kind of price ceiling for the land market in lowland areas. Moreover, the average land sale price of 6.3 million ALL/ha reported from the Immovable Property Registration Office does not represent agricultural land, as most of the land bought, although currently agricultural, may be converted to development land.

2.3.5 Land reform and gender issues

Albania is a country where gender inequality exists in various areas of political, economic and social life. The gap in various gender indicators, such as access to labor, education, policymaking and other basic opportunities, is also reflected in rural living. Moreover, compared with the urban population, rural inhabitants are much more linked to customary rules, particularly systems of laws dating back to the 15th–16th centuries known as the Canun. The Canun is a common system of written rules for social and economic livelihood. Though these rules became obsolete in the communist years, during the transition, with the weakening of the rule of law, the practices were reactivated in rural areas, and still coexist within the population, and shape the way that the population deals with the everyday management of assets. It seems that the more formal institutions and regulations are distant from and appear weak to these villages, systems or rules, the lower the level of acceptance of the legal institutions by the peasantry.

When Albania initiated the redistribution of land, the initial land titles were given in the name of the male head of household. Though the ownership certificates distributed to farmers during 1995–1998 assured that every member of the family has equal rights toward land use and land sale, by ranking the names of family members, the land transactions could not be implemented without the consent of all the members of the family in front of the notary. However, formal rights given to farmers after the land reform have not de facto been acceded equally between men and women (Guri and Jouve, 2009; World Bank, 2011). The way that peasants manage and possess land within a family and village is strongly related to the customary rules, and inheritance and land distribution are strongly influenced by the Canun.

The revival of customary rights after 1991 has reinforced the role of the male head of the family, with the male assuming the role of decision-maker, and membership in a family group being identified through the male line (Lemel, 2000; Zhllima and Imami, 2012). Females leaving their birth family via marriage are reluctant to claim any share of land, though the head of the household may make attempts to compensate the daughters for the lack of land with other in kind or monetary rewards or dowries (Voell, 2004). Land from their own families is judged to be unnecessary for women because they will have access to their husband's family's land as his family responsible for providing for her even in the event of his death (Zhllima *et al*., 2010; Wehrmann, 2010). Such rules, regulating access to assets such as land, make females more vulnerable in cases of strong high economic fluctuations, and they would not benefit in the case that the land increases in value over years.

Evidence from a 1998 survey in Albania found that 62% of farmers declared that sons would inherit the farmland (Wheeler and Laha, 1995; Lastarria-Cornhiel and Wheeler, 1998; Wheeler, 1998). Such inequality makes women and the rest of family very vulnerable in cases of distressed sales of land by the male head of the household, which occurred frequently among rural people during the first years of land reform. Although agreement of each family member in front of a notary official is required on the approval of land sale, anecdotal evidence suggests that what commonly happens is that females are obliged (forced) by the male head of household to obey the decision-making of the males of family head of household.

Widowhood is an issue of further concern in rural Albania. The Civil Code gives widows rights of inheritance equal to that of other members of the husband's family. In customary practice, when the male head of the household passes away, most widows remain in the husband's home, do not remarry and may enjoy the rights of the husband's land. This is not always the case though, as some widows may return to their own family, thereby losing their rights to their husband's land. This is especially true of widows with young children, who may wish to remarry. A widow without children cedes her rights to her husband's land and leaves all property to her husband's closest male relative (Lastarria-Cornhiel and Wheeler, 1998). In the majority of cases, a divorced woman also leaves her rights to the husband, although by law she may enjoy her share of rights (on half of the land), if no pre-marriage contract has been established.

There is a common understanding that the information that females have as to their formal rights on land management, inheritance and sale rights is at a lower level than that of men in Albania (World Bank, 2011). Therefore, campaigning on information, and the provision of assistance, and promotion of women's advocacy on land rights in local areas, are of great importance to reduce the friction between the customary rights derived from the Canun and the legal rights deriving from land ownership certificates.

2.4 Findings and Conclusions

The economic, institutional and agricultural land reforms of the 20th century in Albania gave rise to the establishment of new property rights which in most cases were followed, but produced friction among groups of society. The 1945 reform was the beginning phase of class struggle, when the large landowners were pressed to became members of cooperatives or obliged to have their lands expropriated and leave the villages to move toward more remote areas (Prifti, 2004). Within the first year of transition to a market economy in the early 1990s, the Albanian political class embraced rapid redistribution of land from state to private ownership. In the plain areas, agricultural land was redistributed to farmers according to an 'equally per capita' principle in most cases, creating conflicts between the pre-collectivization and the post-collectivization owners who gained land from the transitional reforms. However, in some parts of the country, land was distributed according to community consensus, respecting the pre-collectivization land boundaries. The land reform of 1991 has been, in many ways, one of the most important land reforms of the 20th century in Albania. It remains one of the most radical reforms in the former communist countries of Central and Eastern Europe, with the highest level of de-collectivization and individual privatization, including the distribution of state farms, giving modest attention to pre-1945 owner's claims (Deininger, 2003; World Bank, 2011).

The land reform of 1991 was also accompanied by (and created) clashes among groups of society, with serious consequences in terms of security of tenure. Studies by Zhllima and Imami (2012), Civici (2001), and Guri and Jouve (2003), have found that perceived tenure insecurity has been, and still is, common in rural Albania. Insecurity is lower for plots acquired through inheritance or other customary rights, where there are no post-collectivization/pre-1945 owners' claim conflicts, compared with plots acquired through state reform, in which such conflicts are commonplace. Furthermore, societal groups such as females have unequal access to and information on land rights.

Low security of land rights is an ongoing concern for the development of an agricultural sector in Albania. Investments are positively associated with the possession of plot titles and are negatively associated with prevalence of ownership conflict, and with plot distance from the home. Furthermore, low land security, together with other factors, such as the lack of capital flow in villages, the high prices of land, and a low willingness to sell land among villagers, have created a very weak land sale market. Perceived land rights insecurity negatively affects willingness to buy land. Recent surveys show that farmers prefer to purchase and invest in land plots that are based on ancestral rights, which are perceived as more secure than official titles that are not combined with ancestral rights.

Taking these findings into consideration, several land security enhancement measures should be considered. The current challenges for Albanian land policymakers remain the same as those of a decade ago: to reinforce land rights and re-establish a functional land policy framework to support a more active land market and more sustainable use of Albanian agricultural land. The Albanian government should achieve the finalization of the compensation process for 'pre-1945' owners, in order to diminish possible claims, and reduce the threat perceived by the post-collectivization farmers. Another step would be the finalization of agricultural land registration, and a review of the registration for areas with property rights problems and overlaps. The new practices of registration, and partial redistribution of land, should be accompanied with a bottom-up approach whereby community participation and transparency are the underlying principles.

Information campaigns are essential to raise the awareness of farmers to accomplish and support land registration and reduce the information gap between males and females.

Rental market activities are more frequent, and immune to land insecurity. Therefore, rental land markets seem to be a more active, helpful, and easily accessible instrument in consolidating land for agricultural purposes in areas of higher land fragmentation. While enhancing the functioning of the land sales market can be an extended process, and requires reforms and increased trust and awareness related to public institutions, the land rental market can be more easily promoted and developed. So the rental market should be the subject of land advisory programs and taxation fiscal incentives. The need for reviving the land rental market becomes more critical if farmers' rental agreements are to be eligible to benefit from national and EU pre-accession based subsidies. The rental market may also be helpful toward the majority of farms meeting the minimal size criteria described in the national and EU pre-accession based subsidy scheme. However, relying entirely on the rental market greatly constrains agricultural development, as rental activity does not promote long-term investments. This conclusion is supported by the previously mentioned data, which revealed that none of the farms renting land were being used for perennial crops. Moreover, it remains important to find out how land taxation could motivate land use, taking into consideration the weak role that the land market took in redistributing land.

There are still many important issues uncovered by researchers on the challenges created after the ending of land reform in Albania. There are no specific studies evaluating the influence of land rights security on farm productivity. The hypotheses that have resulted from the literature findings are that clearer land rights result in an increase of long-term interventions of farmers to the land, and, thereby, in an increase of land productivity.

Agricultural land administration problems in Albania may hinder the impact of EU pre-accession funding on farm competitiveness and rural development. The lack of clear documentation of land in some regions of Albania, in terms of location, possession or lease, may reduce the eligibility of many farmers. Further research may be oriented to assessing the possible obstacles on fund absorption by Albanian farms due to mismanagement of the land administration system.

Socioeconomic studies should also put into focus the gender differences in the perception of land rights, and the intra-family division of the land rights, in various areas of rural Albania. These analyses should scrutinize the gender patterns of land ownership in terms of use, sale, rent, and inheritance actions, and the main factors that affect negatively (or positively) gender-balanced property rights. Customary rights should also be investigated in order to reduce the gap between the formal rules and the de facto practices implemented in rural Albania. These findings would help policymakers to make adjustments to the institutional and legal base of land rights management and exchange practices in agricultural land, in order to defend the position of females in rural areas—whose roles are very important for maintaining a stable rural livelihood.

References

ARD (2007) *Land Tenure and Property Rights Regional Report. Volume 2.7: The Balkans and the Caucasus*. Produced for review by the US Agency for International Development (USAID) by ARD, Inc., Burlington, Vermont.

Bardhoshi, N. (2004) *E Drejta Kanunore dhe Shteti i së Drejtës [The Customary Right and the Legitimacy of the State]*. POLIS University, Tirana, Albania.

Binswanger, H.P., Deininger, K. and Feder, G. (1995) Power, distortions, revolt and reform in agricultural land relations. In: Behrman, J. and Srinivasan, T.N. (eds) *Handbook of Development Economics, Volume IIIA*. North Holland (Elsevier), Oxford, UK and Amsterdam, Netherlands, pp. 2661–2772.

Bouquet, E. and Colin, J.-P. (2009) L'etat, l'ejido et les droits fonciers. Ruptures et continuités du cadre institutionnel formel au Mexique. In: Colin, J.-P., Le Meur, P.-Y. and Léonard, E. (eds) *Les Politiques d'Enregistrement des Droits Fonciers: Du Cadre Légal aux Pratiques Locales*. Karthala, Paris, pp. 299–332.

Civici, A. (1997) Evolution des politiques économiques et agricoles de 1945–1995. *Options Méditerranéennes: Série B, Etudes et Recherches* No. 15, 15–50. Available at: http://ressources.ciheam.org/om/pdf/b15/CI970319.pdf (accessed 17 September 2012).

Civici, A. (2001) Evolution des politiques foncières et dynamique des espaces ruraux en Albanie [Evolution of the land policies and dynamics of the rural areas in Albania]. In: Jouve, A.-M. and Hervieu, B. (eds) *Terres Méditerranéennes. Le Morcellement, Richesse ou Danger?* Published for CIHEAM-IAMM, Montpellier, by Karthala, Paris, pp. 127–146.

Civici, A. (2003) Dé-colletivisation et réconstruction de l'agriculture Albanaise 1989–2002. Une transition spécifique? PhD thesis, Ecole Nationale Supérieure Agronomique de Montpellier, France.

Cungu, A. and Swinnen, F.-J. (1999) Albanian's radical agrarian reform. *Economic Development and Cultural Change* 47, 605–619.

de Janvry, A., Gordillo, G., Platteau, J.R. and Sadoulet, E. (eds) (2001) *Access to Land, Rural Poverty, and Public Action*. Oxford University Press, Oxford, UK and New York.

de Soto, H. (2001) *The Mystery of Capital: Why Capitalism Triumphs in the West and Fails Everywhere Else*. Black Swan, London.

de Waal, C. (2004) Post-socialist property rights and wrongs in Albania: an ethnography of agrarian change. *Conservation and Society* 2, 19–50.

Deininger, K. (2003) *Land Policies for Growth and Poverty Reduction. A World Bank Policy Research Report*. World Bank, Washington DC and Oxford University Press, New York.

Demsetz, H. (1967) Toward a theory of property rights. *The American Economic Review* 57, 347–359.

EBRD (1996) *Transition Report Update*. European Bank for Reconstruction and Development, London.

EC (2010) *Communication from the Commission to the European Parliament and the Council. Commission Opinion on Albania's Application for Membership of the European Union, {SEC(2010) 1335}, Brussels, 9.11. 2010, COM(2010) 680*. European Commission, Brussels. Available at: http://ec.europa.eu/enlargement/pdf/key_documents/2010/package/al_opinion_2010_en.pdf (accessed 29 August 2012).

EC (2011) *Commission Staff Working Paper: Albania 2011 Progress Report Accompanying the Document Communication from the Commission to the European Parliament and the Council, Enlargement Strategy and Main Challenges 2011–2012*. Available at: http://ec.europa.eu/enlargement/pdf/key_documents/2011/package/al_rapport_2011_en.pdf (accessed 29 August 2012).

Erebara, G. (2011) The demographic component in Albanian food and economic crisis of 1980s. Master's thesis, Department of Economic History, University of Lund, Sweden.

Feder, G. and Feeny, D. (1991) Land tenure and property rights: theory and implications for development policy. *World Bank Economic Review* 5, 135–153.

Feder, G. and Nishio, A. (1996) The benefits of land registration and titling: economic and social perspectives. Paper presented at: *International Conference of Land Tenure and Administration. Orlando, Florida in November 1996* [Later version available in *Land Use Policy* (1999) 15, 25–43].

Fishta, I. and Toçi, V. (1984) Ekonomia e shqipërisë në vilet e para të ndërtimit të söcialist 1944–1948. Tirana, 8 Nëntori.

Furubotn, E.G. and Richter, R. (1997) *Institutions and Economic Theory: The Contribution of the New Institutional Economics*. University of Michigan Press, Ann Arbor, Michigan.

Gambeta, V. (1999) *Pronesia mbi Token dhe Historia e saj [Ownership of Land and its History]*. Mesonjetorja e Pare, Tirana, Albania.

Giovarelli, R. and Bledsoe, D. (2001) *Land Reform in Eastern Europe: Western CIS, Transcaucuses, Balkans, and EU Accession Countries*. Prepared for FAO (Food and Agriculture Organization) by the Rural Development Institute (RDI), Seattle, Washington.

Guri, F. (2007) Study of the land market in the Albanian coastal area (case study of Durres-Kavaje area): the importance of tourism and the land management policies. PhD Thesis, Agricultural University of Tirana, Albania.

Guri, F. and Jouve, A.-M. (2003) Formation d'un marché foncier sur le littoral albanais. In: Elloumi, M. and Jouve, A.-M. (eds) *Bouleversements Fonciers en Méditerranée: Des Agricultures sous le Choc de l'Urbanisation et des Privatisations*. Published for CIHEAM-IAMM, Montpellier, by Karthala, Paris, pp. 213–231.

Guri, F. and A.-M. Jouve (2009) Réorganisations foncières et transmission de la propriété dans les exploitations agricoles du littoral albanais. Vers un abandon de l'agriculture? In: Jouve A.-M. (ed.) *Transitions Foncières dans les Balkans: Roumanie, Albanie, Grèce*. CIHEAM-IAMM, Montpellier, France. *Options Méditerranéennes: Série A, Séminaires Méditerranéens* No. 82, pp. 39–54.

ILD (2008) *Diagnosis for Extralegality in Albania*. Institute of Liberty and Democracy, Lima, Peru. Available at: http://lgi.osi.hu/cimg/0/0/9/1/3/Albania_FR_1.pdf (accessed 30 August 2012).

INSTAT (1991) *Statistical Yearbook 1990*. Instituti i Statistikës (INSTAT), Tirana, Albania.

Kareco, T. (2008) Monarkia – organizim shtetëror që garantoi reforma demokratike ekonomiko – shoqërore: reforma agrare demokratike nga një mon. *Tema* 14(2507), 20–21.

Kodderitzsch, S. (1999) *Reforms in Albanian Agriculture: Assessing a Sector in Transition*. Technical Paper, World Bank, Washington, DC.

Lastarria-Cornhiel, S. and Wheeler, R. (1998) *Gender, Ethnicity and Landed Property in Albania*. Working Paper No. 18, Land Tenure Center, University of Wisconsin, Madison, Wisconsin.

Lemel, H. (2000) *Rural Property and Economy in Post-Communist Albania*. Berghahn Books, Oxford, UK and New York.

Lerman, Z. (2000) Agriculture in transition economies: from common heritage to divergence. *Agricultural Economics* 1481(2000), 1–20.

Lorenzoni, G. (1930) *La Questione Agraria Albanese*. Gius Laterza and Figli Edizione, Bari, Italy.

Lusho, S. and Papa, D. (1998) *Land Fragmentation and Consolidation in Albania*. Working Paper No. 25, Albania Series, Land Tenure Center, University of Wisconsin, Madison, Wisconsin. Available at: http://ageconsearch.umn.edu/bitstream/12792/1/ltcwp25.pdf (accessed 30 August 2012).

MAFCP (2003) *Agriculture Annual Report 2003*. Ministry of Agriculture, Food and Consumer Protection, Tirana, Albania.

MAFCP (2007a) *Agriculture and Food Sector Strategy 2007–2013*. Ministry of Agriculture, Food and Consumer Protection, Tirana, Albania.

MAFCP (2007b) *Inter-sectoral Rural Development Strategy of Albania, ISRDSA, 2007–2013*. Ministry of Agriculture, Food and Consumer Protection, Tirana, Albania. Available at: http://www.dsdc.gov.al/dsdc/pub/rural_development_strategy_173_1.pdf (accessed 29 August 2012).

MAFCP (2010) *Statistical Yearbook 2009*. Ministry of Agriculture, Food and Consumer Protection, Tirana, Albania.

Majko, P. (2012) Speech of Pandely Majko, MP of the Socialist Party. *Koja Jone*, 1 Shtator [September] 2012. Tirana, Albania. Available at: http://kohajone.com/ (accessed 14 September 2012).

Moor, M. and Flyn, N. (1997) *Real Estate Professionals and the Immovable Property Market in Albania*. Working Paper No. 11, Albania Series, Land Tenure Center, University of Wisconsin, Madison, Wisconsin.

Morone, G. (1997) *Sviluppo Umano e Sostenibile in Albania, 1997–1999*. Ricerca cofinanziata MURST-CNR. Franco Angeli Ed., Milan, Italy.

OSCE (2004) *Legal Sector Report for Albania*. Organization for Security and Co-operation in Europe, Tirana Office, Tirana, Albania.

Pashko, G. (1993) Obstacles to economic reform in Albania. *Europe–Asia Studies* 45, 907–927.

Platteau, J.-P. (1996) The evolutionary theory of land rights as applied to sub-Saharan Africa: a critical assessment. *Development and Change* 27, 29–86.

Prifti, P. (2004) *Remote Albania: The Politics of Isolationism*. Onufri, Tirana, Albania.

Roselle, S. and Swinnen, J.F.M. (2004) Success and failure of a reform: insights from the transition of agriculture. *Journal of Economic Literature* 42, 404–456.

Roucek, J.-S. (1933) Economic conditions of Albania. *Economic Geography* 9, 256–264.

Sabates-Wheeler, R. and Waite, M. (2003) *Albania Country Brief: Property Rights and Land Market*. Land Tenure Center, University of Wisconsin, Madison, Wisconsin. Available at: http://www.nelson.wisc.edu/ltc/docs/albaniabrief.pdf (accessed 30 August 2012).

Sandstrom, P. and Sjoberg, O. (1991) Albanian economic performance: stagnation in the 1980s. *Soviet Studies* 43, 931–947.

Sikor, T. and Müller, D. (2009) Land fragmentation and cropland abandonment in Albania: implications for the roles of state and community in post-socialist land consolidation. *World Development* 37, 1411–1423.

Stahl, J. (2007) *The Political Ecology of Postsocialist Land Use Change: Case Studies from Three Villages in Southeastern Albania*. Landwirtschaftlich-Gärtnerischen Fakultät der Humboldt-Universität, Berlin.

Stahl, J., Sikor, T. and Dorondel, S. (2009) Transparence de l'administration foncière en Albanie et en Roumanie. In: Jouve, A.-M. (ed.) *Transitions Foncières dans les Balkans: Roumanie, Albanie, Grèce*. CIHEAM-IAMM, Montpellier, France. *Options Méditerranéennes: Série A, Séminaires Méditerranéens* No. 82, pp. 55–64.

Stanfield, D., Jazoj, A., Sherko, R. and Marquardt, M. (2002) *Institutional and Technological Strategies for the Management of Immovable Property Registration Systems: Lessons from the Albanian Land Market Action Plan*. Land Tenure Center, University of Wisconsin, Madison, Wisconsin.

Swinnen, J.F.M. and Mathijs, E. (1997) Agricultural privatisation, land reform and farm restructuring in Central and Eastern Europe: a comparative analysis. In: Swinnen, J.F.M., Buckwell, A. and Mathijs, E. (eds) *Agricultural Privatisation, Land Reform and Farm Restructuring in Central and Eastern Europe*. Ashgate, Aldershot, UK, pp. 333–373.

Swinnen, J., Vranken, L. and Stanley, V. (2006) *Emerging Challenges of Land Rental Markets – A Review of Available Evidence for the Europe and Central Asia Region*. Chief Economist's Regional Working Paper Series 1(4), World Bank, Washington, DC.

Vlora, E. (1973) *Kujtime 1885–1925*. Shtypshkronja Pegi, Tirana, Albania.

Voell, S. (2004) North Albanian customary law kanun as a habitus. Presented to: *8th Biennial EASA Conference, Face to Face: Connecting Distance and Proximity, Vienna, Austria, 8–12 September 2004*. European Association of Social Anthropologists, London.

Wehrmann, B. (2010) *Governance of Land Tenure in Eastern Europe and Commonwealth of Independent States (CIS)*. Land Tenure Working Paper No. 16. Produced for GTZ (Deutsche Gesellschaft für Internationale Zusammenarbeit), Federal Ministry of Economic Cooperation and Development and FAO (Food and Agriculture Organization). FAO, Rome. Available at: http://www.fao.org/docrep/012/al378e/al378e00.pdf (accessed 30 August 2012).

Wheeler, R. (1998) *Past and Present Land Tenure in Albania – Patrilineal, Patriarchal, Family Centered*. Working Paper No. 13, Albania Series, Land Tenure Center, University of Wisconsin, Madison, Wisconsin. Available at: http://landportal.info/sites/default/files/ltcwp13.pdf (accessed 30 August 2012).

Wheeler, R. and Laha, M. (1995) *Property Rights, Women and the Family in Contemporary Albania*. Internal report of project of the Land Tenure Center University in Albania, under subcontract to Development Alternatives, Inc. Financed by US Agency for International Development, BASIS IQC LAG-I-00-98-0026-0. Land Tenure Center, University of Wisconsin, Madison, Wisconsin.

World Bank (2006) *Status of Land Reform and Real Property Markets in Albania*. World Bank Office, Tirana, Albania.

World Bank (2007) *Albania: Strategic Policies for a More Competitive Agriculture Sector*. Report No. AAA18-AL, World Bank Sustainable Development Sector Unit, Europe and Central Asia, Tirana, Albania. Available at: http://siteresources.worldbank.org/INTALBANIA/Resources/Strategic_Policies_for_more_competitive_Agriculture_sector.pdf (accessed 3 September 2012).

World Bank (2011) *Governance in the Protection of Immovable Property Rights in Albania: A Continuing Challenge*. World Bank Issue Brief, Washington, DC.

Zhllima, E. and Imami, D. (2012) The Albanian land rights security perception and factors influencing it. *Albanian Journal of Agricultural Sciences* 11, 43–52.

Zhllima, E., Viaggi, D. and Müller, D. (2010) Property rights of land and its perception in rural part of central Albania. *New Medit* 9(3), 56–64. Available at: http://dev.iamb.it/v2/share/img_new_medit_articoli/311_56-64zhllima.pdf (accessed 30 August 2012).

3 Agricultural Extension in Transition

Halina M. Zaleski
University of Hawai'i at Mānoa, USA

3.1 Introduction

Agricultural extension is a system that links the development of agricultural information and technology, and its application by agricultural producers. Advice has been provided to farmers for thousands of years, with the earliest known example being a tablet from Mesopotamia from 1800 BC, which contained information on the control of rats, and on the watering of crops (Ahmed, 1982, as quoted in Bne Saad, 1990, and Jones and Garforth, 1997). During the Han dynasty (AD 25–200) in China, agricultural research was being conducted and information was being disseminated to landowners (Jones and Garforth, 1997). Scientific research, including agricultural research, greatly expanded following the Renaissance in Europe, and landowner interest in the application of scientific farming methods developed. Science also benefited from the accumulated knowledge of farmers; Darwin, for example, corresponded extensively with plant and animal breeders while developing the theory of evolution (Stone, 1980).

Rapid development of agricultural science, and the establishment of agricultural societies and agricultural schools, were seen in mid-19th century Europe and North America (Jones and Garforth, 1997). In the USA land-grant colleges were established in 1862 by the Morrill Act to provide education in agriculture and the mechanical arts (USDA NIFA, 2011). Colleges of veterinary medicine and agriculture were established in the 19th century throughout Europe, including Eastern Europe. Some examples include the University of Veterinary Medicine in Budapest, Hungary in 1787 (Kozári, 2012), the University of Agronomic Science and Veterinary Medicine in Bucharest, Romania in 1852 (University of Agronomic Science and Veterinary Medicine, 2012), and the Sofia University St. Kliment Ohridski in Sofia, Bulgaria in 1888 (Sofia University St. Kliment Ohridski, 2012).

The first steps toward organized extension systems came with the hiring of itinerant agricultural lecturers in the northeastern USA in the 1820s, in France in the 1830s, and in German states in the 1850s (Jones and Garforth, 1997). In Europe, these teachers reported to central agricultural societies, but by the 1860s, the majority of their funding came from state funds. By the end of the century, several countries had made the agricultural lecturers a part of the civil service (Jones and Garforth, 1997). In the UK, a board of agriculture was established at the end of the 1880s, and funding was provided for local governments to provide agricultural extension. The local governments sometimes hired their own extension officers, but many used instructors from agricultural departments in colleges and universities (Jones and Garforth, 1997).

The development of agricultural extension in the USA started by following a similar path, through farmers' institutes using the expertise of land-grant universities, but the Smith-Lever Act of 1914 placed agricultural extension in the land-grant system rather than under local governments (Jones and Garforth, 1997; USDA NIFA, 2011).

Throughout the development of public agricultural extension, the major players have been farmer organizations, agricultural universities and government agencies, but these have been differently integrated and balanced in different countries. The pros and cons of current models for organizing and delivering extension services may be affected by the developmental or transitional stage of a country. Developing countries, transitional countries, and developed countries in North America and Europe—with their differing needs and resources—may not all find the same system of extension to be the most effective. In all cases, there is a need to integrate extension and research, and to develop the coordination and communication that are vital in providing effective support to farmers.

In the US model, the primary responsibility for agricultural extension, research, and teaching rests with the land-grant universities, although government institutes also conduct research and some outreach activities, often related to their regulatory functions. Funding for agricultural research, teaching, and extension is provided by the government and by private sources, including farmer organizations. On one hand, the advantages of the land-grant system include separation of extension from regulatory functions, integration of extension with research and teaching, and the involvement of students in research and sometimes extension. On the other hand, extension, research and teaching sometimes find themselves competing for limited resources. As state and federal support of universities covers less and less of their costs, and student tuition, private support, and competitive research grants provide a larger share of university budgets, priorities can shift.

In a common European model, the responsibility for extension rests with the ministry of agriculture, as in Albania, while the universities are responsible for teaching, and research is conducted by both government institutes and universities. Advantages of housing extension in the ministry of agriculture include less competition for resources among extension, research, and teaching, and the development of government programs based on farmer needs. Applied research may also be housed in the ministry, allowing for integration of research and extension. However, separating regulatory and extension functions may be difficult, and farmers may be less likely to trust an agency that is seen as both regulatory and advisory.

A third model for the organization and delivery of extension places the primary responsibility with farmer organizations. Agricultural societies and farmer institutes often played important roles in the development of agricultural extension, and in some countries they continue to play an important role. Examples of this system are provided by chambers of agriculture in European countries such as Poland. The chambers are created by government statute, but are democratically controlled by farmer members. Funding may be provided through ministry of agriculture support, member dues, member services such as agricultural loans, or investments. There are similarities between this model and the commodity boards supported by check-off funds in the USA, for example, the National Pork Board which promotes pork, funds research, and provides producer education. Unlike chambers of agriculture, however, the US commodity boards are not permitted to lobby government or advocate for farmer priorities. The advantages of producer-controlled extension include responsiveness to farmer needs and priorities, and an understanding of practical concerns; but in this system, extension is not closely integrated with agricultural research and tending. The success of the chambers depends on educated agricultural producer members.

One of the important differences among extension systems is in the approach to the role of farmers. Sometimes, extension systems may consider farmers as the recipients of extension services, and be organized on a top-down model, with researchers developing technologies, and extension workers delivering the results to producers. The most extreme example of this was provided by 'forced extension' in Japan, where in 1910 the law required farmers to adopt the recommendations of village society extension workers (Jones and Garforth, 1997). A top-down model may be preferred in developing and transitional countries, especially if farmers are poorly educated and there is limited experience with democratic institutions. Other extension systems may consider farmers as contributors or partners, especially in developing priorities for research and extension, but potentially also in the development of technologies and in communication with fellow producers. Such systems are more likely to work with active farmer organizations, but may also include advisory committees to public extension institutions. In systems such as those providing extension through chambers of agriculture, farmers are not just partners, but leaders, in extension. For such a system to work, the farmers must be able to work effectively with the ministry of agriculture and with university researchers.

In transitional countries, the balance among farmer organizations, agricultural universities, and government agencies in the organization and delivery of extension may shift as the country develops its economy and democratic institutions, and as effective farm leaders emerge. Here, Albania provides an example of a country in transition from a centrally planned economy and institutions to a market-driven and democratic system. This chapter will include a discussion and comparison of extension in Albania with extension systems in other eastern European countries.

Albania's transition from centrally planned to private market-driven agriculture, and the development of institutions to support agriculture, are particularly interesting because of the country's unique history. Throughout most of its history, Albania had been subject to other countries, most recently the Ottoman Empire until 1912 and the Axis countries during World War II (Zickel and Iwaskiw, 1994). Because of these limited stretches of national independence, Albania had developed no agricultural research or extension. Under communist rule after the war, Albania developed institutions to support centrally planned agriculture. During the transition to a market economy, Albania has redesigned and redeveloped institutions, including an extension system managed by the Ministry of Agriculture, to support market-driven agriculture. This process has been supported by a variety of projects led by non-governmental organizations (NGOs), and European and American agencies, with somewhat different approaches and emphases. This patchwork has created some gaps and challenges, even while agricultural production has made significant gains.

During the transition, Albanian farms have faced major challenges, as they are typically small with low capital investment and limited access to credit. Farms often consist of unconnected parcels. With the dissolution of collective farms, the former employees received parcels of land and became independent farmers. These farmers generally had little or no previous experience in production and business management, and faced the challenge of developing the knowledge and skills needed to operate successful farms. Many are subsistence farmers, but commercial farming is growing. The growth of agricultural production in recent years is a compliment to the farmers and the support they have received. The marketing infrastructure is growing too. While increasing commercial production is important to meet national needs and generate exports, providing effective support to subsistence farms may be a key component of rural development policy, especially at a time when the cities are already challenged by a rapid influx of people from the rural areas.

3.2 Historical Background of Extension in Albania and Eastern Europe

3.2.1 Pre-war

Although Albania declared independence from the Ottoman Empire in 1912, independence did not lead to changes in agriculture (Zickel and Iwaskiw, 1994). Due to the political influence of large landowners, Albania was the only Balkan country where there was no land reform between the two World Wars (Zickel and Iwaskiw, 1994). Land ownership and agriculture before 1945 were governed by semi-feudal relationships, with seven families owning on average 2079 ha, 4713 families owning on average 19 ha, and 128,961 peasants with holdings averaging 1.8 ha; 14% of the rural population was landless (Daku, 1997). About 96% of the arable land was under crops, and about 4% in orchards (Daku, 1977). The peasants generally practiced subsistence agriculture, often with ancient tools such as wooden plows (Zickel and Iwaskiw, 1994). Much of the best farmland was covered by coastal marshland (Zickel and Iwaskiw, 1994).

During this period, educational opportunities were limited. Before independence, there were no schools operating in the Albanian language. Although the Albanian government opened elementary and secondary schools, in 1946 about 85% of the population was illiterate. There were no universities in the country (Zickel and Iwaskiw, 1994), and no agricultural research or extension. One exception was the establishment in the 1930s of a large experimental farm (which later became the first state farm) by Italian colonists (Zickel and Iwaskiw, 1994).

Like Albania, Poland was partitioned among Austria, Germany and Russia, and did not regain independence until 1918. Unlike Albania, however, Poland had developed strong national institutions during long periods of independence dating back to the 10th century. Agricultural education, extension and research in Poland already had a long tradition. Serfs had been emancipated and feudal relationships ended in Poland during the first half of the 19th century, and many peasants owned their land (Davies, 1982). Land reforms from 1919 to 1938 passed 2,655,000 ha from large state, private, and church estates into peasant hands (Davies, 1982). In many ways, Poland and Albania anchor two ends of a spectrum, with other Eastern European countries falling somewhere between these two extremes.

Farmer organizations were well established in Poland, and agricultural cooperatives in Poland date to the 1860s. The cooperatives included credit banks making short-term loans to peasants, and *rolniki* or farmers' circles enabling bulk purchases of farm supplies and effective marketing of farm products by members (Davies, 1982). As in many other countries in the 19th century where agricultural societies played important roles in the development of extension (Jones and Garforth, 1997), the *rolniki* in Poland also contributed to rural education, and to the improvement of farming methods (Davies, 1982). Agricultural extension in Poland was formally established in 1883, when a paid speaker was employed to give educational lectures and conduct farm visits. By 1908, agricultural companies were employing agricultural instructors. Starting in 1918, self-governing district agricultural chambers were established in Poland, and provided extension education as well as representing farmer needs to the government. Unlike Albania, Poland was one of the few former Eastern Bloc countries with a relatively well-developed extension system and structures in place to disseminate agricultural progress (Kania, 2010). Although agricultural teaching and research were well established in Poland's universities, agricultural extension developed as a separate system.

While organized agricultural extension programs were established in many Western European countries and in Poland, there was no organized extension in most Eastern European countries. Even without a formal extension system, agriculturists within the universities provided information to farmers. In Romania, for example, Ion Ionescu de la Brad (1818–1891), a prominent agronomist, was active in agrarian reform, agricultural education, and economic and statistical research (Ion Ionescu de la Brad University of Agricultural Sciences and Veterinary Medicine Iaşi, 2012). In the 19th century agricultural teaching, research and the development of new technologies were established in universities in most European countries; including in Eastern European countries such as Poland, Hungary, Bulgaria, and Romania. Albania was an exception in not having any universities.

3.2.2 1945–1990

Following World War II, communist governments instituted central planning in Eastern European countries. The agricultural landscape was changed by land reform, by the expropriation of the land of large landowners, and by pressure to create cooperative or state farms. Support continued for teaching and research in agricultural universities. In post-World War II Albania, state and collective farms were created, the available arable land was almost doubled by marsh reclamation, and land was improved through desalination, irrigation, and fertilization. Universities were founded. Albania was closed to foreign assistance from the 1980s to 1991, but a good base was created for future development. In other countries, changes in agriculture were less extensive, and there was more engagement with the rest of the world.

Immediately after World War II, the Albanian government was reorganized under the communist party. In 1944, the provisional government expropriated all foreign-held lands (Zickel and Iwaskiw, 1994). The government launched Albania's first agricultural reform in 1945, confiscating land held by absentee landlords and non-farmers, and redistributing about half of the arable land to the peasantry, although the state became the sole proprietor of the land (Zickel and Iwaskiw, 1994; Daku, 1997). Forest and pasture land was nationalized. Agricultural debts were annulled, and access was provided to low-cost irrigation water (Zickel and Iwaskiw, 1994). By 1946, peasants were cultivating more land and producing more corn and wheat than ever before, and the specter of famine had receded (Zickel and Iwaskiw, 1994).

Another set of reforms was initiated by the Albanian government in 1946 (Daku, 1997). Small peasant holdings were pooled to create cooperative farms, resulting by the 1950s in the incorporation of most of these small holdings into cooperatives (World Bank/ European Community, 1992, quoted in Daku, 1997). Large tracts of land held by religious groups, foreign companies and large landowners were converted into state farms (Daku, 1997). The state farms were provided with machinery and equipment, and were eligible for state subsidies (Zickel and Iwaskiw, 1994; Daku, 1997). State farm workers earned fixed salaries (Zickel and Iwaskiw, 1994; Daku, 1997). In 1967, First Secretary Enver Hoxha proclaimed that all of Albania's private farmland had been collectivized, and that the only legal forms of agricultural production were by state farms, cooperative farms, and personal plots granted to members of cooperative farms and limited to 200 m^2 (Zickel and Iwaskiw, 1994).

Starting in the late 1970s, Albania developed a policy of self-reliance and tried to create an independent economy. The government no longer sought foreign aid, credits or investment (Zickel and Iwaskiw, 1994). With a goal of food self-sufficiency, the number of state farms was gradually increased, primarily by developing new farms on reclaimed land (Daku, 1997), but also by amalgamating small cooperative farms (Zickel and Iwaskiw, 1994); the number of state farms leveled off by 1980 (World Bank/ European Community, 1992, quoted in Daku, 1997). The government also took gradual steps to reorganize cooperative farms into 'higher-type' farms that more closely resembled state farms (Zickel and Iwaskiw, 1994). Collectivization of livestock was targeted in a campaign in the late 1970s and early 1980s, but mismanagement resulted in the large-scale slaughter of livestock and production shortfalls (Zickel and Iwaskiw, 1994). As a result, increasing animal husbandry became a focus of the *Eighth Five Year Plan* from 1986 to 1990 (Zickel and Iwaskiw, 1994). During the period 1985–1989, there were 550 farms with an average size of 1060 ha (Kodderitzsch, 1999).

By the second half of the 1980s, a shift of government funding from agriculture to industry, and rapid population growth, were resulting in food shortages (Zickel and Iwaskiw, 1994). The trend to consolidation of farms was reversed with the breakup of some larger cooperatives (Daku, 1997). In an attempt to alleviate dire food shortages, in 1990 the regime removed the 200 m^2 limitation on the size of the personal plots of collective farm members (Zickel and Iwaskiw, 1994). In the same year, the government decided to allow private production of cattle, and collective farms transferred part of their stock to their members. Members of collective farms in mountain areas were to receive an additional 0.2 ha of private land to raise livestock (Zickel and Iwaskiw, 1994). In 1991, 216 state farms held 24% of the arable land (Zickel and Iwaskiw, 1994), cooperative farms held 72%, and 4% was in small individual plots (Daku, 1997).The area of agricultural land per person active in the agricultural sector in Albania was 1.5 ha, compared with 4.4 ha in Poland, and 10.5 ha in Bulgaria (Daku, 1997).

Under the centrally planned agricultural system, institutions for agricultural research and education were established, but no separate extension system was developed (Qafoku, 1995). Technologies were developed by research institutes under the Ministry of Agriculture, and directly communicated by agricultural specialists to the state and cooperative farms (Dishnica, 2007). For livestock, for example, the Husbandry Institute conducted breeding programs and the Bull Station and Tirana Biologic Materials Production Station supported the national artificial insemination program in bringing highly productive animal genetics on to Albanian farms (Tahiraj, 2002). Agriculture High Schools trained specialists, for example in animal husbandry, who worked on state and cooperative farms. These schools also provided the required training for farmers working with livestock (Tahiraj, 2002). The Higher Agricultural Institute (later renamed the Agricultural University of Tirana) was established at Tirana in 1951 (Agricultural University of Tirana, 2012). The Institute was composed of four faculties: Agronomy, Agrarian Economy, Forestry Engineering and Veterinary Medicine; by 1990, it had grown to 6000 students and 480 professors (Agricultural University at Tirana, 2012).

Although Albania did not achieve its goal of self-sufficiency during the period of central planning, agriculture had been transformed and the stage set for subsequent rapid development. As already mentioned, the control of the large landowners had been ended. Agriculture had been greatly expanded and the number of arable acres almost doubled, from 391,000 to 704,000 ha, through land reclamation and improvement programs (World Bank/European Community, 1992, quoted in Daku, 1997; Daku, 1997). Land had been reclaimed by draining the coastal marshes, and improved through irrigation, desalination, terracing, and fertilization (Zickel and Iwaskiw, 1994; Britannica Online, 2012). Livestock had been improved through the use of modern technologies such as artificial insemination and the introduction of more productive genetics. Institutions had been developed for agricultural research and education, education greatly expanded, and the first universities founded.

In Poland, in contrast, under the communist government after World War II attempts at collectivization of land were short-lived. Land reform in 1946 had redistributed most of the remaining large estates in eastern Poland among the peasantry (Davies, 1982). A campaign of forced collectivization into State Agricultural Enterprises and voluntary production cooperatives was carried out from 1950 to 1955, but its progress was slow and was reversed in 1956 to 1957 by decollectivization and the return of land to private ownership by peasants (Davies, 1982). Most land in Poland was in private hands; more than 75% of arable land was privately owned by more than 2 million private agricultural farms. Agricultural organization was left to cooperatives (Davies, 1982). Unlike Albania, Poland did not need to redistribute agricultural land from cooperative farms during the transition to a market economy (Kania, 2010).

Agricultural education, research and extension continued in Poland, but with more government involvement. In 1947, Poland dissolved the agricultural chambers and moved their activities into peasant self-help unions. In 1970, regional agricultural research centers were established to modernize agriculture through research and specialized extension. Further reorganization in 1975 transformed the regional centers into provincial centers of agricultural progress, which included the state farms, and which were responsible for professional development and agricultural services (Kania, 2010). Unlike Albania, Poland had a formal extension system in place during the communist years.

3.3 Extension in Transitional Countries

3.3.1 Albania 1991 to present

In 1991, the government of Albania abandoned central planning, gradually adopted market reforms for agricultural production, abolished fixed pricing, and rescinded restrictions on foreign aid and investment (Dishnica, 2007).

In July 1991, the government redistributed—without requiring payment—all the land in cooperative farms to those who had worked on them (Zickel and Iwaskiw, 1994; Daku, 1997; Dishnica, 1997). Other qualifying residents of villages attached to cooperative farms obtained up to 0.4 ha. The National Land Commission, chaired by the Minister of Agriculture, oversaw the work of district and village land commissions, which demarcated and distributed the land (Zickel and Iwaskiw, 1994). Property could be inherited, but not sold or leased, making consolidation of tiny landholdings difficult, and limiting the use of land as collateral for credit (Zickel and Iwaskiw, 1994). In 1992 and 1993, the state farms were similarly redistributed (Daku, 1997). From 1991 to 1994, 531,000 ha was distributed to 460,000 farms, each consisting of three to seven plots, including the land under the house, and operating on a subsistence or semi-commercial level (Daku, 1997). Farms averaged 1.2 ha/family, 0.7 cows/family, and 7 head of sheep and goats/family (Tahiraj, 2002). Pasture lands were not redistributed at this time, but were managed by the Ministry of Agriculture, which charged users a seasonal fee to graze their livestock on eight district pastures (Zickel and Iwaskiw, 1994). Breeding centers for cattle and other livestock were privatized and their genetic resources were no longer available to small livestock producers (Tahiraj, 2002). In 1998, almost 50% of the population was active in agriculture.

Albania also changed its policies limiting cooperation with other nations. In 1991, the country joined the World Bank (World Bank, 2007). It also became a member of the International Development Association (IDA), International Bank for Reconstruction and Development (IBRD), the International Finance Corporation (IFC), and the Multilateral Investment Guarantee Agency (MIGA). In 2003, Albania entered into negotiations for a Stabilization and Association Agreement with the European Union (EU). This agreement was signed in 2006, and has furthered political and economic reform and provided financial assistance with the goal of further integration into the EU. In 2008, Albania graduated from IDA concessional flows to IBRD lending. The World Bank lent Albania US$60 million toward agriculture, environment, and community development in 2007 (World Bank, 2007).

Many foreign players and projects have been involved in the development of agricultural institutions in Albania. In the 1990s, the Support for Agriculture Restructuring in Albania (SARA) project, financed by the United States Agency for International Development (USAID), focused on strengthening policy analysis capacity and restructuring agricultural research at the Ministry of Agriculture, and restructuring the two agricultural economics departments at the Agricultural University at Tirana. During the same time period, an Extension Project and Alliance for Vocational Agricultural Training in the Albanian Region (AVATAR) project, both financed by the EU, worked to establish an extension service and to train extension specialists. Other projects financed by different European countries supported the restructuring of research, extension, and vocational education during the 1990s (Daku, 1997).

In addition to strengthening the Ministry of Agriculture, Food and Consumer Protection (MAFCP) and the Agricultural University at Tirana, international projects have directly addressed agricultural challenges. Land O'Lakes carried out major USAID-funded development projects, such as the Dairy Campaign Development from 1993 to 1999, and the Livestock Service Centers Development from 2000 to 2003. In 1993 and 1994, Land O'Lakes conducted a broad extension effort reaching 3800 women dairy producers to improve cow health, and the quality and quantity of milk produced. In the following two years, the extension program was expanded to reach an additional 1200 women (USAID, 2000). Livestock service centers were developed to provide farmer customers with both supplies and service. The International Fertilizer Development Centre (IFDA), funded by USAID, conducted projects in Albania starting in 1992, with a focus on the fertilizer market. Following a survey of the members of the Albanian Fertilizer and Agri-Business Dealers' Association (AFADA) in 1998, a two year project was conducted to train AFADA dealers in extension methods to create an alternative to the government extension system (Androulidikas *et al.*, 2002). With the support of the Irish League of

Credit Unions, the first credit unions were registered in Albania in 1997 (USAID, 2000). The World Bank Albania Agricultural Services Project from 2001 to 2008 addressed agricultural production, trade and land constraints in order to increase smallholder incomes (World Bank, 2001). More recently, USAID projects have focused on the supply and marketing of high-value agricultural products, such as watermelons, apples, mandarins, and olives (USAID Albania, 2012).

In recent years, with continuing democratic and market reforms, Albania has been maintaining a relatively stable market economy with a modest annual agricultural growth of 2.5%, but growth has been as high as 10% for vegetables and fruits (Dishnica, 2007). Even with annual agricultural growth, however, Albania is still heavily reliant on imported products. Productivity of agriculture is low, with the agricultural sector employing about half of the workforce, but contributing only about a fifth of the GDP (gross domestic product) (CIA, 2012). Entry into the EU will challenge the competitiveness of Albanian agriculture. Meanwhile, the economy has gradually shifted toward services and construction, with less reliance on agriculture (23% of GDP in 2007; Dishnica, 2007) and industry (World Bank, 2007). Aid projects continue to be important in assisting development in Albania. The World Bank Status of Projects in Execution for FY10 (World Bank, 2010) lists 15 ongoing development projects in Albania. USAID continues to fund projects, such as the 2009 Albanian Agricultural Competitiveness Project (Development Alternatives, Inc., 2009).

Extension in the Ministry of Agriculture, Food and Consumer Protection

The Ministry of Agriculture, Food, and Consumer Protection (MAFCP) started the 1990s conducting research and providing farmer support through 18 district agricultural research institutes (Dishnica, 2007). These institutes had 342 researchers and extended resources, including facilities, and plant and animal seedstock (Dishnica, 2007). To support newly independent farmers, in 1993 the Extension Service was started with a pilot project financed by the EU that covered six districts (EBRD and FAO, 2007). The pilot was expanded in the following years to cover 18 districts, including olive-producing districts. In 1996, the Extension Service was incorporated into the Ministry's Science Directorate, which became the Extension and Science Directorate, separating extension from the Ministry's regulatory and monitoring functions (Dishnica, 2007). The initiative was temporarily halted in 1997 during a time of social turmoil, following the collapse of a financial gambling pyramid scheme (EBRD and FAO, 2007).

Starting in 1997, restructuring of research reduced the number of institutes to nine, of which five were in Tirana and four in other regions. These institutes included research, economic and administrative units, and laboratories, experiment stations, libraries and production units (Dishnica, 2007). Projects financed by the EU, and the Dutch and US governments, enabled the restructuring, as well as the upgrading of laboratory equipment and other resources (Dishnica, 2007).

A further major restructuring of extension and scientific research occurred in 2006. The existing nine institutes were reorganized into five regional Agricultural Technology Transfer Centers (ATTCs) and the Food Safety and Veterinary Institute in Tirana. The ATTCs and other extension activities were consolidated under the Ministry's Directorate of Extension Service, Science and Agricultural Information (Dishnica, 2007). The new centers are primarily focused on providing service and technology transfer, while responsibility for basic and original research was shifted to the Agricultural University at Tirana (Dishnica, 2007). The ATTCs, however, continue to have excellent facilities for applied research, and laboratories for soil and other testing. Considerable practical expertise appears to be available at the ATTCs in some areas, although much of the information available to extension is based on translations of material from other countries. Such information is useful in informing production management, but locally developed information is needed in areas such as marketing.

Information developed at the ATTCs is transferred through 12 regional Departments of Agriculture, Food and Consumer Protection (DAFCP), which house specialists in horticulture, livestock and economics, and provide regional coordination and training of agents. Regional and district offices organize workshops and provide information to farmers. The 12 DAFCP support 250 extension agents working through 120 district Agricultural Information Centers (AIC), where farmers and farmer groups can seek technical assistance and consultation. The AIC often have one agent each, and may or may not have computer and Internet access. However, the AIC are dispersed throughout the country to provide easy access for small farmers (Dishnica, 2007). More recently, efforts are being made to increase the use of mass media to provide information to farmers and the general public (Dishnica, 2007). Information on a wide range of topics and issues is being provided, including the development of local organizations, small enterprises, environmental issues, rural infrastructure, and social programs (Dishnica, 2007).

The MAFCP collects data intended to inform and evaluate extension. Data are collected on requests for information and workshop attendance, so extension activities can be designed to meet demand. It is not clear how the data is used to evaluate programs. Of direct interest to agricultural economics, detailed data are collected from 200 representative farms on crop and livestock production, and on income and expenses. The data appear to be a valuable but underutilized resource.

Research in the Agricultural University of Tirana

The Higher Agricultural Institute was renamed the Agricultural University of Tirana (abbreviated AUT) in 1991 (Agricultural University of Tirana, 2012). The University worked to develop its academic programs, and carried out academic restructuring from 1992 to 1994 with the assistance of the SARA project, financed by the US government (USAID), and by the Gesellschaft für Technische Zusammenarbeit (GTZ, now GIZ) project, financed by the German government (Agricultural University of Tirana, 2012). Relationships and exchanges were established with universities in Europe and the USA, leading to recognition of AUT diplomas. In the academic year 2005/6, AUT started organizing studies in conformity with the European 3+2+3 system (Bachelor, Master and PhD) for its growing instructional programs (Agricultural University of Tirana, 2012). Education is important for extension because, as Diamond (2000) points out, well-prepared extension officers are a key component of an effective extension system.

The responsibility for scientific research was transferred from the MAFCP to the University in 2006 (Dishnica, 2007), leading to an expansion of research and training at AUT. Agricultural research at AUT has been supported by both national and international projects, and is seen as providing a 'concrete contribution to the progress of Albanian agriculture' (Agricultural University of Tirana, 2012). Mechanisms to facilitate research, such as clarifying workload and evaluation expectations for any faculty conducting research, funding research through faculty research accounts within the university, and consideration of professional liability still need to be developed. Applied research and outreach activities have been carried out by individual faculty members as private consultants on personal time, but there is some interest in developing an integrated research and extension center at AUT. Better coordination between the MAFCP and the University will be important in developing research priorities, in making effective use of the resources within the MAFCP, and in avoiding duplication of extension and applied research efforts.

Role of farmer organizations

Farmer organizations are relatively new in Albania, and are often organized by commodity. Examples of commodity associations include the Albanian Dairy and Meat Association (ADAMA, formerly the Dairy Entrepreneurs National Association, formed in 1997) and the Livestock Entrepreneurs Albanian Association (LEAA) formed in 1999; these were both developed with assistance from Land O'Lakes (USAID, 2000). Farmer organizations vary in size and in the range of their activities.

The Albanian Agribusiness Council (Keshilli Agrobiznesit Shqiptar, KASH) was founded in 1999–2000 with the help of USAID, and held its First National Congress in

2001. KASH was formed to unify the various existing farmer and commodity organizations, to reach all sectors of agriculture, and to advocate for and promote agribusiness. Unlike the chambers of agriculture in other Eastern European countries, KASH is not a direct membership organization. At its Second National Congress in 2005, KASH developed a national advocacy program and invited the Albanian government and political parties to respond by describing their approach to agricultural development. By its Third National Congress in 2010, KASH consisted of 22 agribusiness trade associations. In addition to representing farmers on political questions, KASH has organized regional and national agribusiness fairs to help farmers market their products more widely, and has encouraged member involvement in conferences and training (Albanian Agribusiness Council, 2012).

KASH plays an important role in advocating for farmer needs in the political arena. It works with the government of Albania and is represented on various councils and committees, such as the Business Advisory Council, and the Fair Board. KASH participated in the negotiations of free trade agreements with other Balkan countries. The MAFCP provides office space to KASH's regional councils (Albanian Agribusiness Council, 2012). These regional councils work with the district governments just as the national council works with the national government. In addition to working with various levels of government in Albania, KASH sees importance in its role in advising international donors and projects to ensure that they effectively meet the needs of Albanian agriculture (Albanian Agribusiness Council, 2012).

3.3.2 Extension in transition: Eastern Europe

Each country in Eastern Europe is developing its own approach to extension, with variations in the relationships among extension, research and farmers, and in the roles of key institutions, including the ministry of agriculture, agricultural universities, private industry and farmer organizations. In Bulgaria, as in Albania, extension is housed in the Ministry of Agriculture. In Poland and in Hungary, farmer-controlled chambers of agriculture play an important role. Delivery of extension in Romania has fluctuated, shifting between central and local offices of the Ministry of Agriculture, and culminating in the development of chambers of agriculture.

Development aid is a common thread in Eastern Europe. The Program of Community aid to the countries of Central and Eastern Europe (PHARE) was designed to prepare countries for accession to the EU, starting in 1989 in Poland and Hungary, and expanding in 1994 to other Central and Eastern European Counties (CEEC), including Bulgaria and Romania, among others. In 2001, assistance to Balkan countries, including Albania, was transferred to the Community Assistance for Reconstruction, Development and Stability in the Balkans (CARDS) program (EC, 2012). Pre-accession programs include a Special Accession Programme for Agriculture and Rural Development (SAPARD) (EC, 2012). In 2012, European Union assistance to candidate countries was combined into the Instrument for Pre-accession Assistance (IPA) (EC, 2012). USAID has also supported many projects in Eastern Europe to assist in the transition from centrally planned to market economies, including completed projects in Poland (USAID Poland, 2012), Hungary (USAID Hungary, 2012), Bulgaria (USAID Bulgaria, 2012), and Romania (USAID Romania, 2012), and ongoing projects in Albania (USAID Albania, 2012).

Extension and research in Bulgaria

In Bulgaria, the majority of farmers are traditional farmers with limited resources, and growing a variety of crops, primarily for family consumption. Commercial farmers make up a minority of farms, but have more resources (land, animals, money, knowledge), generally specialize in particular areas of production (dairy, horticulture, fruticulture), and market regionally or nationally. There are also semi-commercial farmers that are in transition from traditional to commercial farming (Marinov, 2011). The average size of farms in 2007 was 24.3 ha (Eurostat, 2012).

In Bulgaria, the National Agricultural Advisory Service, NAAS, was created in 1999 within the Ministry of Agriculture and Food through the assistance of the PHARE Project, implemented by ADAS Consulting Ltd, England (National Agricultural Advisory Service, 2012). NAAS includes a Centre for Vocational Education, 28 Regional Agricultural Advisory Services (RAAS), and 105 extension officers (Marinov, 2011). It is currently (2007–2013) receiving support for rural development through the European Agricultural Fund for Rural Development (EAFRD), under measure 143 'Provision of advice and consultation in agriculture in Bulgaria and Romania' (Marinov, 2011). Extension staff receive information, monitoring of farmer advisees, and professional backup and training from regional offices of the Ministry of Agriculture and Food (Marinov, 2011).

In addition to government extension, private agronomists and consultants provide service to farmers. Nongovernmental consultancy includes donor-supported NGOs and international projects. Farmers' associations in Bulgaria include the Association of Agricultural Producers of Bulgaria, the Cooperative 'BioBulgaria', and other, similar organizations (Marinov, 2011). The Bulgarian Ministry of Agriculture and Food includes the Agricultural Academy (AA), and specialized, commodity-oriented (fruit production, vegetable production, grain production, sheep production, etc.) research institutes within the AA that develop new technologies, provide subject matter expertise, and provide analytical (soil, water, feed) laboratory services (Marinov, 2011). The Agricultural University of Plovdiv, founded in 1945, provides education (BS and MS) and research related to agriculture (Agricultural University Plovdiv, 2012). The Ministry funds scientific research at Plovdiv, based on priority topics, such as crop or animal husbandry, apiculture, or viticulture. Informal cooperation exists among the Agricultural University, the Regional Agricultural Advisory Services offices and the regional research institutes of the AA (Marinov, 2011). New technologies tend to benefit larger farmers more than small and medium farmers (Marinov, 2011). Agricultural research is also conducted in agricultural units in general universities (Standing Committee on Agricultural Research, 2012a).

Extension and research in Hungary

In Hungary, the primary agricultural products are meat, maize, wheat, sunflower seeds, potatoes, sugar beets and dairy products (US Department of State, 2012). The average farm size was 29 ha in 2007, but the majority of farms are small farms producing for their own consumption, while about 35% of farms are commercial producers (Eurostat, 2012). Hungary approached extension by encouraging private organizations (Bárczi *et al*., 2000). In 1994, the Hungarian Chamber of Agriculture (HCA) was founded by the Hungarian Parliament as a national body representing agricultural producers. The law was renewed in 2000 and 2011. The purpose of the Hungarian Chamber of Agriculture was to reduce the role of the state by having the Chamber act in the areas of development and EU subsidies, as well as represent the interests of agriculture to the Hungarian government. HCA operates a Farm Information Service with 202 advisors throughout Hungary (Hungarian Chamber of Agriculture, 2012).

The HCA is a federation of 20 territorial chambers. Members of the territorial chambers are involved in crop and livestock agriculture, forestry, fishing, hunting, and related processing, trading and service industries. Membership is voluntary. In 2012, the HCA had 11,000 members representing 60% of Hungary's agricultural output (Hungarian Chamber of Agriculture, 2012). The HCA connects farmers with the EU's Common Agricultural Policy (CAP) and rural development (Hungarian Chamber of Agriculture, 2012). It is involved with other agricultural chambers and international organizations, such as the EU's Committee of Professional Agricultural Organizations (Hungarian Chamber of Agriculture, 2012).

Agricultural research in Hungary is conducted by a well-developed system that includes government research institutions, including those under the Ministry of Agriculture and Rural Development, institutions of the Hungarian Academy of Sciences, and universities with an agricultural focus. Research is funded through government and private projects, including funding from the Ministry of Agriculture and Rural Development (Standing Committee on Agricultural Research, 2012b).

Extension and research in Poland

In Poland, most farms are small, usually under 5 ha, and grow products for home consumption, while their owners depend on off-farm income. Medium-size farms over 15 ha and large farms up to 200 ha are much more efficient, and produce 90% of the food that is consumed or exported (US Department of State, 2012). Average farm size in 2007 was 12.3 ha (Eurostat, 2012). While Poland is a net exporter of agricultural products, its trade surplus has been reduced by competition from more efficient producers in other EU countries (US Department of State, 2012). Within Europe, Poland is a leading producer of dairy, apples, potatoes and rye, and a significant producer of rapeseed, grains, hogs and cattle (US State Department, 2012).

In Poland in 1990, the Agricultural Advisory Service was reorganized into 49 Agricultural Advisory Offices under the Ministry of Agriculture and Rural Development (Kania, 2010). To increase the level of professional expertise, from 1990 to 1995, the Polish–American Extension Project partnered extension professionals from the USA with their counterparts in the Polish extension service (Place *et al.*, 2000). Polish extension also benefited from foreign aid from other countries, including Denmark and Ireland (Kania, 2010). The agricultural chambers were re-established in 1996 (Kania, 2010). Local Agricultural Chambers and the National Council of Agricultural Chambers are involved in providing information and support to farmers, preparing analyses of agricultural needs, promoting agricultural products, and representing farmer interests with government agencies and administration (National Council of Agricultural Chambers, 2012). Extension services are also available from private consultants, farm supply and marketing firms, and research institutes and universities (Kania, 2010).

Agricultural research in Poland is conducted by government research and development institutions, including those under the Ministry of Agriculture and Rural Development, institutions of the Polish Academy of Sciences, and ten universities that include an agricultural focus. Research is funded through government and private projects, including funding from the Ministry of Agriculture and Rural Development (Standing Committee on Agricultural Research, 2012c).

Extension and research in Romania

As in other Eastern European countries, most farms in Romania produce for their own consumption, while about 35% are commercial farms. Average farm size was 11 ha in 2007 (Eurostat, 2012). Principal agricultural products in Romania are maize, wheat, potatoes, oilseeds, vegetables, livestock, fish and forestry (US State Department, 2012). Extension in Romania has fluctuated between centralization and decentralization, and this provides an interesting contrast. In 1998, the National Agency for Agricultural Consulting (ANCA), a network of 41 County Centers for Agricultural Consulting (OJCA), and Local Centers for Agricultural Consulting (CLCA) were established with the support of an EU-financed project under the PHARE program. In 2001, the system was decentralized and the county and local centers were moved under district and local authorities. Then, in 2004, the system was recentralized under ANCA and the Ministry of Agriculture and Rural Development. Later, in 2010, the county and local centers were again decentralized to district and local authorities, and development started on a system of farmer-managed agricultural chambers (Stefanescu, 2011).

Agricultural research in Romania is conducted by national institutes, by research institutes within ministries, such as the Ministry of Agriculture and Rural Development, and by agricultural universities. There is limited agricultural research conducted by private entities, which are usually privatized former public institutes. Some research is funded by the Ministry, but most is funded through research projects (Standing Committee on Agricultural Research, 2012d).

3.3.3 Chambers of Agriculture

During the transition to market economies, chambers of agriculture were established in many Eastern European countries. Such self-governing chambers of agriculture have a long history in Europe, having been established by statute in Poland in 1918, and in France in 1920. They are organized democratically, and any farmer can join and vote. Support for the activities of the chambers often comes primarily from membership dues, as for the Slovak Agriculture and Food Chamber, or partially from government, as for French Assemblée Permanente des Chambres d'Agriculture, whose budget of 702 million euros comes from investments and agricultural loans (48%), local chambers and members (23%), the state and EU (20%), and other sources (9%). The roles of the chambers also vary from country to country, but always include advocacy for farmer needs and priorities. The chambers provide varying levels of service to their farmer and commodity organization members. These may include vocational training, certification programs, consulting, providing market information, putting on trade fairs, organizing workshops, and many other activities in support of members. In some countries, such as Hungary, the chambers of agriculture are responsible for running the agricultural extension program, in this case the national Farm Information Service.

The chambers of agriculture provide a strong and unified voice for farmers, not only in national policy, but also in defining farmer needs and priorities for extension and research. This European model differs from the US check-off supported commodity boards in two important ways. First, the chambers are very active in addressing farmer issues in the political arena, while American commodity boards are prohibited from engaging in any political activities. Second, the chambers have a united national organization, while the commodity boards function as separate organizations. Similarities include the involvement of all these organizations in helping to define priorities for, and in some cases deliver, agricultural extension and research.

The chambers of agriculture and other farm organizations are represented in the EU through the Committee of Professional Agricultural Organisations (COPA) and the General Confederation of Agricultural Cooperatives (COGECA). This umbrella organization facilitates communication and coordination among its members, and represents farmer interests in the evolution of the CAP (COPA–COGECA, 2012). Albanian farmer organizations are not yet partnered with COPA–COGECA.

3.4 Discussion

Development of agricultural extension services has been an important component of the transition from centrally planned to market-driven economies in Eastern Europe. Even though in Albania, as in other eastern European countries, the goal of integration into the EU is an important driver of structures, a range of approaches is evident. Extension may be housed primarily within the national ministry of agriculture, as in Bulgaria, Poland and Albania. Alternatively, responsibility for extension can be housed in self-governing but government-supported chambers of agriculture, as in Hungary. Extension in Romania has fluctuated between centralized and local units in the Ministry of Agriculture, but Romania is now moving to a chamber of agriculture model similar to Hungary's. Extension activities are also conducted by private agencies and organizations, and by universities.

In Albania, the transition from a centrally planned to an open, competitive economy involved extensive restructuring of some institutions, such as the Ministry of Agriculture and the Agricultural University of Tirana, and the creation of new agencies, such as the extension service within the Ministry, independent Livestock Service Centers, and the Albanian Agribusiness Council (KASH). Extensive assistance was provided by European and US aid agencies and investors to support these changes. With many different contributors to development, change has been rapid and transformative, but not always coordinated. Development of extension, in particular, has not followed a single model. Extension activities are carried out both by the public extension system housed in the Ministry of Agriculture and through private organizations and consultants. Faculty members at the Agricultural University of Tirana are also engaged in private consulting in support of agricultural enterprises.

Effective extension requires links to agricultural research. As with extension, research in Eastern Europe has developed in both governmental institutes and in agricultural universities. In either case, government agencies provide a significant source of funding. Unlike most European countries, in which research has long been established in agricultural universities, in Albania responsibility for research was transferred from the Ministry of Agriculture, Food and Consumer Protection to the Agricultural University of Tirana only in 2006. Although the Ministry continues to have the capacity and funds to perform applied research, the University has developed expertise and the potential for research through its faculty and graduate students. The Ministry has an opportunity to expand research relevant to agricultural enterprise by working with the University, but the mechanisms for funding such research are still being developed at the University, making it difficult for faculty to access research funds. Because of the lack of any mechanism in the University to set up research accounts for individual investigators, the approach envisioned by the University is to set up a private entity for research and extension. The Ministry, however, is not able to fund researchers or research groups involved in private enterprise.

Communication with farmers is another key component of effective extension. While extension is often thought of as communicating agricultural research from the university to the farmers, communication of priorities from the farmers to the university is equally important. Farmer organizations, such as the chambers of agriculture established in many European countries, often play a key role in developing and communicating priorities. The Albanian Agribusiness Council is a still a young organization and is an umbrella organization for commodity groups rather than a direct membership organization like the Chambers of Agriculture. To date, the Council has focused mainly in the political arena, but has the potential for increased involvement in consulting on services, such as research and extension, to benefit farmers. In addition to involvement in setting priorities, feedback from farmers is important in the evaluation of extension and research activities. For example, in a hypothetical instance where a researcher had a developed an improved variety of cabbage, and the farmers growing that variety found that it suffered far more insect damage than other varieties, a mechanism should be available for this feedback to reach the researcher to inform future work.

Some of the challenges that Albania faces in coordinating the efforts of its various institutions arise from the extensive international support, with projects from different nations working from different models of the relationships between research and extension. In Europe, research is seen as primarily the responsibility of the universities, while extension is primarily the responsibility of the ministry of agriculture, although it may be delegated to farmer-controlled chambers of agriculture. In contrast, in the USA, both research and extension are seen as primarily the responsibility of universities in the land-grant system. In most European countries, as in the USA, farmer organizations have a powerful voice and often a strong political presence, irrespective of any formal role in extension. Albania is following this trend in developing the Albanian Agribusiness Council, and a strong voice for farmers will help to ensure that their needs are met.

Other Eastern European countries have evolved differently. The development of institutions to support market-driven agriculture has been built on a relatively strong historical foundation and on existing agricultural research and educational institutions, and in Poland on an already established extension system. The scope of change was thus narrower and many changes were focused on improving competitiveness and marketing in order to smooth entry into the EU. Contrasting the development of agricultural institutions in Albania and in other Eastern European countries, it is immediately clear that, while Albania still has challenges to resolve, it has accomplished a great deal in a relatively short time. Even with more established institutions, Eastern European countries, such as Romania, are still experimenting with the most effective structure for their extension systems.

Extension is not a single unified system in most countries, as is shown by the examples discussed in this chapter. Components of extension include information services and research though the ministry of agriculture, research and outreach through agricultural universities, services provided by agricultural cooperatives and private suppliers, private consultants, and services provided by commodity groups and national farmer organizations. In the USA also, many players are involved in extension, including: the extension system housed in the land-grant universities; agencies within the US Department of Agriculture, such as the Natural Resource Conservation Service, which consult directly with farmers; government research agencies, such as the Agricultural Research Service; commodity boards supported by check-offs on product sales; cooperatives, such as Land O'Lakes/Purina Feeds; farm supply distributors; and private consultants. While the balance of effort among the various agencies is different in different countries, in each case, coordination rather than duplication of efforts is important for the effective use of resources. Farmer organizations, such as chambers of agriculture, are an important component in providing coordination and ensuring that farmer needs are met.

References

Agricultural University of Tirana (2012) Details available at: http://www.ubt.edu.al//index.php?lang=english (accessed 13 January 2012).

Agricultural University Plovdiv (2012) Details available at: http://www.au-plovdiv.bg/en/ (accessed 4 April 2012).

Ahmed, A. (1982) *The Role of the Information System in Development*. Studies Series, No. 314. Ministry of Culture and Information, Baghdad, Iraq.

Albanian Agribusiness Council (2012) Details available at: http://www.kash.org.al/ (accessed 31 May 2012).

Androulidakis, S., Freeman, C.C. III, Peqini, I., Agolli, S. and Korra, L. (2002) Private extension in Albania: impact of Albanian Fertilizers and Agri-Business Dealers Association on farmers adoption of technology. *Journal of International Agricultural and Extension Education* 9 (1) 47–55. Available at: http://www.aiaee.org/attachments/254_Androulidakis-Vol-9.1-6.pdf (accessed 30 August 2012).

Bárczi, J., Kozári, J. and Tóth, K. (2000) *Principles of Agricultural Extension Applicable in Hungary*. Gödöllő University of Agricultural Sciences, Hungary. Available at: ifsa.boku.ac.at/cms/fileadmin/Proceeding2000/2000_V_01_Barczi.pdf (accessed 12 April 2012).

Bne Saad, M.H.A.Al-h. (1990) An analysis of the needs and problems of Iraqi farm women: implications for agricultural extension services. Unpublished doctoral thesis, University College, Dublin, Ireland.

Britannica Online (2012) Albania: Agriculture, forestry and fishing. Available at: http://www.britannica.com/EBchecked/topic/12472/Albania/42629/Agriculture-forestry-and-fishing (accessed 27 March 2012).

CIA (2012) The World Factbook. Europe: Albania. Central Intelligence Agency, Washington, DC. Available at: https://www.cia.gov/library/publications/the-world-factbook/geos/al.html (accessed 18 March 2012).

COPA–COGECA (2012) Committee of Professional Agricultural Organisations–General Confederation of Agricultural Cooperatives. Details available at: http://www.copa-cogeca.be/Main.aspx?page=HomePage&lang=en (accessed 10 April 2012).

Daku, L. (1997) Investing in agricultural extension: the case of Albania. Research project submitted to the Faculty of the Virginia Polytechnic Institute and State University in partial fulfillment of the requirements for the degree of Master of Science in Agricultural and Applied Economics. MSc thesis, Virginia Polytechnic Institute and State University, Virginia. Available at: http://scholar.lib.vt.edu/theses/available/etd-02898-154928/unrestricted/ETD.PDF (accessed 30 August 2012).

Davies, N. (1982) *God's Playground: A History of Poland, Vol. 2, 1795 to the Present*. Columbia University Press, New York.

Development Alternatives, Inc. (2009) *Albanian Agricultural Competitiveness Program: Annual Report Fiscal Year 2009*. Produced by Development Alternatives, Inc. for USAID. Available at: http://pdf.usaid.gov/pdf_docs/PDACP170.pdf (accessed 28 March 2012).

Diamond, J.E. (2000) Agricultural and extension education in Albania. *Journal of International Agricultural and Extension Education* 7(1), article 6. Abstract available at: http://www.aiaee.org/vol-71-spring-00/318-agricultural-and-extension-education-in-albania.html (accessed 2 May 2012).

Dishnica, T. (2007) *Country Report: Albania. The National Agricultural Research and Extension Service*. Department of Extension Service, Research and Agricultural Information, Ministry of Agriculture, Food and Consumer Protection, Tirana, Albania. Available at: http://www.fao.org/fileadmin/templates/nr/images/research/skopje/ws1_report_Albania_en.pdf (accessed 10 February 2012).

EBRD and FAO (2007) *A Systematic Analysis of the Agribusiness Sector in Transition Economies. The Albanian Olive Oil Sector*. Commissioned by the European Bank for Reconstruction and Development (London) and the Investment Centre of the Food and Agriculture Organization of the United Nations (FAO), Rome. Available at: http://www.eastagri.org/files/ALB_Oliveoil_VC_final.pdf (accessed 10 February 2012).

EC (2012) Enlargement: Overview. European Commission, Brussels. Available at: http://ec.europa.eu/enlargement/instruments/overview/index_en.htm (accessed 5 September).

Eurostat (2012) Farm Structure in Hungary. European Commission, Brussels. Available at: http://epp.eurostat.ec.europa.eu/statistics_explained/index.php/Farm_structure_in_Hungary (accessed 31 May 2012).

Hungarian Chamber of Agriculture (2012) Welcome to the Hungarian Chamber of Agriculture! Available at: http://www.agrarkamara.hu/Bemutatkoz%C3%A1s/WelcometotheHungarianChamberofAgriculture.aspx (accessed 10 April 2012).

Ion Ionescu de la Brad University of Agricultural Sciences and Veterinary Medicine Iaşi (2012) Details available at: http://www.uaiasi.ro/index.php?lang=en&pagina=pagini/home.html (accessed 11 April 2012).

Jones, G.E. and Garforth, C. (1997) Chapter 1 – The history, development, and future of agricultural extension. In: Swanson, B.E., Bentz, R.P. and Sofranko, A.J. (eds) *Improving Agricultural Extension: A Reference Manual*. Food and Agriculture Organization Corporate Document Repository, Rome. http://www.fao.org/docrep/W5830E/W5830E00.htm (accessed 30 August 2012).

Kania, J. (2010) Organization of Agricultural Extension in Poland. In: IFPRI (International Food Policy Research Institute) (facilitator) *Agricultural Extension and Advisory Services Worldwide*. Available at: http://www.worldwide-extension.org/europe/poland (accessed 27 March 2012).

Kodderitzsch, S. (1999) Reforms in Albanian agriculture: assessing a sector in transition, World Bank Publications. Technical Paper No. 431.

Kozári, J. (2012) Szent István University Gödöllő. Details available at: http://www.agroporadenstvi.cz/poradenstvi/AAC/AAC_HU_SZIU.pdf (accessed 10 April 2012).

Marinov, L. (2011) Analytical Report – Bulgaria. In: IFPRI (International Food Policy Research Institute) (facilitator) *Agricultural Extension and Advisory Services Worldwide*. Available at: http://www.worldwide-extension.org/europe/bulgaria (accessed 7 April 2012).

National Agricultural Advisory Service (2012) National Agricultural Advisory Service, Ministry of Agriculture and Food, Bulgaria. Available at: www.naas.government.bg/en/ (accessed 12 April 2012).

National Council of Agricultural Chambers (2012) National Council of Agricultural Chambers (KRIR), Warsaw, Poland – general information. Available at: http://krir.pl/index.php?option=com_content&task=view&id=889&Itemid=100 (accessed 7 April 2012).

Place, N.T., Evans, D.E., Andrews, M.P. and Crago, N.E. (2000) Implications and impact among American extension professionals and near-associates resulting from the Polish–American Extension Project. *Journal of International Agricultural and Extension Education* 7(1), 5–16. Available at: http://www.aiaee.org/attachments/314_Place-Vol-7.1.pdf (accessed 31 August 2012).

Qafoku, N. (1995) Passé, présent et avenir du système de vulgarisation en Albanie. In: *La Vulgarisation Agricole dans les Pays de la Méditerranée du Nord-est: Synthèse des Actes du Séminaire de Volos/CIHEAM-IAMM/FPH. Cahiers Options Méditerranéennes* 2(2), 149–150. Available at: http://ressources.ciheam.org/om/pdf/c02-2/CI011005.pdf (accessed 31 August 2012).

Sofia University St. Kliment Ohridski (2012) Details available at: http://www.uni-sofia.bg/index.php/eng/ (accessed 7 April 2012).

Standing Committee on Agricultural Research (2012a) SCAR – National Research Systems – Bulgaria, Standing Committee on Agricultural Research (SCAR), European Commission, Brussels. Available at: http://ec.europa.eu/research/agriculture/scar/bg_en.htm (accessed 12 April 2012).

Standing Committee on Agricultural Research (2012b) SCAR – National Research Systems – Hungary, Standing Committee on Agricultural Research (SCAR), European Commission, Brussels. Available at: http://ec.europa.eu/research/agriculture/scar/hu_en.htm (accessed 12 April 2012).

Standing Committee on Agricultural Research (2012c) SCAR – National Research Systems – Poland, Standing Committee on Agricultural Research (SCAR), European Commission, Brussels. Available at: http://ec.europa.eu/research/agriculture/scar/pl_en.htm (accessed 12 April 2012).

Standing Committee on Agricultural Research: (2012d) SCAR – National Research Systems – Romania, Standing Committee on Agricultural Research (SCAR), European Commission, Brussels. Available at: http://ec.europa.eu/research/agriculture/scar/ro_en.htm (accessed 12 April 2012).

Stefanescu, S.L. (2011) Current status of the agriculture advisory and extension system in Romania. In: IFPRI (International Food Policy Research Institute) (facilitator) *Agricultural Extension and Advisory Services Worldwide*. Available at: http://www.worldwide-extension.org/europe/romania (accessed 7 April 2012).

Stone, I (1980) *The Origin: A Biographical Novel of Charles Darwin*. Doubleday, New York.

Tahiraj, N. (2002) Status Report for Animal's Genetic Resources. Ministry of Agriculture and Food, Republic of Albania, Tirana, Albania. Available at: ftp://ftp.fao.org/docrep/fao/010/a1250e/annexes/CountryReports/Albania.pdf (accessed 31 August 2012).

University of Agronomic Science and Veterinary Medicine (2012) Details available at: http://en.usamv.ro/ (accessed 11 April 2012).

US Department of State (2012) Background Notes/Country Fact Sheets. Available at: http://www.state.gov/r/pa/ei/bgn/index.htm (accessed 31 May 2012).

USAID (2000) *Albania Dairy Improvement Campaign, USAID CA# DHR-A-00-95-00040-00, Final Report, 1995–2000*. Available at: http://pdf.usaid.gov/pdf_docs/PDABS359.pdf (accessed 26 March 2012).

USAID Albania (2012) Details available at: http://albania.usaid.gov/ (accessed 31 May 2012).

USAID Bulgaria (2012) Details available at: http://bulgaria.usaid.gov/ (accessed 31 May 2012).

USAID Hungary (2012) USAID Europe and Asia, Regional Services Center/Budapest. Available at: http://transition.usaid.gov/locations/europe_eurasia/countries/rsc/index.html (accessed 31 August 2012).

USAID Poland (2012) USAID Mission to Poland. Available at: http://www1.usaid.gov/pl/ (accessed 31 August 2012).

USAID Romania (2012) USAID Europe and Asia, Romania, Overview. Available at: http://transition.usaid.gov/locations/europe_eurasia/countries/ro/ (accessed 31 August 2012).

USDA NIFA (2011) US Department of Agriculture National Institute for Food and Agriculture: About Us, Extension. Washington, DC. Available at: http://www.csrees.usda.gov/qlinks/extension.html#yesterday (accessed 24 April 2012).

World Bank (2001) The World Bank, Albania – Agricultural Services Project. World Bank Office, Tirana, Albania. Available at: http://www-wds.worldbank.org/external/default/main?pagePK=64193027&piPK=64187937&theSitePK=523679&menuPK=64187510&searchMenuPK=64187283&siteName=WDS&entityID=000094946_01060204005784 (accessed 27 March 2012).

World Bank (2006) Sourcebook: Agricultural Investment. Module 3 – Investments in Agricultural Extension and Information Services. World Bank, Washington, DC. Available at: http://go.worldbank.org/3CO4VK9ON0 (accessed 31 August 2012).

World Bank (2007) *Albania: Strategic Policies for a More Competitive Agriculture Sector*. Report No. AAA18-AL, World Bank Sustainable Sector Unit, Europe and Central Asia, World Bank, Tirana, Albania. Available at: https://openknowledge.worldbank.org/bitstream/handle/10986/7563/AAA18optmzd0AL.pdf?sequence=1 (accessed 5 September 2012).

World Bank (2010) *Status of Projects in Execution FY10, SOPE: Europe and Central Asia Region, Albania*. World Bank, Washington, DC. Available at: http://siteresources.worldbank.org/INTSOPE/Resources/5929468-1286307702807/7453146-1286464746323/AlbaniaFinal.pdf (accessed 28 March 2012).

World Bank and European Community (1992) *An Agricultural Strategy for Albania*. World Bank/European Community, Washington, DC.

Zickel, R. and Iwaskiw, W.R. (eds) (1994) *Albania: A Country Study*. GPO [Government Printing Office] for the Library of Congress, Washington, DC. Available at: http://countrystudies.us/albania/index.htm (accessed 26 March 2012).

4 Albania's Agriculture Industry

Brinton Foy Reed[1] and Engjell Skreli[2]
[1]*University of Hawai'i at Mānoa, USA;*
[2]*Agricultural University of Tirana, Albania*

4.1 Introduction

The fall of communism in the former Soviet Union primed many Eastern European countries for a shift from centralized leadership to pluralistic democracy, and was accompanied by wholesale economic change in the form of privatization and trade liberalization. However, the transition to a market from a planned economy can be tumultuous and is marked by obstacles to economic growth, such as persistent or increased income inequality, expansion of informal markets, and lack of exposure to the inner workings of market exchange. Despite such obstacles to growth and development, economies in Eastern Europe have enjoyed higher overall standards of living following transition (Bonneau, 2011). Albania is no exception. Despite economic collapse shortly before and during the shift toward a democratic government, Albania has achieved a measure of stable economic growth relative to its regional and economic peers, and its economic performance in the near-term is expected to maintain an upward trajectory.

Before transition, Albania's economy was based on large-scale, collectivist agricultural production, with centralized economic and financial governance (Lerman, 2001). Albania remained an agriculture-based society after transition and to this day agriculture plays a vital role in Albania's economy. Despite its relative stability and performance, especially during the global economic downturn in 2008, its gross domestic product (GDP) and total agriculture production trail the average of other upper middle income countries and Eastern Europe (World Bank, 2012). The factors limiting production and growth in Albania's agriculture sector, especially lack of rural land markets and the presence of a substantial informal economy, are complex issues with roots reaching back to the transition period itself. These form the basis for a web of interrelated constraints that, together, inhibit vertical integration and efficiencies of scale, particularly at the base of agricultural value chains. At this level, producers and processors lack, among other things, access to modern equipment and know-how, credit and financial services, storage facilities, permanent market places, and irrigation infrastructure. The consequence of these limitations is to prevent the majority of producers from escaping subsistence-based livelihoods, and to give processors little incentive to adhere to food quality standards or operate within the law. These conditions ultimately discourage investment and impede Albania's economic development.

Leaving aside this introduction and the conclusion, this chapter consists of four main sections. The first begins with an overview of Albania's economy, agricultural sector performance, and the sector's role in economic growth. It then includes a description of Albania's production and processing capabilities and ends with a discussion of the major characteristics that define Albania's current agriculture sector. The next section gives an account of the major historical events from the transition period to present, and discusses various aspects of ownership and privatization in the context of constraints to growth. The following section provides a review of several value chain studies, and identifies key elements within those chains that should be considered as policy targets (challenges). Finally, the chapter closes with an analysis of the sector's strengths and weaknesses, as well as potential opportunities and threats going forward.

4.2. Agriculture Sector Overview

4.2.1 Albania's economy

Agriculture, as the principal component of primary production, is a main driver for most of the world's developing economies. This is especially true of preindustrial and transition economies in which service and industry sectors are fledgling, or are stagnant at low levels. In the case of transition economies, the relationship between economic growth and agriculture forms a positive feedback loop; agricultural production feeds into GDP growth, and GDP growth, in turn, boosts agricultural productivity by stimulating the development of market services, supply chains, distribution channels, and access to financial services (Lerman, 2001). It is thus reasonable to assume that long-term trends in GDP growth affect the development of the agriculture sector, and that constraints to growth in the agriculture sector also constrain GDP. For this reason, it is important to consider post-transition GDP trends when evaluating the performance of Albania's agriculture sector over similar time periods.

At the launch of Albania's economic transition in 1992, GDP per capita in purchase power parity (PPP), or GDP adjusted for currency exchange rates, had bottomed out at US$1963 (World Bank, 2012) and it was the poorest and most isolated country in Europe (Jarvis, 2000). In the three years preceding the transition, Albania experienced a serious economic contraction, resulting in grinding poverty, failing infrastructure, and high inflation (Bonneau, 2011). From 1989 to 1990, Albania's GDP growth rate fell from 9.8% to −9.6%, and in 1991, it shrunk to −29.6%. In the year following the transition, however, GDP growth rates reached 9.6%, and they remained high (above 5%) until 2002, with the exception of a drop in 1997 of over 19 percentage points to about −9.6%. This economic shock was a result of the social and financial upheaval caused by the collapse of massive pyramid schemes, the liabilities of which approached half of the country's GDP (Jarvis, 2000).

Despite economic upsets, Albania's economy continues to grow and has stabilized in the last decade. Its average GDP growth rate of 5.1% from 2007 to 2010 indicates that Albania has, in recent years, outpaced all other countries in the region in terms of development (World Bank, 2012; see Fig. 4.1). Albania, like other lower middle income countries (its economic peers until 2010 as defined by the World Bank, 2011), showed great resilience during the global economic downturn, maintaining positive growth while the average for all developing nations in Europe and Central Asia declined in step with the world economy. During that time period, other Balkan nations averaged less than half of Albania's growth rate, except for Kosovo, which grew at a similar rate. In 2009, Albania's growth exceeded that of all other Balkan nations, and in 2010, Albania reached 'upper middle income' status, which indicates a gross national income (gross national product less business taxes) per capita of US$3976–12,275 (World Bank, 2011).

Other economic performance indicators, such as poverty levels and unemployment rates, have improved in step with Albania's increased GDP performance. According to World Bank (2012) estimates, total poverty decreased from 25.4% to 12.4% in the nine years between 2002 and 2011. Although slowing rates of poverty reduction have been noted, especially in urban areas, poverty levels are expected to continue decreasing as Albania's economy grows. From 2002 to 2008, unemployment rates also improved, dropping from 17% to

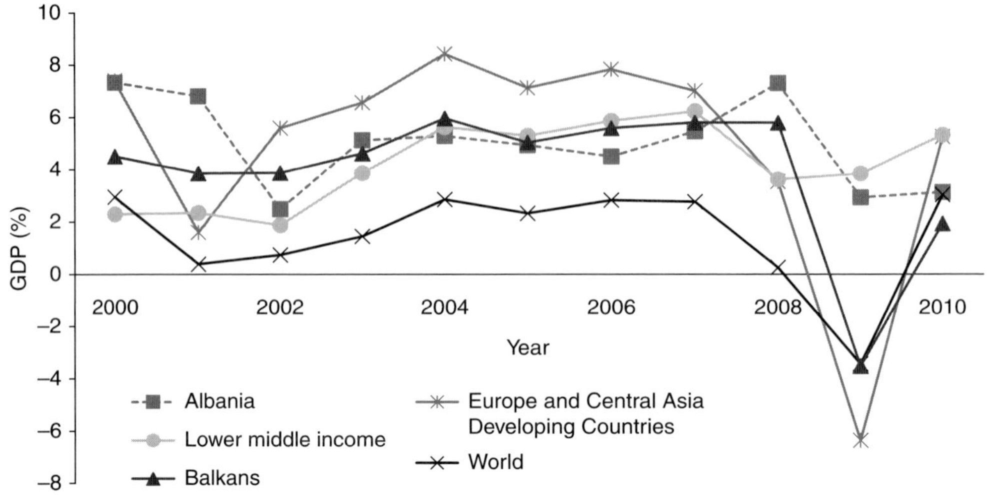

Fig. 4.1. Gross domestic product (GDP) per capita annual growth in Albania compared with other countries/ regions, 2000–2010. Source: World Bank (2012).

12.8% (World Bank, 2011). These improvements are attributed primarily to relatively stable and increasing economic development.

Despite encouraging economic growth in recent years, Albania's GDP per capita (PPP) remains well below the average for upper middle income nations, the world average, and all but Moldova and Bosnia and Herzegovina in the Balkans. In addition, Albania is still an import-oriented economy despite a strong record of aggressive trade liberalization following transition. Its export base is small and undiversified. Its imports in 2010 were close to 39% of GDP, while exports were just 13% of GDP. Export volume was a third of that of imports, despite an increase of trade volume in 2010 by 20% and a remarkable year-on-year increase of exports by 56%. This trend of growth in exports is expected to continue as the effects of Albania's ascension to the WTO (World Trade Organization) in 2000, and subsequent free trade agreements, set in (US State Department, 2011).

4.2.2 The place of agriculture in Albania's economy

As mentioned previously, economic growth and growth in the agriculture sector of a nation are often strongly linked. This is particularly true when agricultural contributions to GDP are high, as is the case with many post-communist transition economies. Much of the labor force, infrastructure, and political administration of these nations were geared for large-scale agricultural production during communist rule, and afterwards, many transition economies retained a strong agricultural economic base. Such is the case for Albania (Lerman, 2001). Following transition in 1991, agriculture was the main source of economic growth, contributing over 50% of total GDP, although foreign direct investment (FDI), and privatization of businesses stimulated rapid growth of the services sector from the mid-1990s onward (Fig. 4.2).

Even so, Albania's agriculture value added as a percentage of GDP remains much higher than Eurozone and world averages, slightly higher than that of lower middle income countries (World Bank, 2011; Fig. 4.3), and consistently higher than that of all other Balkan nations (World Bank, 2012). This suggests that Albania still struggles to industrialize and depends heavily on the agriculture sector for economic growth.

Nearly half (47.8%) of Albania's 1.053 million-strong labor force works in agriculture, although agricultural production accounts for

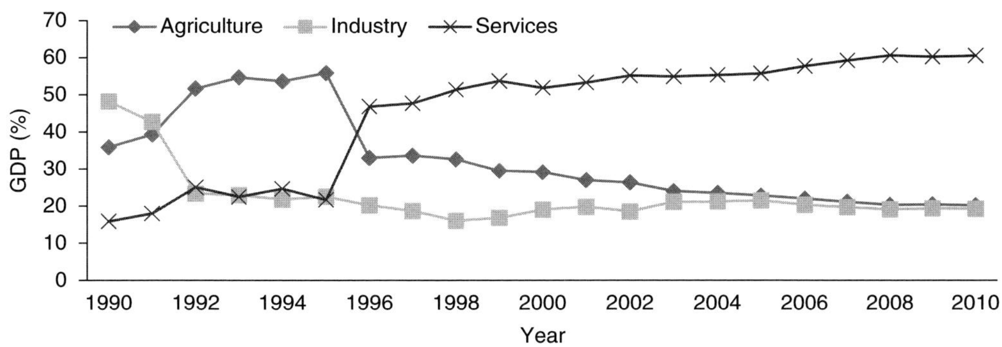

Fig. 4.2. Gross domestic product (GDP) composition by sector in Albania, 1990–2010. Source: World Bank (2012).

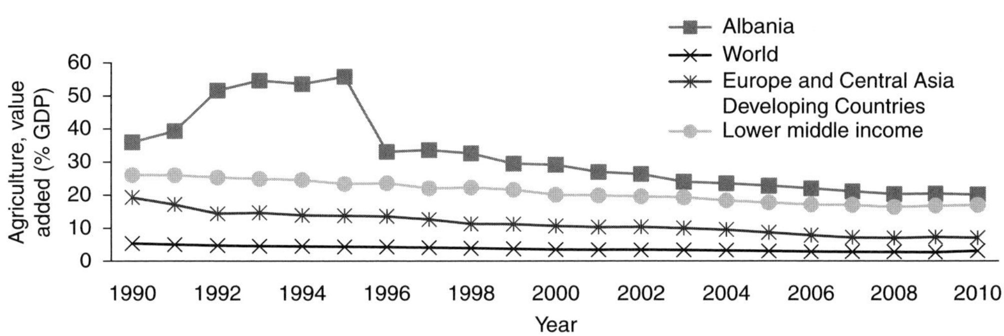

Fig. 4.3. Agriculture value added as % gross domestic product (GDP). Source: World Bank (2012).

less than 21% of GDP (CIA, 2011). For rural Albanians, agriculture is the main source of income and employment, and can provide an employment safety net if off-farm employment is abandoned or unavailable. The productivity, in terms of value added per worker for the agriculture sector, is only 32% of that of other sectors and around 20% of the European Union (EU) average for agriculture (World Bank, 2011). This is about two thirds of the average for other developing countries in the region (World Bank, 2012). Albania's low agricultural productivity per capita is typically attributed to the prevalence of small, inefficient family-owned farms with a lack of modern farming equipment and access to markets (CIA, 2011).

Despite low per capita productivity, the agriculture sector grew by about 40% from 2001 to 2008, and continues on an upward trend (World Bank, 2011). In the years following transition, growth in agricultural productivity was spurred by a shift away from staple crops toward more profitable products, such as livestock and horticulture. This growth was a result of allocation realignment—that is, moving resources toward producing alternative goods—and thus required little investment. However, Albania's agricultural product mix has stabilized and further growth is constrained by lack of investment in human capital and physical assets, such as modern farm machinery (World Bank, 2007a).

Growth is also constrained by lack of reinvestment at the farm level. In 2009, only 5% of farms reported reinvestment of profits. Reasons for low reinvestment rates include low profitability of farming compared with off-farm opportunities, unclear land tenure rights, and constraints related to poor irrigation and roads. While credit sources are scarce for small farmers, some commercial farms have access to commercial loans and investment from donor projects. Finding sources of investment for the agriculture sector remains one of Albania's primary economic objectives (MAFCP, 2011).

4.2.3 Production and processing

In 2008, agricultural output reached 1509 billion euros, which contributed just over 17% to GDP that year. Of that, crop production contributed 41.2%, livestock contributed 47.8%, and agricultural services contributed 2.8%. The main agricultural products, in terms of contribution to total output, were milk (19.8%) followed by vegetables (13.3%), fruit (11.8%), cereals (11.4%), and beef (10.3%), (MAFCP, 2011).

Permanent crops, those that produce yields for multiple seasons, steadily increased their share of total annual production from 10% in 2000 to 15% in 2009. The most important permanent crops (by output)—grapes (36%), olives (34%), and other fruit (28%)—experienced growth in both planted area and yield. This growth is attributed to higher quality cultivars, a shift toward higher value crops, and increased demand for processed products such as wine and olive oil. From 2000 to 2009, fruit trees and vineyards both experienced production growth rates exceeding 100%, and production from citrus products grew 284.6%. Olive tree production was more modest, experiencing a growth rate of only 32.6%.

According to Albania's Ministry of Agriculture, Food and Consumer Protection (MAFCP), in 2009 the crop production sector comprised, by volume, potatoes (31%), fruit (29%), cereals (28%), potatoes (7%), and industrial crops (5%). From 2000 to 2009, production rates for most crops increased despite net decreases in cultivated area. This is especially true for cereals production, which increased 11% despite a decrease in planted area of 20%. Cereal and vegetable yields increased by a third, and potato and bean yields increased by about a quarter. Albanian farmers, like others around the world, have been able to grow more with less land owing to increased availability of inorganic fertilizers, and the diffusion of knowledge of modern farming techniques and technology (MAFCP, 2011).

Albania's yields were impressive for certain crops in 2009. Yields of grapes, chestnuts, cauliflowers, figs, cherries, broccoli, and dry beans were all higher than the average rate for EU-27 (the 27 countries of the EU as of 1 January 2007). Yields of grapes were nearly three times as high; and those of chestnuts and figs were 2.5 and 2.0 times as high, respectively. Albania's Mediterranean weather gives it an agricultural advantage over EU-27 countries in terms of yields for these crops. In the future, this climatic advantage could be leveraged to build strong trade ties with the EU nations.

Livestock production is an important component of Albania's agriculture sector, in terms both of heritage and contribution to agricultural production. It contributed to over 50% of total agricultural production in 2007 and over 47% in 2008 (Meço *et al.*, 2010). The livestock sector grew at an annual rate of 3–5% until 2007, before experiencing negative growth rates in 2008 and 2009. There is strong domestic demand for livestock products, especially from small ruminants. Within the livestock sector, the primary output is milk (42% by volume). This is followed by beef (21%), meat from other ruminants (13%), and eggs (11%). Poultry and honey production is small (5% and 3%, respectively), but is growing very fast (MAFCP, 2011). There is much anticipation for the expansion of export channels for meat and meat products, although the lack of quality standards remains an obstacle to meeting demand abroad.

Much of Albania's agricultural production for market is sent to food processors, and despite recent growth in this industry, several important issues continue to hinder development of the food processing industry. In general, its growth is constrained by the same factors that affect primary production—lack of vertical integration, difficulties in meeting environmental and quality standards, and unavailability of credit. Most product lines in this industry have low asset utilization rates due to obsolete or unrepaired equipment, and to the seasonality of some agricultural products. In addition, processors are subject to the entire value added tax burden (20% in 2010) for final products (that is, the tax burden for production and processing falls squarely on processors), as most producers operate in an unofficial capacity by not registering their businesses and do not pay taxes (Meço *et al.*, 2010). This inequitable tax distribution incentivizes processors to work in the informal, or shadow, economy, even if they are officially recognized businesses.

In 2008, the food processing industry contributed only 1.4% to total gross value added but, from 2000 to 2005, this grew 278% for food and 205% for beverages (Ruli and Belortaja, 2003). Bread and flour production contributed to nearly half of the output for the entire industry. This was followed by soft drink production and milk and meat processing. From 2000 to 2008, employment in the industry grew by 20%, although the numbers of employees per company have remained low, according to government statistics (MAFCP, 2011).

The food processing segment is dominated by small, inefficient operations. Over 95% of officially registered companies employ ten workers or fewer. Many processing companies, especially the smaller ones, operate on the margins of the informal economy. There is, however, a small segment of well-developed food processing companies in beef, fish, milk, wine, and beer processing, who have been able to modernize and meet recent legislation requirements for food safety and quality. In order to prime the sector for growth, the Albanian government has provided subsidies to legal processors for interest rate payments, and some support for extension and other services.

4.2.4 Sector characteristics

Albania's agriculture sector is characterized by smallholder subsistence farming on fragmented plots, an aging and decreasingly educated rural labor force, and a substantial informal economy. These characteristics contribute to low agriculture productivity, inconsistent supply, and uncompetitive products. Such factors are often identified as key challenges to the continued development of Albania's agriculture industries.

Like other transition economies in central Europe, Albania's agricultural production base is typified by a few large farms organized as corporations or cooperatives, and a large number of small, household farms that produce for their own consumption (Mathijs, 2002). The predominance of small operations is primarily due to the lack of farmland consolidation following the transition to private ownership in the early 1990s. Farm plots awarded to households during the privatization process were small and often scattered over long distances. The lack of markets for rural land, which is discussed in more detail later, has prevented farmers from reaching economies of scale, and made procurement of mechanized equipment out of reach for the vast majority of rural producers (Fortuzi *et al.*, 2008a).

In 2009, average farm size ranged from about 0.7 to 1.6 ha, depending on the region. Nearly 90% of all farms were less than 2 ha and 46% of those were less than 1 ha in size. Around 11% were 2–5 ha, and only around 1300 farms were larger than 5 ha. The majority of farms over 5 ha were industrial sized, reaching up to 50 ha, according to MAFCP (2011).

Fragmentation of plots is extremely high. On average, parcels per farm were 3.8 in 2009, and average parcel size was 0.3 ha. This fragmentation and the distance between fragmented parcels greatly reduce the efficiency of farm labor and resources. The problem is compounded in many areas by dilapidated roads and, for some farmers, lack of transportation options. This characteristic of Albania's farmland ownership structure has resulted in a high number of small, abandoned plots, and is considered a major impediment to growth in the sector.

Subsistence farming dominates domestic production. Only 18% of all crops and 30% of all livestock products reach markets. According to a study conducted by the Agricultural University of Tirana, major factors hindering the production of for-market agriculture products are, in order of severity: lack of permanent marketplaces, lack of transportation, price collusion, and lack of sources for processing services. The hurdles facing farmers who wish to sell products outside their districts, or abroad, include transportation problems, inferior product quality compared with imports, and lack of time available for travel to markets (Fortuzi *et al.*, 2008b).

Another trend of Albania's agriculture sector is the aging of the agricultural labor force due to migration of young people to urban from rural areas. Many farms are left to parents or abandoned as the younger generation of would-be farmers move to cities or emigrate. From 2004 to 2008, the total number of farm owners under 40 years of age nearly halved, from 11.4% to 6.7%, and 50% of farmers in 2008 were over 55 years old (Petrick and Weingarten, 2004). Aging and migration have contributed to ongoing decreases in available labor and specifically, skilled agricultural labor. This is compounded by lack of training and education for new farmers. According to some estimates, only 30% of farmers have any formal agricultural education. Those who do have some agricultural education have often received it informally from older farmers who attended vocational agriculture schools (MAFCP, 2011). The lack of knowledge and training concerning cultivar selection, soil nutrient cycling, and conservation agriculture practices inhibit yield quality and quantity at the production level.

Finally, Albania has a sizeable, unreported informal economy that could be as large as 50% of GDP (CIA, 2012). In the agriculture sector, organizations in the informal economy typically take two forms—small, household farms, and unregistered businesses. Few household farms pay tax on production or property. The government forgoes such taxation as farming households are typically impoverished and collection costs are high. There are many new and small to medium-sized agriculture businesses that are not registered. In addition, some businesses that are registered participate in the informal economy (Bode *et al.*, 2009). Even though the size and total economic effects of shadow economies are notoriously difficult to measure, even a small presence of these in any sector can present a serious investment risk. Without transaction records, accounting registers, or even official recognition as a concern, arbitration and litigation is difficult or impossible. Consequently, Albania's informal economy is a major barrier to the introduction of FDI and private investment (Ruli and Belortaja, 2003; World Bank, 2007b).

These, and other features of Albania's agriculture sector, are generally the result of communist legacy, or of the policies undertaken during the transformation from a planned to market economy. In order to fully understand how each has affected Albania's current economic situation, it is important to consider the major historical events that caused them. Putting these key characteristics into a historical context also helps in the formulation of policies aimed at alleviating constraints to growth. In the next section, we will look at the historical cause of such constraints and identify major milestones in the development of Albania's agriculture sector.

4.3 Albania's Agriculture Following Transition

4.3.1 Milestones

Like other post-communist countries in the Central and Eastern Europe (CEE) and the Commonwealth of Independent States (CIS) nations, Albania's pre-transition agricultural activities were dominated by collective, large-scale farms that cultivated thousands of hectares of farmland. These institutions were operated by hundreds of member workers and, alongside large cooperatives and state farms, formed the backbone of agriculture production. Rural subsistence households working less than 1 ha made up the difference in production quantities, and markets, input, and budgets were controlled centrally. Price controls, profitability accounting, and other market factors were excluded from agricultural systems, resulting in an agriculture sector that was economically less efficient than its market-oriented counterparts (Lerman, 2001).

Beginning on 17 July 1991, with the signing of Law 7501: *On Land*, land privatization began in Albania (see Table 4.1). Of the Baltic and other CEE countries, Albania was the only one in which all agricultural land had been nationalized, and so Law 7501 also set the legislative framework for ownership (Fortuzi *et al.*, 2008a). Albania's nationalization of all land in 1976, combined with a lack of complete ownership records, made redistribution to previous owners difficult, and thus was a major impetus for the per capita calculation of post-transition land entitlements to workers (Lerman, 2001). In about half of the rural areas, distribution followed the legislative requirement of equal per capita shares. Ownership was issued to heads of household, who were then responsible for

Table 4.1. Important milestones in the development of Albania's agriculture.

Date	Milestone
17 July 1991	Law 7501: *On Land* enacted, launching Albania's land privatization effort
8 September 2000	Albania joins the World Trade Organization (WTO)
29 March 2002	First free trade agreement signed, indicating a policy trend toward preferential trade agreements
12 June 2006	Stabilization and Association Agreement (SAA) signed by Albania and the EU
19 December 2006	Central Europe Free Trade Agreement (CEFTA) signed
11 July 2008	The SAA between the Republic of Albania and the European Communities (EC) is ratified
28 April 2009	EU membership application submitted

distributing shares to family members. The remaining rural land, which was located in the northern foothill and mountain regions, was distributed by local leaders according to pre-1945 land ownership records. Such distributions recognized ancestral land rights, were viewed with legitimacy, and were not seriously challenged despite being inconsistent with official policy (USAID, 2011).

This method of per capita privatization was similar to those adopted by other post-Soviet states that wished to disperse cooperative farms, had little or no private land, and no clear ownership records. Consequently, Albania's agriculture sector, like that of many CEE countries, is characterized by small, family-owned farms. All CEE countries, including Albania, are showing a gradual shift toward farm ownership structures that are similar to those found in market economies; that is, the bottom 50% of farms (by size) utilize around 10% of total farmland, while the top 10% utilize approximately 40% (Lerman, 2001). However, Albania is behind its peers in this regard. It is expected that small farm consolidation will increase once challenges, such as developing healthy land markets, are met.

Following transition, Albania implemented a fairly successful trade liberalization policy in which it liberalized prices and foreign trade, and allowed for convertibility of domestic currency. By the end of 1991, foreign trade was freed from state monopoly with the ratification of the Law on State Enterprises, and in 1992 Albania introduced currency convertibility and a floating exchange rate policy. In mid 1996, price controls for nearly all commodities were abolished. A major milestone of the liberalization process was reached on 8 September 2000, when Albania was accepted into the WTO. As a result, it simplified its tariff structure and established uniform, low rates. Albania's aggressive trade liberalization policy and ascension into the WTO paved the way for a series of free trade agreements, which increased regional integration and the capacity for development of Albania's export base (Michalopoulos, 1998).

On 29 March 2002, Albania signed its first free trade agreement with the Republic of Macedonia. This marked the beginning of an effort to supplement WTO membership with EU and European Commission (EC) agreements at the regional level, and bilateral trade agreements at a local level. Of these agreements, the most notable is perhaps the Stabilization and Association Agreement (SAA), which was signed by Albania and the EU on 12 June 2006. This marked the first major step in Albania's EU accession process (US State Department, 2011). The *Interim Agreement on Trade and Trade-related Matters* was signed by Albania and the EC on 1 December 2006 and, on 11 July 2008, the SAA between the Republic of Albania and the EC was ratified. Following that, the Central Europe Free Trade Agreement (CEFTA) was signed on 19 December 2006 by Albania, the Republic of Bosnia and Herzegovina, the Republic of Croatia, the Republic of Macedonia, the Republic of Moldova, the Republic of Montenegro, the Republic of Serbia and UNMIK (UN Interim Administration Mission in Kosovo)/Kosovo. This was followed by free trade agreements put into effect with Turkey on 5 May 2008, and the European Free Trade Association (EFTA) agreement with Iceland, Liechtenstein, Norway, and Switzerland.

The EU continues to be Albania's primary trading partner; in 2010, the EU contributed around 65% of Albania's imports and received about 70% of Albania's exports. Also, because of CEFTA, signatory countries have increased bilateral trade. Although these trade agreements have helped to prime Albania for the development of an export market, the full benefits of the agreements have yet to be realized, as evidenced by persistent trade deficits with regional partners (US State Department, 2011). Despite the swift and largely successful trade liberalization process that Albania has followed, it still suffers from low competitiveness of agriculture products, and other barriers to reaching international markets.

4.3.2 Ownership, privatization, and constraints to growth

At the time of transition, reforms for the agriculture industry were focused on equitable privatization of arable land. Market-based productivity, however, is based not only on private ownership, but also on the awarding of resources to efficient producers. This basic principle necessitates the development of asset markets in which efficient producers can invest earned profits to achieve operational growth. Perhaps the most important of assets for agricultural producers is land and equipment. Without transferability of key assets such as these, farmers have little hope of expanding operations, no matter how productive their farms are. In this situation, efficiency is weakly rewarded and economic growth is stunted. Although privatization of land in Albania was conducted equitably and in good faith, property markets for farmland never fully materialized owing to unclear property rights, weak enforceability of contracts, and mistrust among producers. The lack of available land continues to limit producers to low-income subsistence farming. Furthermore, by limiting growth in farm operations (income and farm size), it effectively limits the development of markets for other key assets such as mechanized farm equipment or modern processing machinery. Barriers to land market development are among the most important constraints to the development of Albania's agriculture sector.

More than 80% of Albania's arable land is privately owned. Of the remaining publicly owned land, around 16% was offered to Albanian households during the privatization process, but was subsequently refused due to long distances between parcels or lack of fertility. Despite this high proportion of privately owned land, Albania's fledgling property market is largely limited to urban areas; less than 2% of rural households sold property and less than 3.6% of them had rented out land as of 2007. There is evidence that informal renting and property sales occur in rural areas, although the extent of these activities is unknown (MAFCP, 2011). The development of a formal land market continues to be hindered by land ownership conflicts, lack of available credit, and transaction costs associated with formal sales and rental arrangements.

In recent years, the Albanian government, with assistance from the World Bank, developed the legal and administrative capacity to manage land transactions, and in the last decade, land consolidation rates have grown steadily. From 2000 to 2008, the total number of smallholder farms decreased from 418,000 to 353,000. In an effort to promote land consolidation, transaction costs were subsidized and technical assistance was offered to buyers and sellers in select communities. The result, after a six-year period leading up to 2007, was 189 transactions and a net consolidation of 95 ha. Self-sustained land consolidation remained elusive, however. In 2010, a new national strategy to consolidate land was launched with the help of the Food and Agriculture Organization (FAO), the results of which remain to be seen (MAFCP, 2011).

4.4 Albanian Agricultural Value Chains

4.4.1 Value chain approach

In order to reach a more comprehensive understanding of Albania's agriculture sector, it is important to consider in detail the operations and interactions of individual participants that produce or add value to products or services. For the reason that a specific product or service ultimately binds all participants and their activities together, it is reasonable to treat that product or service as a point of reference for analysis. Furthermore, each entity that influences the final product or service can be thought of as a link in a chain that, at every step, adds value to that product or service. Together, these links form a conceptual model called a value chain.

A value chain can be defined as the full range of activities that are required to bring a product from its conception to its end use, and the value chain approach is a product-based conceptualization of business activities that focuses on the relationships among economic components and industry participants. This approach is often used to identify opportunities for improvement, as well as barriers to efficiency for a product line or an entire industry of similar products. These activities can include research and development, production, distribution, marketing, and even support from indirect participants, such as advertising and marketing firms, banks, or research extension services.

Value chain analysis provides a framework for understanding how various components of Albania's agriculture sector work together. It gives a real-world view of how the products discussed reach final users. Most importantly, it helps to identify specific challenges and opportunities at every interaction and process involved in bringing the products to market. While the scale and scope of value chains varies considerably depending on the nature of analysis, a systematic value chain analysis is typically composed of three integral component types: actors, value chain supporters, and environments. These interacting components are shown in Fig. 4.4 and are discussed below (Roduner, 2007).

Value chain actors (the small lightly shaded boxes) are the organizations that directly add value to products as they move up through the value chain. Their actions and interactions form the framework of the value chain, and the quality of the products or services that they produce is usually taken as the primary indicator of value chain health. Typical value chain actors are suppliers, producers, processors, distributors, wholesalers, retailers, and exporters.

As their name suggests, value chain supporters (shown in Fig. 4.4 as small darker shaded boxes) facilitate the actions and interactions of the value chain actors. These service-based components may enhance production capacities, provide financial management,

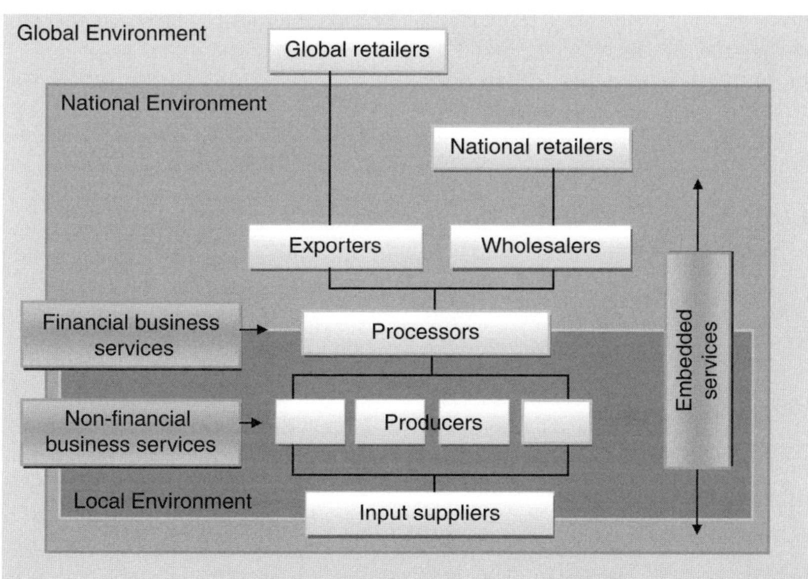

Fig. 4.4. Value chain framework. Source: Roduner (2007).

ensure equitable access to information, promote knowledge sharing, or any other services that value chain actors may need.

Value chain environments (the larger boxed areas in the figure) are typically divided into three geopolitical levels within a value chain network to identify macroeconomic, political, regional, and local factors affecting the overall functioning of a value chain. At the lowest level, the local environment, such factors could include municipal taxation, zoning laws, and the existence of cooperatives and local commerce groups. At the national level, factors such as national policy, regulatory frameworks, government services, and food regulation and inspection come into play. Factors at the global level include tariffs, trade agreements, currency markets, and international aid organizations. It should be mentioned that there are many factors that can be attributed to the local and national environment, such as infrastructure, distribution networks, and organizations such as national unions and lobbying groups that have semi-autonomous local chapters.

Value chain analysis is a useful tool when comparing similar industries, identifying constraints or opportunities, or studying entire product lines. This section reviews the findings of five recent value chain studies focused on various industries within Albania's agriculture sector, and uses the value chain approach to identify common challenges and differences among them.

4.4.2 Review of selected value chain studies

The following value chain studies have been selected with the objective of extracting common problems or challenges across the industries analyzed. Each study was selected according to three criteria. First, each industry studied must be a priority for the Albanian government. Priority industries include grapes, olives, meat and milk, and food processing (MAFCP, 2011). The second selection criterion was publishing date—recent studies were preferred owing to the dynamic nature of Albania's agriculture sector. The third criterion was that each study must use a value chain approach so as to provide an analysis of shared resources, opportunities for integration and savings, and capacity bottlenecks across industries. Based on these criteria, the industries reviewed here are fruits and vegetables, small ruminants for meat and milk production, and olive oil and wine production. The results of each study are segmented into the value chain components they are most strongly associated with.

The first study is from Skreli *et al.* (2009), and is entitled '*Competitiveness of Albanian agriculture: value chain study for fruit and vegetable sector in Fier region*'. It focuses on the fruit and vegetable value chains in Fier, as this region is the most important contributor to Albanian fruit and vegetable production, producing approximately a quarter of Albania's vegetables, and nearly as much fruit. The main fresh vegetables and fruits considered were watermelons, melons, tomatoes, cucumbers, intensively grown apples, and table grapes. The main processed products were paprika and cucumber. Albania's major fruit and vegetable processors are located near the center of the Fier region and were thus considered in this study. In addition, raw produce from farms in the Fier region are shipped to processors inexpensively, so some processors that are outside the Fier region were also considered if their inputs come from Fier.

To better reflect the actual value chains of Albania's fruit and vegetable industries in the Fier region, we offer a modified version of the standard value chain shown in Fig. 4.4 (Fig. 4.5). This simplified value chain depiction is based on information taken from the Skreli *et al.* (2009) case study, but can be used as a general reference for the review of the other value chains discussed in this section. Even though each of the five value chains has individual characteristics that set it apart from others in the sector, the last four that are discussed are fairly similar to the value chain for fruit and vegetables industry for the Fier region. Like the case study it is based on, Fig. 4.5 focuses on local- and national-level value chain segments.

Producers are divided into two groups: large, commercial farmers and small subsistence farmers. Small farms typically perform little or no soil analysis and know little about

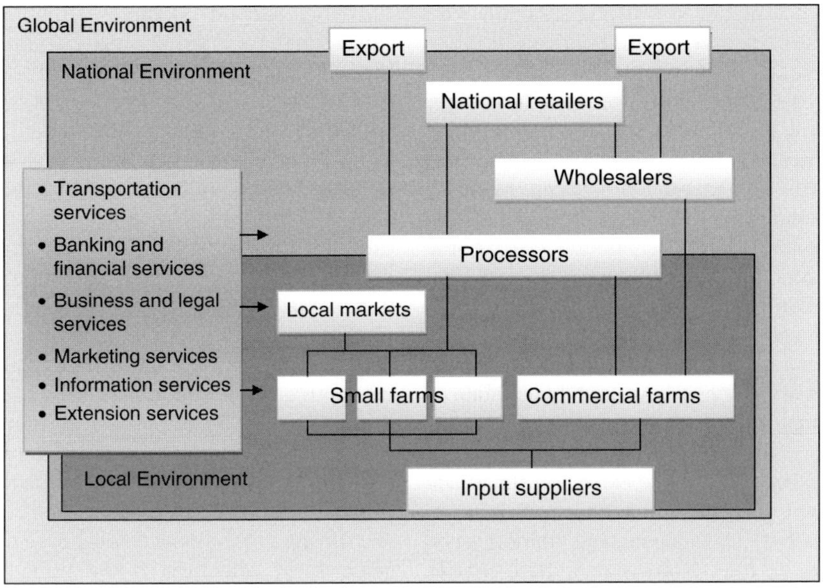

Fig. 4.5. Fruits and vegetables value chain. Based on information from Skreli *et al.* (2009).

the seeds and cultivars they purchase. They rely heavily on the advice of suppliers when selecting inputs, which often results in the use of inappropriate cultivars for the plot conditions. Subsistence farms consume most of their own produce, only selling it when there is a marketable surplus. Part of the produce is home processed. Both processed and fresh produce is sold directly to consumers via local markets, although household farms sometimes also sell to processors. Local markets are often informal, and little or no transaction records are kept.

Among producers, quality and quantity of output is limited by the lack of standardized production methods, such as storage and pruning techniques, the poor quality of seeds and of other inputs, misuse of pesticides and fertilizers, and unsophisticated management. In addition, problems with irrigation infrastructure result in water shortages and represent a major factor in increased production costs at the farm level. Although government-organized subsidies to provide drip irrigation and freshwater wells have been successful, the effects were too small to significantly affect national production. There is still a strong need for large-scale rehabilitation of the irrigation infrastructure that was virtually destroyed during Albania's economic transition. Lastly, absence of vertical cooperation among actors results in an inconsistent and unreliable supply chain for these products. There are reported cases of processors or traders coordinating with other chain levels through the development of ordering systems from processed or export goods, but these remain anecdotal and isolated.

Fresh produce grown by commercial farms goes directly to wholesalers who sell to national retailers or export products. Produce destined for processing goes to processors, who, as a rule, act as wholesalers for national retailers and export directly. Commercial farms generally rely less on suppliers, have economies of scale, and produce higher quality products in greater quantities.

Lack of post-harvest facilities plagues vegetable and fruit processors and has a negative effect on the competitiveness of Albanian produce overall. Post-harvest capacity for cold storage, cleaning and sorting, and packaging facilities is also limited. Loss of yields due to lack of transportation infrastructure and storage facilities varies throughout the year. Losses are heaviest during the harvest season when capacity fills. These bottlenecks can depress prices to the point that sometimes even harvesting vegetables is unprofitable. These factors contribute to unused processor capacity and are often cited as obstacles to processor competitiveness.

The value chain supporters identified for this study are public and private extension services, banking and financial services, business and legal services, information services, transportation services, and marketing services. Many of these services are available at both local and national levels. Access to credit due to unreliable collateral and high interest rates was identified as a major barrier to technology improvement. Though there have been private initiatives to provide pricing and other market data to farmers and processors, these have not been able to reach a large enough scale to affect most value chain actors. Technological, economic, and marketing advice given through extension services is insufficient; the agricultural and marketing economists involved have inadequate training and, even if they did have such training, there are not enough of them. At the regional level, one agricultural economist provides farm management consultation, and one horticulture specialist provides marketing consultation and post-harvest advice to farmers. There is a consensus that marketing consultation and, particularly, farm management training must be further developed, and that the entire agricultural research system must be overhauled. The function of agricultural research has recently been transferred to the Agricultural University of Tirana (AUT), but the transfer process has not been completed yet and no public research is being carried out. However, some business and marketing workshops for extension staff and faculty at AUT have been conducted through the HED/USAID (Higher Education for Development/United States Agency for International Development) funded AHEED (Albania/Hawaii Higher Education and Economic Development) project.

In terms of environmental factors, food safety and inspection policy plays an important role in industry competitiveness, especially considering the stringent quality assurance processes that must be in place before entering the EU. In addition, it is a vital component of developing a sustainable export base. Albania has approved a large number of food safety regulations, but compliance is insufficient. One reason for this is the ubiquitous presence of informal arrangements, and lack of regulation and contract enforcement.

The second study is from Deutsche Gesellschaft für Internationale Zusammenarbeit (GIZ, 2010), and is entitled '*Fruits and vegetables sector study prepared in the framework of IPARD Program – Albania*'. This study found that actors in the fruits and vegetables industry face many of the same challenges discussed both in the first study and previously in this chapter. According to the GIZ study, farmers (or 'producers' in Fig. 4.4), face a combination of challenges for growth: small, fragmented plots, an aging workforce, lack of post-harvest storage options, high input costs, and unreliable input quality. In addition, they have limited knowledge of agronomic and management systems. As a result, they suffer from low average productivity (except for melons) and high post-harvest losses of produce.

Processors, the next level of actors in the value chain, also face serious difficulties—the most important of which are lack of proper storage bins and refrigerated trucks, the use of obsolete equipment, lack of food quality enforcement, and the limited skills of marketing and sales personnel. Perhaps the most significant challenge processors face in terms of maintaining product quality is that farmers, who have difficulty in maintaining yield quality, send the freshest and highest quality produce directly to markets and older, lower quality produce to processors. In addition, the mixing of fruit and vegetable varieties to minimize shipping costs at the farm level makes processing difficult for food processors, especially when combining feedstock from different suppliers. This problem is systemic and demonstrates how the interactions among actors can determine the overall health of the value chain.

Value chain support issues include weak public extension services, lack of qualified processing consultants in rural areas, and lack of access to credit. Value chain environment issues include limited market segmentation, unreliable land ownership rights and taxation, lack of food safety and quality enforcement, prevalence of informal (shadow economy) processors, high competition from imported products for high-end market segments, and lack of irrigation infrastructure.

These constraints to growth prevent many vegetable and fruit producers from bridging the production gap from subsistence farming to commercial farming, and present a challenge to commercial farmers and producers who do not participate in the informal economy. Such challenges will persist until steps are taken to reduce the competitiveness of value chain participants operating in Albania's informal economy. These issues, which are discussed in more detail later, are the subject of ongoing research and projects initiated by the Albanian government and nongovernmental organizations (NGOs).

The third study, from the European Bank for Reconstruction and Development (EBRD) and FAO (EBRD and FAO, 2007), is entitled '*A systematic analysis of the agribusiness sector in transition economies: The Albanian olive oil sector*'. This study reveals that producers of olives and olive oil have low education profiles and are also subject to small, fragmented, and inefficient plots. In addition, their lack of farming equipment and pest management contribute to low yields. Processors in this industry are constrained by lack of professional and technical skills, including marketing, have little or no quality management systems in place, and are unable to retain olive oil quality during the bottling and packaging processes. However, there have been significant investments in many cases, and thus technology for extracting the oil from olives is typically good. Distributors, which typically receive and repackage products before transporting them to retailers, also suffer from lack of professional and technical skills, including modern managerial techniques. Most distributors are family owned and operated, and only 20% of all distribution channels are 'formal', or operate legally. It is likely that they face stiff competition from informal competitors which are not likely to pay taxes or adhere to food quality or production standards. There is little vertical integration among actors in this industry, although upstream partnerships and buyouts with farmers are being introduced.

Service issues for the olive oil industry include a lack of credit, quality certification process, and technical and managerial consultants. In addition, transportation service costs are high because of the small and fragmented nature of olive producers and olive oil processors. As for value chain environment challenges, quality standards are rarely enforced, food safety regulations are unclear, bribery and corruption are widespread, and there is little access to wholesale markets for olive oils.

The fourth study, from Kipi *et al.* (2010), is entitled '*Study of small ruminants subsector in Albania*'. This explored the value chain for milk and meat. It identified the following as major challenges for producers: shortage of labor for shepherding and milking; insufficient sources for feed during winter; poor quality breeds; inappropriate milk handling, high cost of transportation for both milk and meat; and income loss due to low lamb weight before slaughter. Processors struggle with milk and meat spoilage resulting from improper handling, lack of equipment and knowledge to ensure safe milk and meat processing, lack of branding, and the high costs of cheese production. In addition, processors for fresh meat from small ruminants work mostly in the informal economy where lack of safety regulations results in product quality that is too inconsistent for exportation.

Value chain support issues for the meat and milk industry are lack of extension services, veterinarians, credit, and reliable sources of market information. Environmental factors affecting this industry include high unmet domestic demand for small ruminants due to low consumer purchasing power and low export tariffs, lack of food safety legislation, and difficulties in competing with the informal sector (Bello *et al.*, 2011). However, growth in milk and meat production is a strong national priority and producers enjoy direct support in the form of subsidies.

The last study, from Skreli (2008), is entitled '*Study on Agrobusiness Market – Fier Qark*'. This focused on a number of value chains, including grapes, wine, and fruits and vegetables. For this review, however, we focus on grape and wine production. Farmers in these industries, like those of the other industries discussed above, face difficulties related to small farms, lack of modern equipment, and a dearth of technical expertise. These farmers have an especially difficult time ensuring quality, and most products are substandard in terms of variety and food safety. Importantly, the findings of this study indicate that difficulties in farmer and producer interactions have led to some vertical integration in the wine value chain—wine producers prefer to manage their own vineyards. However, lack of intellectual property (IP) protection and lack of marketing expertise in the industry has resulted in a lack of trademarks and branding for wine producers. Trademark enforcement has led to frequent infringement cases and a loss of market share for popular wine producers. As with other industries, credit availability is low but vineyard production is subsidized. Poor irrigation infrastructure is cited as another major issue for producers.

Even though these studies, with the exception of the first two, looked at different value chains, they share many characteristics. While each has its own set of difficulties, it is the challenges they have in common that are perhaps the most important to consider. Any mitigation of these challenges is likely to benefit many value chains at once, and thus have a greater impact on the development of the agriculture sector overall. For this reason, a comparison of similarities can be a helpful step in identifying such problems. Table 4.2 summarizes some of the commonalities between the four types of value chains, as well as the notable differences, which were identified in the five studies.

Several important conclusions can be drawn from these studies. First, agriculture value chains face similar, fundamental challenges despite the type of products they sell. These challenges, some of which have been discussed previously in this chapter, affect the overall competiveness of Albania's agriculture sector and likely have secondary effects on other economic sectors. Second, many of these problems directly affect the main actors of the value chain—the producers and processors. For this reason, it is unlikely that the value chains reviewed will experience significant growth unless the systemic problems are addressed at the lowest levels of production. Section 4.3.3 discusses these common problems within the context of value chain actors, supporters, and environments.

4.4.3 Summary of findings

Analysis of the five value chain studies discussed above revealed many challenges for

Table 4.2. Similarities and differences for the value chain studies reviewed.

Value chain	Differences	Similarities
Fruits and vegetables	• Mixing of fruit and vegetable varieties makes processing difficult • Lack of standardized production methods	• Small farm size and aging workforce with low educational profile represent the most important structural problems and result in lack of modern technologies and processes, inconsistent supply from farms in terms of variety and quality, unreliable input quality and scheduling from suppliers, and high production costs
Olives and olive oil	• Recent investment has enabled some producers to use efficient, modern technology	• Poor post-harvest knowledge and lack of transportation and storage capacity
Meat and milk	• Shortage of labor for shepherding and milking • Informality of fresh meat processing industry has resulted in lax safety standards that result in low quality and exportability of products	• Low managerial and technical skills at processing and distribution levels
Grapes and wine	• Lack of trademark and intellectual property rights enforcement • Wine producers have started to integrate vertically by investing in their own vineyards	• Poor product marketing • Limited horizontal cooperation and vertical integration • Lack of support services, especially credit and extension

Albania's agriculture sector going forward. Many of these challenges likely plague other value chains in the sector as well. Table 4.3 summarizes the major difficulties identified in the Section 4.4.2 and organizes each challenge by the value chain element it affects the most. It should be noted than many of these challenges negatively impact actors and competitiveness on multiple levels of the value chain.

As mentioned earlier, the central challenge that farmers face is small and fragmented plots. In addition, a lack of a young and educated labor force leads to inefficient operations, and reliance on input suppliers with misaligned incentives and selling products with questionable quality. This, combined with weak infrastructure and a dearth of modern farm equipment, is the major factor contributing to the production of low quality products with unreliable quantities at high costs. Because these producers form the foundation of the value chain, any problems they experience are passed on as products travel up the chain toward end users. Thus, these problems, which are common across all value chains in the sector, and are discussed in Section 4.5,

Table 4.3. Value chain findings.

Value chain level	Main challenges
Actors	
Producers	• Small and fragmented plots • Aging workforce and shortage of skilled labor • High input costs, unreliable input quantity and quality • Limited agronomic and management knowledge, overreliance on input suppliers • Few irrigation options at the farm level • Lack of available farming equipment • Lack of proper storage bins and post-harvest storage
Processors	• Obsolete or unrepaired equipment • Weak enforcement of food quality regulations and little or no quality management or certification systems • Limited marketing and sales skills or knowledge • Lack of branding and trademarks
Wholesalers and retailers	• Low impact of wholesale and especially large retail outlets on value chains • Lack of retail outlet branding
Supporters	
Transportation services	• Lack of refrigerated trucks • Bad road conditions which contribute to high transportation costs
Banking and financial services	• Limited access to credit due to unreliable collateral and high interest rates
Information services	• Lack of access to current market data
Extension services	• Training for agricultural economists and production economists is inadequate • Little public and private research
Consulting services	• Few qualified technical and management consultants in rural areas
Environments	
Local	• Limited market segmentation • Bribery and corruption
National	• Unreliable land ownership rights and taxation • Prevalent shadow economy • High competition from imported products in high-end market segments • Weak infrastructure (roads and regional irrigation) • Lack of trademark and intellectual property rights enforcement

are arguably the most critical constraints to future growth and, unfortunately, and are likely to be the most difficult to mitigate.

For the same reasons, processors receive inadequate and unreliable supply from farms. This presents a major challenge to vertical integration, and the post-processed product quality suffers as a result. Processors are characterized by lack of social capital and operable, modern equipment. Lack of storage options, and little or no enforceable quality assurance programs both contribute to low quality output. The problem is compounded for processors who operate legally—competition with informal processors is especially intense at this level of the value chain.

In terms of value chain supporters, the most common problems faced at all levels are lack of available credit, proper transportation storage and refrigeration, market information sources, and qualified individuals offering consultation and extension services. Of these, the unavailability of credit is perhaps the most pervasive problem, and limits growth of most actors within the value chain. This is a particularly acute problem for producers and processors, who cannot provide collateral for credit without first having credit to purchase assets and expand production. With little or no refrigerated storage and transportation capacity, agricultural product losses are high, especially for those that are produced seasonally. Fortunately, much work is now being done to develop human capital in order to provide high quality extension and consulting services to actors all along the value chain.

The challenges associated with the value chain environment, especially at the national level, are, by their nature, the most defining challenges all economic segments face. Unclear land ownership rights and contract enforceability are the chief contributors to the stagnation of rural land markets and prevalence of fragmented farmland. It is possible to achieve some measure of consolidation through targeted programs, such as those discussed in previous sections, but so far the success of these programs has been quite limited. Finally, bribery and corruption are a hallmark of informal economies, and such activity can strongly limit agricultural development (Bello *et al.*, 2011). It is likely that Albania's vision of a market economy will be fully realized until its shadow economy is contained and converted.

These common problems are by no means limited to the value chain studies that have been reviewed in this section (Section 4.4). Nearly all of them limit, in some fashion, the development of other value chains in the agriculture sector, and even may inhibit growth in the industry and service sectors as well. For this reason, they should not be considered value chain problems per se, but as problems affecting the entire sector. Thus, the following analysis incorporates the findings of this section with those of the entire chapter to provide a vision of Albania's agriculture sector as it currently stands, and give insight into its future development.

4.5 Strengths, Weaknesses, Opportunities, and Threats Analysis

This section provides an overview of the elements most affecting Albania's agriculture sector, and addresses potential challenges and prospects for future growth. These elements are organized into strengths and weaknesses, which provide a current snapshot of the sector, and opportunities and threats, which are forward-looking. This method, called a SWOT analysis, is useful for understanding the essential characteristics of an industry, and also as a starting point for strategic planning. The results of this analysis are presented in Table 4.4.

Albania's agriculture sector has several important strengths in addition to recent economic stability, including a Mediterranean climate and soil conditions that are well suited for growing high value crops such as olives and medicinal and aromatic plants (Imami *et al.*, 2009). This comparative advantage could prove lucrative as farmers shift production towards high demand products and Albania's export base continues to develop (MAFCP, 2011). In addition, high remittance levels from Albania's large migrant workforce constitute a strength, as well as an opportunity for future growth. Albania has one of the highest levels of remittance as a percentage of GDP received in Europe (over 20% of households), and this inflow of cash represents a significant resource

Table 4.4. SWOT (strengths, weaknesses, opportunities, and threats) analysis.

Strengths	Weaknesses
• Steady and growing gross domestic product (GDP) • Fertile land and Mediterranean climate • High levels of remittance • Large labor force • Shift toward producing agriculture products with higher profitability and export potential	• Small farm holdings and land fragmentation, weak land markets • Low availability of agricultural machinery • Lack of vertical integration • Weak enforcement of quality assurance and health standards • Poor infrastructure • Lack of credit, extension and market information sources • Informal market and taxation issues • Decreasing education profile among rural workforce
Opportunities	**Threats**
• Potential integration of European Union provides incentives and guidelines for sector improvement • Government focus on agribusiness development and land consolidation • High and growing domestic demand for high value, labor intensive products such as fruits and vegetables and medicinal plants • High demand abroad for agricultural products • Increased interest in non-traditional tourism development, such as farm and countryside tours • Access to high quality agriculture-related technical and management skills via overseas training in Greece and Italy • Agricultural and economic research capacity growing	• Pervasive shadow economy and corruption on local and national levels • Absence of land markets, especially in rural areas • Urbanization and cross-border migration of farm labor • Aging of rural workforce • Competition from high quality imports

of external financing (Duval and Wolff, 2009). Not only do remittances increase investment in household operations and improve livelihoods, they are associated with consumption of high value added agricultural products.

There are several important opportunities that value chain participants may take advantage of in the near-term as Albania's agriculture sector develops. First, the Albanian government has implemented policies aimed at agribusiness development, including subsidies for new fruit tree plantations and reduced interest rates for investment. Moreover, EU support in the form of the Instrument for Pre-Accession in Rural Development (IPARD) is expected to provide similar financial incentives, as well as guidelines and recommendations for further development. High demand for agricultural products from foreign and domestic consumers provides incentive to introduce governmental measures to increase productivity and enforce quality and health standards. Opportunities exist to learn modern management and production techniques from abroad, and returning Albanian farmers have used this knowledge to their advantage in recent years. Finally, Albania's diverse landscape, rich cultural heritage, traditional food (lamb and goat meat, organic fruits and vegetables, traditional cooking), and its hospitality may lead to the development of rural tourism once human capital and business management capacities are radically improved.

Perhaps the biggest challenges Albania's agriculture sector currently faces stem from the prevalence of both an informal economy and small, inefficient household farms. These characteristics contribute to the lack of vertical integration as well as weak adherence to quality assurance and health standards. In addition, they inhibit further economic growth that would eventually lead to improvement of infrastructure and availability of modern technology. These challenges will, in part, be mitigated indirectly by increased profitability in the agriculture sector, but governmental initiatives could also help to reduce their impact in the near-term. Eventually, economic growth in this sector should lead to the establishment of credit and information sources, although the government and NGOs can get a head start by providing micro loans, expansion of extension services, and low-cost electronic means of distributing information such as market prices, best practices, and economic forecasts.

Looking forward, the most serious threats to growth in this sector stem from the lack of property transferability and widespread graft, on both local and national levels. Without the ability to increase assets, economies of scale cannot be reached by producers or processors, and there exists little incentive to innovate or improve efficiency. Of major concern is Albania's large informal economy, which reduces the competiveness of legal entities and inhibits the enforceability of contracts and standards. These structural problems are not likely to be mitigated in the short-term and are thus major concerns going forward. Other threats that the sector faces are an aging agricultural workforce, with a decreasing educational profile. This is due in large part to the urbanization and international migration of young people in Albania's rural areas. This labor and human capital outflow is an increasing threat to future agricultural production and is directly related to the lack of profitability in the sector. Decision-makers should consider increasing incentives for farm labor until the profitability of such work is high enough to stabilize rural labor force populations and attract skilled labor. The increased incentives and, particularly, the opportunity for profits that follow economic development in this sector will attract the experienced and educated into existing value chains.

4.5.1 Policy recommendations

The problems identified in this chapter are among the most prevalent and challenging faced by Albania's government today. It is no coincidence that the most insidious problems, such as the informal economy and the lack of land markets, are systemic and directly affect the economy at the national level. The likelihood that they have a causative relationship with many of the other difficulties experienced by actors on all levels of the sector's value chains cannot be ignored. It is these problems that should be targeted by policy. However, the ubiquitous nature of these problems makes mitigating them in a holistic fashion unrealistic, and at this stage in development such wholesale policy overhauls could jeopardize economic stability. Therefore, these central problems are best addressed in a piecemeal fashion, and this is the approach that we suggest. In this section, we make five policy recommendations that we believe will promote the healthy and stable growth of the sector.

First, the Albanian government should continue focusing on programs designed to improve production at the farm level, but increase scope to include processors. For farmers and livestock producers, consistency in quality, price, and other standards, such as size, is the most important factor affecting the competiveness of their products. These inconsistencies also present challenges for processors, and ultimately reduce the quality of products all along the value chain. Current programs do not explicitly consider links between farmers and processors. Improving relationships and communications at these low levels could help improve storage and transportation issues, and would promote vertical integration.

Second, government and businesses should partner together to improve business conditions and competitiveness of existing product lines. The government could consider designing a competitive grants program for industry participants. Such grants could be awarded directly to groups of farmers, processors, and wholesalers, and should focus on improving production, technology, processing, marketing, food safety and quality management, and land consolidation through leasing. In addition, the government could issue grants to well-established organizations, primarily groups of processors and wholesalers, to create 'industry leaders' that would develop programs to improve coordination, cooperation and efficiency among value chain actors.

Third, the government should increase efforts to strengthen services and improve infrastructure. Importantly, extension and research services need to be strengthened. There is a lack of research in many key areas, including industry and farm profitability, although the capacity for such research is growing, most notably through AUT. The government should initiate a competitive grants program to fund applied research in order to promote sustained development in the sector. Also, steps should be taken to develop information networks that provide product quality standards for targeted export markets, current prices, and customized advice to farmers and processors. These networks could be established directly by the government or indirectly through support to market information providers, consultancies, research universities, and extension services. In terms of infrastructure, irrigation structures and roads are dysfunctional and need substantial reconditioning, especially in rural areas. Such infrastructure issues are frequently cited as major constraints to economic growth in the agriculture sector.

Fourth, priority should be given to projects that stimulate further development of rural land markets. Increased enforcement of contractual agreements and policies aimed at strengthening land tenure rights and nurturing mutual trust among land owners will likely be key steps in this process. Once rural land markets are established, successful producers will have a means by which to consolidate land for farming or pasture, and reach economies of scale which will drive rapid growth throughout the sector.

Finally, the government must invest and commit to improving food safety and quality assurance programs and their enforcement. It should also take steps to ensure stronger contract enforcement and increase the competiveness of those industry participants that operate legally, that is, in the formal economy. Such steps could include tax breaks for accountable businesses, improvement of public institutions—especially tax collection agencies, efforts to educate the public about consequences of participation in the informal economy, and, perhaps most importantly, increased efforts to fight corruption at all levels of government (Bello *et al*., 2011).

These foci for national policy are important steps on the path toward the maturation of Albania's market-based agriculture sector. As these fundamental elements of Albania's agricultural industries improve, the positive effects will reverberate up each value chain and foster the establishment of a profitable exportation industry and create synergies with other sectors, resulting in national economic development.

4.6 Conclusion

In the last decade, Albania has achieved a high level of stable economic growth and has outperformed many of its regional and economic peers. This is a laudable accomplishment, especially considering the magnitude of its economic collapse in the early 1990s. This positive trend is reflected in other internal measures of development and economic well-being, such as employment and poverty alleviation, and is evidenced on the international stage by its growing trade presence and reclassification as an upper middle income nation. Like other transitioning economies, however, continued growth is highly dependent on an agriculture sector that is plagued with low productivity and a high dependency on imports. While both of these trends are improving, major hurdles must be overcome before Albania can reach its full potential.

Perhaps the two most fundamental impediments to growth have roots in the transition period, and are defining characteristics of Albania's agriculture sector. The first is the prevalence of small, subsistence-oriented household producers that cultivate small and often non-contiguous plots. This fractional and inefficient production base is a direct result of the per capita privatization process during transition, and a subsequent lack of rural land markets. The second is a burgeoning informal economy that persists on every level of Albania's agricultural value chains, but is especially predominant among small farms and food processors. While the establishment or growth of informal economies during economic transition is common, the extent of

Albania's informal economy is quite high—nearly 50% of GDP by some estimates. These systemic problems impede growth by preventing consolidation of resources, discouraging investment, and hindering the operation of quality assurance systems. These conditions, in turn, limit the expansion of value chain services such as availability of credit, information sources, and extension and consultation. Without a means to reach scales of efficiency or ensure a level playing field for all market participants, modernizations in the form of mechanized equipment or quality management systems are out of reach or uneconomical for the majority of operators. As a result, production costs remain high while product quality is low or inconsistent. Low profitability in this sector also drives away an able-bodied and trained rural workforce—a workforce that is key to ensuring production in the long-term. On a macroeconomic level, low agricultural production constrains GDP growth, especially in agrarian societies, and contributes to the trade deficit for agricultural products.

If Albania is able to leverage its strengths, such as fertile land, a good growing climate, a large agriculture labor force, and the availability of high-margin agriculture products, while reducing the effects of the challenges mentioned above, it will be in a better position to attract private and foreign direct investment. This investment will, in turn, open markets and prime economic development in the agriculture sector. With improved access to markets and product quality, Albania can establish an agricultural export base and take advantage of domestic demand, which will lay the foundation for increased industrialization. Once Albania is able to reach markets in Northern Europe, the export of quality agricultural products could serve as a primer for rapid growth and might eventually facilitate the incorporation of Albania into the EU.

Recent moves by the Albanian government to improve the competitiveness of the entire sector, especially at the producer and processor levels, are a step in the right direction, as is its aggressive trade liberalization policy. Some problems, such as infrastructure and the development of extension resources, are fairly straightforward and continue to be targeted by policymakers for improvement. Others, such as unclear property rights, weak contract enforceability, and the informal economy, are well embedded within Albania's economic systems, and their causes and mitigation will likely prove more complicated. A multi-pronged approach to solving these systemic issues, such as providing grants for industry leaders, promoting the development of land markets, and incentivizing companies to operate within the formal economy, will be crucial for tackling such problems. Achieving the transformation of Albania's agriculture sector from smallholder subsistence farming to an internationally competitive industry remains a challenge. Improvement of value chains at the local and national levels has promising implications for the continued development of Albania's agriculture and, ultimately, for the livelihood of all Albanian citizens.

References

Bello, M., Çuçllari, F. and Cini, M. (2011) On the development of the Albanian informal economy. *Management and Marketing* 9, 315–322. Available from: http://www.mnmk.ro/documents/2011-2/14_Albania_FFF.pdf (accessed 19 March 2012).

Bode, R., Canco, G. and Muca, B. (2009) The informality in agriculture Albania's [Albania's agriculture] and the ways for its reduction. In: *Banat's University of Agronomical Sciences and Veterinary Medicine International Scientific Symposium, Timisoara, Romania, 14–15 May 2009. Lucrări Ştiinţifice* 11(2), 135–140. Available at: http://www.usab-tm.ro/pdf/2009/xi2/management09_18.pdf (accessed 20 February 2012).

Bonneau, A. (2011) The effect of privatization method on inequality in Eastern European transition economies. MSc thesis, Stanford University, California.

CIA (2011) The World Factbook. Europe: Albania. Central Intelligence Agency, Washington, DC. Available at: https://www.cia.gov/library/publications/the-world-factbook/geos/al.html (accessed 31 August 2012).

Duval, L. and Wolff, F.C. (2009) *Remittances Matter: Longitudinal Evidence from Albania*. Laboratoire d'Economie et de Management Nantes-Atlantique (LEMNA), Université de Nantes, France. Working Paper EA4272, Document No. hal-00421234, version 1. Available at: http://hal.archives-ouvertes.fr/docs/00/42/12/34/PDF/LEMNA_WP_200920.pdf (Accessed 31 August 2012).

EBRD and FAO (2007) *A Systematic Analysis of the Agribusiness Sector in Transition Economies. The Albanian Olive Oil Sector*. Commissioned by the European Bank for Reconstruction and Development (London) and the Investment Centre of the Food and Agriculture Organization of the United Nations (FAO), Rome. Available at: http://www.eastagri.org/files/ALB_Oliveoil_VC_final.pdf (accessed 31 August 2012).

Fortuzi, S., Tota, O., Sallaku, F. and Musbelliu, B. (2008a) The impact of land reforms in land use changes and environmental situation in Albania. *Research Journal of Agricultural Science* (Romania) 40(1), 679–688.

Fortuzi, S., Pagria, I., Musabelliu, B. and Sallaku, F. (2008b) Role of marketing cooperatives in increasing farmers' partnership on Albanian markets. *Research Journal of Agricultural Science* (Romania) 40(3), 183–190.

GIZ (2010) *Fruits and Vegetables Sector Study. Research report prepared in the framework of IPARD Program – Albania*. Deutsche Gesellschaft für Internationale Zusammenarbeit, Tirana, Albania.

Jarvis, C. (2000) The rise and fall of Albania's pyramid schemes. *Finance and Development* 37(1). Available at: http://www.imf.org/external/pubs/ft/fandd/2000/03/jarvis.htm (accessed 31 August 2012).

Imami, D., Leonetti, L. and Stefanllari, A. (2009) Value chain structure and governance changes in medicinal and aromatic plants sector in Albania – Implications for poverty reduction and rural development. Paper presented to: *19th IAMA [IFAMA] Annual World Forum and Symposium Budapest, 20–23 June 2009, Budapest, Hungary*. International Food and Agribusiness Management Association. Presentation available at: http://www.eoq.hu/iama/conf/imamileoneistfanllari.pdf (accessed 31 August 2012).

Kipi, A., Skreli, E., Uruçi, M. and Mersini, K. (2010) *Study of Small Ruminants Sub-sector in Albania. Research Report Prepared for SNV-Albania in the Framework of ProMali Programme*. SNV (Netherlands Development Organisation)-Fier, Fier, Albania.

Lerman, Z. (2001) Agriculture in transition economies: from common heritage to divergence. *Agricultural Economics* 26, 95–114.

MAFCP (2011) *Draft IPARD Programme for Albania, Second Draft (Version 2.1)*. Ministry of Agriculture, Food and Consumer Protection, Tirana, Albania.

Mathijs, E. (2002) Commercialization and subsistence in transaction agriculture: empirical evidence from Albania, Bulgaria, Hungary, and Romania. Presented to: *Xth EAAE Congress. Zaragoza, Spain, 28–31 August 2002*.

Meço, M., Sula, R., Keco, R. and Çerava, A. (2010) Assessment of profitability of dairy sector in Albania. *Annals. Food Science and Technology (AFST)* 11(2), 154–161. Available at: http://www.afst.valahia.ro/docs/issues/2010/issue%202/2010-2-5-1-Meco.pdf (accessed 31 August 2012).

Michalopoulos, C. (1998) *WTO Accession for Countries in Transition*. World Bank Policy Research Working Paper No. 1934. Available at: http://papers.ssrn.com/sol3/papers.cfm?abstract_id=620568 (accessed 31 August 2012).

Petrick, M. and Weingarten, P. (eds) (2004) *The Role of Agriculture in Central and Eastern European Rural Development: Engine of Change or Social Buffer? Studies on the Agricultural and Food Sector in Central and Eastern Europe, Volume 25*. Institüt fur Agrarentwicklung in Mittel- und Osteuropa (Institute of Agricultural Development in Central and Eastern Europe, IAMO), Halle (Saale), Germany. Available at: http://www.iamo.de/dok/sr_vol25.pdf (accessed 31 August 2012).

Roduner, D. (2007) *Donor Interventions in Value Chain Development*. Working Paper, Community of Practice on Value Chains in Rural Development. VCRD CoP, Berne, Switzerland. Available at: http://www.networklearning.org/index.php?option=com_docman&task=doc_download&gid=115&Itemid=52 (accessed 31 August 2012).

Ruli, G. and Belortaja, S. (2003) Albania – the difficult road towards sustainable growth. In: *Monitoring Albania's Path to European Integration*, 1st edn. Friedrich Ebert Foundation, Tirana, Albania. Available at: http://pdc.ceu.hu/archive/00004737/01/SB_GR_Feb_2004_en.pdf (accessed 31 August 2012).

Skreli, E. (2008) Study on Agrobusiness Market – Fier Qark. SNV (Netherlands Development Organisation)-Fier, Fier, Albania.

Skreli, E., Chan-Halbrendt, C. and Balliu, A. (2009) Competitiveness of Albanian agriculture: value chain analysis for fruits and vegetables sub-sector in Fier Region. Paper presented to: *19th IAMA [IFAMA] Annual World Forum and Symposium Budapest, 20–23 June 2009, Budapest, Hungary*.

International Food and Agribusiness Management Association. Available at: http://www.eoq.hu/iama/conf/1195_paper.pdf (accessed 31 August 2012).
US State Department (2011) Background Note: Albania. Available at: http://www.state.gov/r/pa/ei/bgn/3235.htm (accessed 31 August 2012).
USAID (2011) *USAID Country Profile, Property Rights and Resource Governance, Albania*. US Agency for International Development, Washington, DC. Available at: http://usaidlandtenure.net/sites/default/files/country-profiles/full-reports/USAID_Land_Tenure_Albania_Profile_0.pdf (accessed 17 September 2012).
World Bank (2007a) *Albania: Urban Growth, Migration and Poverty Reduction: A Poverty Assessment*. Report No. 40071- AL, Poverty Reduction and Economic Management Unit, Europe and Central Asia Region, World Bank, Washington, DC. Available at: http://siteresources.worldbank.org/PGLP/Resources/Albania_PA_Grey_Cover_2007_FINAL.pdf (accessed 31 August 2012).
World Bank (2007b) *Albania: Strategic Policies for a More Competitive Agriculture Sector*. Report No. AAA18-Al, World Bank Sustainable Development Sector Unit, Europe and Central Asia, World Bank, Tirana, Albania. Available at: http://siteresources.worldbank.org/INTALBANIA/Resources/Strategic_Policies_for_more_competitive_Agriculture_sector.pdf (accessed 31 August 2012).
World Bank (2011) *World Bank – Albania Partnership Program Snapshot, September 2011*. World Bank, Washington, DC. Available at: http://siteresources.worldbank.org/INTALBANIA/Resources/Albania_Snapshot.pdf (accessed 31 August 2012).
World Bank (2012) World Development Indicators (WDI) and Global Development Finance (GDF). World dataBank. Available at: http://databank.worldbank.org/ddp/home.do?Step=12&id=4&CNO=2 (accessed 31 August 2012).

5 Agricultural Production and the Food Processing Industry in Albania

Ilir Kapaj, Remzi Keco, and Ana Kapaj
Agricultural University of Tirana, Albania

5.1 Overview of the Agricultural Sector: from a Centralized To an Open Economy

The agriculture sector is one of the most important in the Albanian economy, representing 19% of GDP (gross domestic product) in 2010 (MAFCP, 2011). This is not only a result of agriculture's contribution to the country's GDP, but also because of the level of population that is concentrated in rural areas (MAFCP, 2011). Viewed in this context, the progress of the agriculture sector in the past, present and future has been, and remains, a determining factor in the development of the country. But some aspects of the sector, such as size, importance, structure, competitiveness, performance, and investment level have undergone dramatic changes as a result of changes in the economic and political system in the early 1990s. Before analyzing the development of the agriculture sector in the past two decades, it is important to understand its position in the socialist years of 1945–1990.

As with all forms of production in communist Albania, before 1990 agricultural production was totally organized and controlled by the government. It should be emphasized that before the 1990s, the production structures of the agriculture sector were not based on free market economic principles. Instead, they were based on the communist system's strongly held political views of the necessity of securing the country's food needs through domestic production. All agricultural products and quantities of produce were controlled by the central government, with all production units organized into state cooperatives. Under this framework, individual farmers owned no land, but rather were hired to work by the cooperatives. As a result, they seldom took part in any decisions on agricultural production. Before the economic and political transitions of the 1990s, there were about 160 state cooperatives (organized in peripheral urban agricultural lands, with workers from urban areas) with a land area of about 170,000 ha, and 492 agricultural cooperatives (organized in rural areas, with peasant workers) with a total land area of about 530,000 ha (World Bank, 2006). During the communist system, the share of agricultural sector in the country's GDP has changed little. It has ranged from 27.5% of the total GDP in 1960 to 25.1% in 1990 (INSTAT, 1992).

After the political and economic changes of 1990, Law No. 7501 was drafted in order to ensure a fair division of land among agricultural families. However, one of the consequences of this policy was that many land plots were fragmented, with families owning several non-contiguous parcels spread over a wide territory. This makes it difficult to take advantage of economies of scale. Albania also has among the lowest amount of agricultural land per capita (0.22 ha) in the region, although as a result of the privatization process, over 90% of agricultural land is now in private ownership. This early transition period was followed by a 21% contraction in agriculture sector production in 1991, though agricultural production started to increase in 1992 in response to land privatization and the elimination of fixed pricing. Generally speaking, trends in agricultural production have been positive over the last 18 years (1992–2010); more specifically, in the last 10 years, there has been an accelerated trend in production. This has been a result of increased demand for agricultural products and is also a result of farmers' responses to markets.

5.2 Trends in Agricultural Production

Since 1991, agricultural production has been driven by market forces. Farmers make their own farm production decisions, producing for markets and also for their own consumption. Farm production first fulfills the family's needs, and the rest goes to the market. Farms in Albania are generally characterized by semi-subsistence and mixed production systems. On the majority of farms (56%), the most dominant mixed production system is currently cereals–forage–livestock–vegetables–orchards, and this follows two growing seasons a year (MAFCP, 2009).This is classified as a 'mixed farming system,' and is more appropriate for subsistence farming. Such production structures tend to be oriented to meet the subsistence needs of rural farm household families first.

From 2000 to 2010, the trend in the agricultural production has been positive, and the value of total agricultural production has increased by 36% (MAFCP, 2011). More specifically, there has been a 25% increase in the value of livestock production, 25% increase in the value of field crop production, and 134% increase in the value of fruit production (Table 5.1). To better understand agricultural production in Albania during these years, each sub-sector will be analyzed separately, looking at past production trends and the reasons behind these trends.

5.2.1 Livestock production

Livestock production represents the largest share of agricultural production, with 52.2% of the total production value in 2010. Animal products—meat, dairy products and eggs—are a major source of food. Most farming families use the majority of livestock they produce for their own consumption, and little is sent to market. Milk and milk products have a long production tradition in Albania. In the lowland areas, cattle are the dominant livestock, while in hilly and mountainous areas, small ruminants (sheep and goats) dominate. Because of the importance of livestock farming in general, and milk production in particular, for rural areas, the Albanian Ministry of Agriculture, Food and Consumer Protection (MAFCP) has selected the dairy sector as a priority in its supporting policies. One of the stated objectives of Ministry policies is to improve the competitiveness of livestock and milk products in order to replace imports with domestic production, and also to potentially grow exports through the reduction of export taxes. However, there are still significant

Table 5.1. Agricultural production in value (million lek (ALL) in Albania, 2000–2010. Source: MAFCP (2011).

Production type	2000	2005	2008	2009	2010
Livestock	71,460	84,276	86,882	87,727	89,770
Field crops	42,483	42,392	45,741	47,352	53,485
Fruits and vegetables	12,173	15,552	23,046	24,001	28,547
Total	126,116	142,220	155,669	159,080	171,802

problems in this sector, such as those of the traceability of the farming families concerned, level of investment and breed improvement.

In terms of production, cow's milk is by far the biggest commodity in livestock production, followed by beef and veal. The total number of cattle has declined slightly—by nearly 10% from 2001 to 2010, whereas over the same period, beef/veal production has increased by 25%, and milk production by 30% (Fig. 5.1).

More than half of the milk produced in the country is consumed on farm or fed to farm animals (i.e. calves, lambs, kids), while the rest is sent to market. Urban consumers are increasing their preferences for processed milk products, such as pasteurized or condensed milk, fruit yogurt, and ice cream. According to Kapaj *et al*. (2011), 52% of Albanian consumers prefer pasteurized processed milk.

Cattle meat production increased by 25% from 2001 to 2010 and reached 68,000 t in 2010. Beef/veal production amounted to approximately 58 million euros in 2010, making it the second most important commodity in Albanian agriculture in terms of production value (MAFCP, 2011). The increase in demand for meat has also increased farmers' interest in meat production. This interest has been reflected in a shift in the breeds used for meat production in the cattle sub-sector to breeds such as Limousin and Simmental. Sheep and goats are important for both milk and meat production.

Sheep meat, sheep's milk, goat meat and goat's milk together accounted for 50 million euros in 2009. Small ruminant production is an activity that mostly dominates in the mountainous and hilly regions of Albania as these geographical areas, which are mostly pastures and grassy lands, are appropriate for sheep and goat rearing. The number of small ruminants decreased slightly from 2000 to 2010. While in 2000 there were approximately 3,043,000 animals, in 2005 there were only 2,701,000, and in 2010 the number decreased further to 2,581,000 head (Table 5.2). During the transition years, Albania experienced a significant migration of its population to urban areas due to the political and economic changes in Albania. This migration has had an impact on the livestock production, resulting in many abandoned animal farms and a declining trend in number of small ruminants.

Total production of cow's milk has increased in recent years and reached some 1.07 million t in 2010, with a value close to 170 million euros (MAFCP, 2011). Looking at the trends of milk production over the years, it is obvious that the main contribution comes from the cow's milk. From 2000 to 2005, total cow's milk production increased by

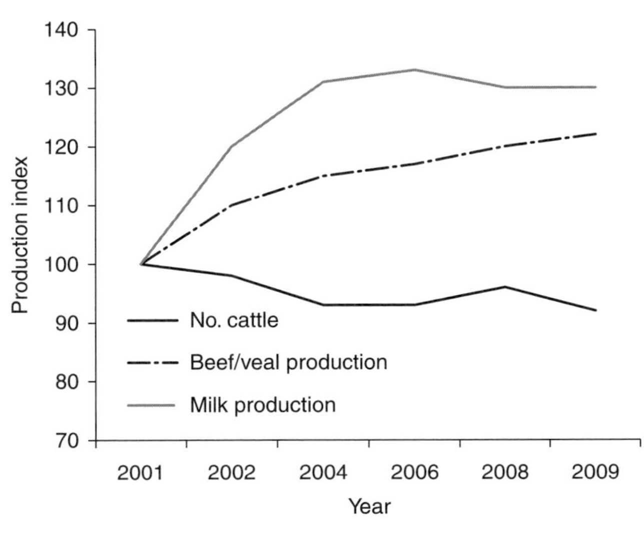

Fig. 5.1. Development of cattle herd size, milk and beef/veal production in Albania, 2001–2009 (index for 2001 = 100). Source: MAFCP (2010).

approximately 13%, with a further increase in 2006, a reduction in 2007 and a return to the 2005 value in 2010 (Table 5.3). The increase in production in earlier years was mostly due to increases in the percentage of highly productive animals. The annual average yield of cow's milk is 2100 l/cow, comparatively lower than in other countries of the region, such as Macedonia, Bulgaria and Romania. The improvements in productivity are due to livestock breeding and the development of milk collection centers with quality standards. The storage facilities of these collection centers has led to improvements in the local marketing systems for milk from small-scale producers (in areas around the centers).

Chicken production saw a steady increase in numbers from 2000 to 2010 owing to increased domestic demand for eggs as well as increased exports to neighboring countries. In 2010, some 8,437,000 animals were recorded, almost 60% more than in 2000. Egg production is, in terms of production value, still more important than the production of poultry meat (respective values 17 million and 5.3 million euros) in 2010.

5.2.2 Field crop production

The second largest production sub-sector of Albanian agriculture is that of field crops (cereals only), which represents 31.1% of the value of the agricultural production sector. The production of vegetables has been higher than that of cereals (Fig. 5.2). This is because of the continuous increase that there has been in greenhouse vegetable production. Vegetable production has increased by 36.7% from 2000 to 2010, while field crop (cereal) production has increased by the smaller value of 22.6%.

To better understand the importance of field crop production in Albania, it is necessary to investigate trends in production over the last decade (2000–2010): according to data from MAFCP (2011), the area under cultivation has decreased for cereals (−17.1%), oilseeds (−34.7%), pulses (−28.4%), and roots and tubers (−21.1%) (Table 5.4). Production of these crops has seen a corresponding decrease, with the production of cereals decreasing by 9.7%, oilseeds by 22.9%, pulses by 6.3%, and roots and tubers by 19.9%. According to the MAFCP, these trends of production reflect a continuous drop in consumer demand for domestic production of these crops (MAFCP, 2010).

As MAFCP data shown in Table 5.4 indicate, the trend in cereal production from 2000 to 2010 has been toward a reduced area under cultivation (−17.1%), which has decreased production (−9.7%), but not at the same rate at which cultivated area has been reduced. As a result, cereal production in Albania results to be economically is not profitable because the high cost of production means that it has no competitive advantage (Agraja, 2007). These changes in

Table 5.2. Number of livestock (1000 head) in Albania, 2000–2010. Source: MAFCP (2011).

Livestock type	2000	2005	2008	2009	2010
Cattle	728	655	541	494	493
Sheep	1939	1760	1800	1768	1806
Goats	1104	941	820	772	775
Pigs	103	147	113	102	98
Poultry	5291	6432	8100	8313	8437

Table 5.3. Milk production in Albania, 2000–2010 (1000 t). Source: MAFCP (2011).

Milk type	2000	2005	2006	2007	2010
Cow's milk	807	930	956	868	930
Sheep's milk	70	75	75	75	72
Goat's milk	71	71	71	73	68
Total milk	948	1076	1102	1016	1070

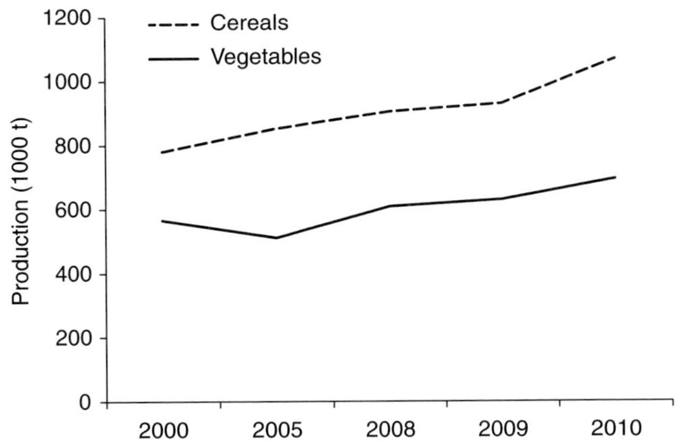

Fig. 5.2. Field crop production (1000 t) in Albania, 2000–2010. Source: MAFCP (2010).

Table 5.4. Crop area (1000 ha), crop production (1000 t) and crop yield trends in Albania, 2000–2010. Source: MAFCP (2011).

Crop area/ production	2000	2003	2005	2008	2009	2010
Crops (1000 ha)						
Cereals	178.2	165.8	157.7	153.3	143.9	147.7
Oilseeds	2.3	2.2	1.9	1.8	3.1	1.5
Pulses	22.5	18.5	17.4	17.4	17.2	16.1
Roots and tubers	12.8	12.5	12.3	10.9	10.7	10.1
Fruit (000 trees)	9,575	9,815	10,291	10,959	11,398	11,934
Vegetables	32.8	33.5	33.6	32.4	32.0	32.5
Production (1000 t)						
Cereals	565.8	502.5	518.9	488.5	498.8	511.2
Oilseeds	3.5	3.5	3.0	1.8	2.6	2.7
Pulses	25.2	22.1	24.5	17.8	22.4	23.6
Roots and tubers	203.0	202.2	193.1	168.2	160.0	169.3
Fruit	183.0	191.3	189.6	215.2	246.0	152.0
Vegetables	620.0	651.0	668.5	675.0	677.4	684.9

cereal production have not significantly affected the massive trade deficit in cereals (MAFCP).

Wheat is cultivated in all regions of Albania, but especially in the Korça region (southeast of the country), where the yields are relatively high (about 4.5 t/ha). A lower yield is achieved in the south coastal areas (about 3.5 t/ha) and an even lower one on the northern hills (about 2.5 t/ha). This difference in productivity is primarily a result of the land reallocation process and land fragmentation. Available gross margin data suggests, on average, that wheat delivers a return of no more than 200 euros/ha.

Maize was the sixth most important commodity in the Albanian crop sector in 2009 (MAFCP, 2009).The total area cultivated with maize is approximately 50,000 ha. The average yield of 4 t/ha is very small in comparison with surrounding countries. Total maize production decreased by 9% in 2009 (around 200,000 t) although the development of the livestock sector demands large quantities of this crop. The lack of advanced technologies and irrigation systems, as well as the low regional price for maize in the years 2000–2009 are some of the reasons why production has not increased.

5.2.3 Fruit and vegetable production

Fruits and vegetables represent the third most important (after livestock and field crops) group of commodities according to production value, and represented 16.6% of the total production value in 2010. The total production of this sub-sector has increased by 134.5% from 2000 to 2010 (Table 5.1). Data for the years 2000–2010 show that the production of vegetables has increased continuously, as demonstrated by both a stable total area under cultivation and a steadily increasing total production.

About 255,000 farms, or more than two thirds of the total number of Albanian farms, produce vegetables and grapes for fresh consumption or for wine production, and 152,000 farms (43%) produce fruit. Most of these farms (about 85%) are small and traditional subsistence farms, but about 10% are market oriented larger traditional holdings, which make greater use of inputs, of hired workers, and of mechanized agronomic services. The remaining 5% of farms are more oriented toward specialized and intensive productions, mainly greenhouse production, which made up 8.2% of the total vegetable production in 2009 (MAFCP, 2010).

In 2009, the production base consisted of about 30,000 ha cultivated with open field vegetables and melon crops, 10,000 ha cultivated with potatoes, 14,000 ha cultivated with beans and about 730 ha of protected crops (MAFCP, 2009, 2011). There were about 10.5 million fruit trees, of which about 86% (9 million) were citrus crops. Total vegetable output included 570,000 million t (Mt) of open field vegetables, 60,000 Mt of greenhouse vegetables, 240,000 Mt of melons and watermelons, 200,000 Mt of potatoes, and 23,000 Mt of beans, while fruit production reached 150,000 Mt, corresponding to 9.9 million fruit trees (2009). In addition, about 163,000 Mt of grapes and 10,000 Mt of citrus fruits are also produced.

The total area cultivated for vegetables has remained more or less stable, while production increased continuously up to 2009, mainly as a result of yield increases. Yields have improved as a result of the introduction of new technologies, including new hybrids, improved cultivation techniques, modern irrigation systems and greenhouse production. Some of the most important problems hampering further development in the sub-sector are limited quality control systems, lack of an effective marketing system and poor quality of the transport infrastructure.

Fruit production, mainly apple and grapes, provided on average more than 20% of the total agricultural output in 2009 (MAFCP, 2009, 2011). Output in value more than doubled from 2000 to 2009. In the same period, the contribution of fruit production to the national agricultural output increased from 9.7% to 20%. Grapes, grown over 8000 ha, were the fourth most important commodity in terms of production value in 2010. Grape production has increased 38% since 2000 (MAFCP, 2011), but has been relatively stable over the last 3 years (2008–2011), with some 80,000–90,000 t/year being produced, at a yield at about 8 t/ha. In 1990, only 7000 ha (36%) of vineyards remained out of more than 19,000 ha that existed in 1989, and only 4.3 million (70%) pergola vines remained out of more than 6.1 million that existed before the 1990s. This is because of the political and economic changes that occurred in 1990, which were followed by a large-scale destruction of state property such as vineyards. Thus, the increase in recent years reflects the gradual recovery of this sector.

5.3 Overview of the Food Processing Industry in Albania

The food processing industry in Albania comprises all enterprises dealing with the processing of fresh agricultural products (livestock products, milk, cereals, vegetables, fruits and fish products). During the communist period, this industry was entirely state owned. However, since the beginning of the transition to a market economy, most state-owned processing enterprises have been partially or totally closed. Privatization has become a first step toward industrial revitalization, and private food processing enterprises began to develop in the early 1990s; at present, the industry has been almost 100% privatized. Despite its rapid development, and 10% annual growth, the processing industry in Albania is still immature and very small compared with the European food processing industry overall. The latter has passed through consolidation processes, a significant part of which has taken the form of cross-border mergers and acquisitions, according to a UNDP (United Nations Development Programme) study in 2005 (UNDP, 2005). This consolidation was stimulated by market pressures, such as increased competitiveness and legislative regulations in terms of quality (driven by arising consumer awareness on food quality and safety), and has led to increased quality goods at lower prices.

According to the UNDP study (UNDP, 2005), consolidation processes have led to major changes in Europe in terms of the structure of the food processing industry. The number of food processing companies in most European Union (EU) Member States has dropped by 50% since 1990, and consolidation has led to further cost efficiency increases in the sector. This process in Western Europe has, in turn, put Albanian companies in a weak competitive position (a situation that still exists) because of the lower quality and higher cost of production in the Albanian processing industry (UNDP, 2005). Food processing enterprises in Albania operate under conditions which are not compliant with EU food safety standards (excepting some large enterprises that do fulfill these safety standards). As a result, products cannot be sent to EU markets. Investments and support by rural development policies, in terms of regulations enforcement bodies and food safety information and guidelines, are needed in order to increase the competitive capacity of the agro-processing sector in Albania.

Based on total employment and total added value in the Albanian agricultural sector, the overall productivity of Albania's food processing sector is about 28,000 euros/worker (UNDP, 2005). This figure is higher, on average, than that of new EU members, but lower for EU members of longer standing. For Estonia, Lithuania, Hungary and Slovakia (newer members of the EU from Eastern Europe) an average productivity of 15,000 euros has been reported, while the average for Belgium, Denmark, Finland and the UK (longer standing members) is 50,000 euros. While there is substantial entrepreneurial interest in the Albanian food processing industry, several factors hinder its development (UNDP, 2005). These include problems of raw materials supply (small-scale primary production and weak infrastructure), lack of access to finance at a time where there is a major need for capital for modernization, and inadequate business management experience and marketing skills.

The food processing industry contributed about 42% of the total value produced by the whole agriculture sector in 2010, when it reached its highest level of 40 million euros—at a time when the Albanian food processing industry was still in its infancy. Before 1990, the food processing industry processed about 50–80% of all agricultural products, as well as marketing the products. After 1990, some commodities, such as the meat and olive subsectors, used primarily imported raw materials as a consequence of insufficient domestic supply. During the last five years, the development of the industry has been rapid, with a mean annual growth rate of 15%. The most rapid growth is evidenced in the milk and meat processing industries. The growth of these two sectors is influenced by the fact that domestically produced meat and milk (especially domestic meat) is preferred over imported meat, and meat and milk are very much components of the traditional Albanian diet (INSTAT, 2011).

The processing industry is concentrated in the Tirana region as well as in other highly populated regions, (Durres, Fieri, Shkodra) owing to the growing demand for food and because of labor availability in these regions. In 2010, there were 2156 enterprises in the food industry sector, of which 60% produced bread and flour, and 16% produced dairy products (MAFCP, 2011).The number of processing enterprises in 2010 increased by 4.6% compared with the period 2005–2006.

The number of employees in processing industry in Albania has increased by 9.5% in 2010 compared with 2006. This trend was driven by increased investments in food processing industry during this period. The investments came mostly from the private entrepreneurs (55% up to 65% of the total investments in the industry) and the rest from other alternative sources (foreign credits, bank credits and from government budget). The availability of labor is an important factor, as the food industry employs more than 10,000 people (INSTAT, 2011). The largest number of employees is involved in bread production (3656 employees, or 34%), followed by those in milk processing units (1122 employees or, 10%) (MAFCP, 2011). Besides the retail and wholesale meat and beverage sub-sectors, most of the food industry companies in Albania are of relatively small size. Over half (57.5%) of the enterprises have 1–10 employees, nearly 36% of them have 11–50 employees and the rest of them, approximately 7% of enterprises, have 51–250 employees; none have more than 250 employees (Kapaj, 2012). Being mostly small in terms of production scale is a reason why the industry faces problems related to production ineffectiveness and lack of financial means for investment in quality improvement.

Investment is another important indicator showing the sustainable development of the food industry in Albania. Investments have increased, achieving the highest level in the year 2005, with about 2.9 billion lek (ALL). For the years 2008, 2009 and 2010, investments were lower, 2.06, 0.611 and 0.524 billion lek, respectively. The reasons for the decrease in investments were the global economic crises and, in particular, the Greek and Italian crises (Greece and Italy are the main trading partners of Albania), the consequences of which are hampering development of the Albanian economy. Two main features can be observed when analyzing these investments. Firstly, investment levels in processing technology improvement (buying new technologies and upgrading existing ones) have increased each year on a level comparable with investments in the construction of processing plants. This indicates that, following the expansion of this industry, food quality improvement has started. Second, from 2006 to 2010, the largest portion of investments has been insured or guaranteed by local private investors and, in terms of loans, only a small portion has come from the banking sector. The government must address this aspect, and support processing enterprises in obtaining both more public and private loans for technology improvement and for promoting vertical integration of the industry by means of more favorable interest rates.

In recent years there has been an increasing awareness of food quality standards. A study by Kapaj (2012) found that 75% of food and beverage enterprises understand the necessity of implementation and certification of quality management systems, which is 15% higher than it was in 2004. Furthermore, the Agricultural and Food Sector Strategy (AFSS) for 2007–2013, drafted by the Albanian government, emphasizes the importance of quality improvement, and establishes governmental institutions and agencies in charge of providing food quality and safety information and training to processing enterprises.

Quality of products is affected by the quality of raw materials. In the case of Albania, the raw materials produced in the country do not meet either the required quality or quantity, and also are not price competitive. This is due to low yield and the lack of modern equipment. The frozen vegetable industry illustrates this, as more than 2000 t of raw vegetables per year (about 90% of raw materials used by that industry), are imported from Greece, Italy and Macedonia. Prices of these raw material imports are lower than those of the raw materials produced domestically. As a consequence, the development of the vegetable sector in Albania is not stimulated. Putting in place custom duties for imported vegetables is one method of making domestic raw materials competitive.

With the intention of investing in the food industry, Deutsche Investitions- und Entwicklungsgesellschaft mbH (DEG) (German Investment and Development Company) conducted a study that summarized information from existing food processing industry studies industry in eight countries in southeast Europe (Albania, Bosnia and Herzegovina (BiH), Croatia, Kosovo, Macedonia, Moldova, Montenegro, and Serbia), with the particular aim of identifying potential larger scale investment opportunities (GFA Consulting Group, 2010).This study developed a SWOT (strengths, weaknesses, opportunities, threats) analysis specifically for Albania's food industry (Table 5.5). The study found the strengths of the processing industry to be its low labor costs, entrepreneurial skills and establishment of a governmental Food Safety Agency for quality and safety assurance. Weaknesses identified included lack of trained managers, high costs of domestic raw material inputs, lack of modern technology and economies of scale, lack of capacities to access financing and the high costs of quality and safety certification for exporting.

The Albanian food processing industry is often mentioned by policymakers and experts as having good potential for growth, owing to the sound growth rates of 15% from 2001 until 2010. Specifically, low labor costs and suitable climate conditions represent a potential for increasing the domestic production of raw material for the industry. Likewise, the recent establishment of the Food Safety Agency will strengthen law enforcement on food quality and safety, and will result in product quality improvements.

While the potential of a number of subsectors (i.e. milk and meat) of the Albanian food processing industry is often referred to as high, the overall current comparative position is low (compared with imported food products). This is due to the high cost of raw materials, particularly domestically produced fresh meat compared with imported meat. One of the many hurdles (weaknesses) faced by the Albanian food processing industry is that it has a relatively short history compared with many other countries, accompanied by a lack of necessary management skills. Most of the former structures and companies disappeared in the early 1990s, or had to be reconstructed to meet the needs of modern open economies. The most pressing problem for the sector is the predominance of small-scale production

Table 5.5. SWOT analysis of the food processing industry in Albania. Source: GFA (2010).

Strengths	Weaknesses
• Low labor costs • Business orientation and entrepreneurship • Food Safety Agency recently established	• Lack of management experience (short life of the enterprises) • Low supply of raw material to processing industry due to VAT (value added tax) exemption of farmers • Low level of technology • Small-scale production systems • Lax law enforcement in food safety • Weak financial position and lack of capacity to access financing • Certification bodies for quality management (hazard analysis and critical control points, HACCP) not accredited, leading to higher certification costs when exporting
Opportunities	**Threats**
• Low labor costs • Consolidation in primary production, processing and trade • Improvement of product quality • Certification with quality management systems and hygiene standards	• Competition from imports • Corruption • High cost of domestic raw materials • Infrastructure (power cuts, road network, waste/wastewater management)

systems in primary production (fresh agricultural production), and corresponding problems with the quality and quantity of domestic supply (UNDP, 2005). Consequently, dependency on imports of raw materials (95% of raw materials are imported) leads to a relatively low cost structure. In addition, non-existent or underdeveloped supply chain management causes poorly developed marketing strategies and weak branding. This is compounded by weak language skills, which cause inappropriate translation of guides and EU regulations on food quality and safety. Lastly, the weak financial position and low capacity to access financing are hampering the opportunities of industry to grow and increase capacities.

Though the industry faces many obstacles, it has several opportunities for expansion as well. One of the first opportunities of which the Albanian agribusiness sector should take advantage is to seek consolidation (UNDP, 2005), particularly in terms of primary production, processing and trade, improvements in product quality, adoption of EU hygiene standards, and certification of production processes. In recent years, there has been a growing trend among processing industry enterprises regarding sources of information about quality and safety management systems (i.e. HACCP, hazard analysis and critical control points; ISO, International Organization for Standardization). Approximately 75% of enterprises operating in the processing sector have the necessary knowledge about EU legislation (Kapaj, 2012), which is higher than the 60% recorded in 2004 (MAFCP, 2009). Consequently, food processing enterprises are more aware of new developments in the field of quality and safety (due to increasing competition in food markets), and of new potentials for export in regional and European markets.

In order to take full advantage of its opportunities, however, the food processing industry will need to deal with a variety of challenges. Competition from imports is one the most serious of these. In most cases, higher quality imports outcompete domestic products. Even when products of high quality are produced, the high cost of domestic raw materials leads to high production costs. Also, weak infrastructure (power cuts, the road network, waste/wastewater management) hamper the competitive position of the industry by increasing costs of production (e.g. by forcing the use of more expensive energy sources), by raising transportation costs and by weakening the image of the firms in the eyes of customers (lack of wastewater treatment facilities). Finally, corruption in the public sector continues to hamper fair competition in the food sector. Enterprises in the food industry operating in non-conformance with safety regulations and the legislative framework, or avoiding paying taxes, can overpass the situation by paying bribes to the public officers whose responsibility it is to monitor the compliance of the food sector with safety regulations and the legislative framework. As consequence, food safety problems arise in the market and there is not fair price competition.

After this general overview of the food processing industry in Albania, the section that follows will go through each of the sub-sectors of the industry in turn. The aim is to go into more detail on the specific characteristics and developments of these sub-sectors. Also, while describing each sub-sector, we will identify their strong and weak points and the main problems they face.

5.4 Sub-sectors of the food processing industry in Albania

The subsections below describe the various sub-sectors of the food processing industry in terms of production and processing. Separation of the sub-sectors is done according to the nomenclature officially used by the MAFCP, i.e. meat and fish products; canned fruit, vegetables and oil production; dairy products; alcoholic beverages production (beer), water and soft drinks production; and other processed products. The number of firms, size of firms, and number of employees, the level of competition with imports, and future of the subsector are the indicators that are considered to characterize the profile of each sub-sector.

5.4.1 Meat and fish products

Meat processing is currently one of the more advanced sub-sectors within the food processing industry in Albania. However, this sector is dominated by small-scale processors with the exception of fish processing. Meat processing starts with slaughter, which takes place mostly on farm, although a few slaughterhouses are located in the main cities (Tirana, Durres, Korça, Vlora, Fieri, Elbasan, and Shkodra). By EU norms slaughterhouses are lacking standards (in terms of product safety). All are operating far below their capacities due to low input supplies. The facilities are outdated and, in most cases, they do not comply with Albanian law. This situation has resulted in poor hygiene and sanitation conditions, and safety-related problems in the product. EU standards on hygiene, public authority, traceability and HACCP are currently not fulfilled.

While the beef, pork and small ruminant meat industry in general may be struggling, the poultry sub-sector looks much more promising, as it has several large investors that are active in the feed production companies and processing companies for the poultry industry. With the highly vertical integration of poultry production, it is better than the rest of the meat industry in terms of quality, management and profit. There is a trend toward establishing a more fully integrated chain of operations, from farms to feed mill and retail outlets. The 'Aiba' and 'Floryhen' enterprises dominate chicken and egg production in Albania. These two enterprises, together with 'AGROTEK ALB', are also the three big feed producers, and control 70–80% of the chicken and egg market in Albania. The market dominance by these enterprises is due to the higher quality product they offer.

The meat processing sub-sector (including poultry) comprises 59 domestic-owned and four foreign-owned firms (Amadori, BIKI, Thrace, and Montenegro) not taking into account a potentially large number (i.e. >100) of artisanal sausage producers who are not registered legally as businesses (Kapaj *et al.*, 2009). The meat processing industry covers the production of a wide variety of salami and sausages from beef, veal and pork meat, and various byproducts, with poultry byproducts occasionally used as filler. The total production value of this sub-sector (poultry, salami and sausage) is estimated to be 50 million euros. The Albanian Meat Processors Association (AMPA) recommends consolidation in the meat processing sector in the next decade (UNDP, 2005). As a result, the number of companies has decreased slightly (to 54 enterprises in 2010) over the years, but from 2000 to 2005 investment in technology tripled and the number of employees doubled. From 2005 until 2011, the number of enterprises, the number of employees, and investments in technology have remained almost steady.

There are currently 14 (out of 59) consolidated, large-scale meat processors (salami and sausage) that operate throughout Albania. This sub-sector of the food industry processes 21,500 t of meat per annum, 85% of which serves the domestic market. Almost all of the raw meat processed in Albania (95%) is imported, primarily from South America. Local meat, as input for processed meat, is too expensive and of limited availability. Processed meat is mainly pork, but also poultry and beef. Meat processing companies use modern technology for processing frozen imported meat but have limited expertise in processing fresh meat due to the more common use of frozen imported meat. The leading companies have implemented quality standards and have sound documentation of their production, including traceability and applied HACCP. Albanian standards according to food law and MAFCP orders are often in place in these companies. Consequently, these companies are almost ready to apply for EU export licenses. Therefore, the best opportunities to produce for export markets are among this relatively few, but dynamic, group of companies. These companies export to adjacent markets—such as Kosovo and Macedonia—up to 10% of their production volume (own calculation based on information from MAFCP, 2010), and can serve as benchmarks for the rest of the food industry enterprises.

Fish processing is another sub-sector of the Albanian meat processing industry. This has been one of the fastest growing sub-sectors of the food processing industry and is oriented toward exports. Canned fish production has been increasing from 2000 to 2010 (Fig. 5.3). The highest level of production was reached in 2009, with 3000 t.

Joint ventures between local and foreign companies (mostly Italian) are the most likely source of new investment. Processing companies for fish products buy raw materials and

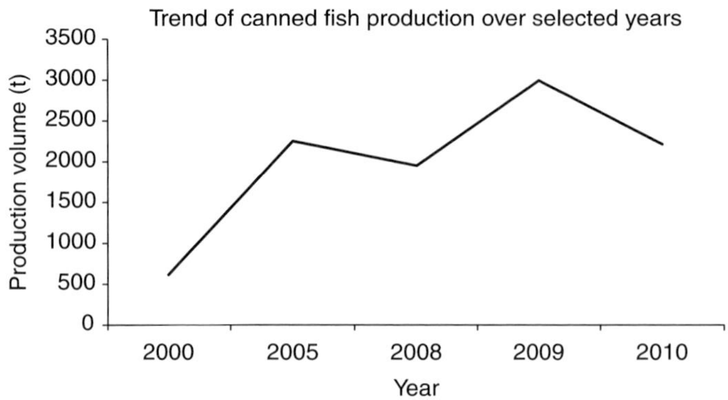

Fig. 5.3. Canned fish production in Albania, 2000–2010. Source: MAFCP (2010).

intermediate products locally for further processing, with the aim of selling these minimally processed fish to the canning manufacturers. Investment opportunities exist in almost every part of the supply chain of the fish processing sector, including harbor facilities, fleet renovation and maintenance facilities, fish collecting/warehousing and fish processing. There are 27 companies in the field, all of which need modernization. Only two of them are major processing companies and these mainly process marine fish. The rest are small size processors which also process freshwater fish and mollusks. The main processing activity is preservation and canning. The main companies are located near the main ports, such as Durres, Shengjin and Vlora (UNDP, 2005). The larger processing companies are Italian, 'Euro Fish' and 'Poseidon'. 'Konservimi Adriatik' and 'Rozafa' are Albanian companies and are well known for their fresh and processed fish. The total number of employees in this sector in 2010 was 912, an increase of more than 100% compared with the year 2000 when the total number was 428. The production volume in this industry has also increased; it was over three and a half times higher in 2010, with 2219 t, compared with 616 t in 2000.

5.4.2 Canned fruit, vegetables and oil production

Fruit processors in Albania produce numerous consumer-ready products. However, most of the raw materials are imported from Greece. The fruit processing industry consists of 22 companies (INSTAT, 2011) which produce canned products (pickles, salads), processed products (sauces, liquids, compotes), and frozen or dried products. The total number of employees in this sector in 2010 was 156 and production was 1844 t.

The vegetable processing industry supplies only 30% of the needs of the domestic market (MAFCP, 2010). It is dominated by micro and small enterprises (with 1–5 employees). There are only two enterprises ('Sejega' and 'Sidnej') with more than 100 employees, and these use both domestic and imported raw materials; they are the only firms producing both for the domestic market and for export. In addition, there are several medium sized (6–10 employees) firms that produce for the domestic market.

Olive oil production is increasing rapidly in Albania, especially along the coastline. With a total production of some 44 000 t in 2010, the production volume is, though, less than 10% of the volume produced by Greece, mainly because of low crop productivity. Low yield is a consequence of the extensive olive cultivation method. Most of the production follows organic production standards, although it is not certified as such. The quality of the raw material is good, but problems arise during the harvesting and storing of olives before they enter into processing activities (Mane-Kapaj *et al.*, 2010). The acidity of oil is up to 3%, which means that it is not always within accepted quality standards (acidity <2% for virgin olive oils and <1% for extra virgin olive oils).The processing of olives and production of olive oil is particularly small-scale, with just a few exceptions. Most small-scale high quality Albanian producers have now reached a level that would make them competitive in any market, but the quantity they produce is very small, and the constraints to their growth are such that their results can be considered as success stories of small individual businesses, rather than a promising trend for the Albanian industry as a whole (Leonetti *et al.*, 2009). Olive oil production was over four times higher in 2010 (7816 t) than in 2000 (1875 t).

The olive oil sub-sector provides potential for small-scale olive oil producers, particularly in the field of organic olive oil. At the same time, there are hurdles to be overcome: product quality should be improved, the value chain is rather weak, and international trade is not yet mature. From the growing export market for organic olive oils in Western Europe, the production conditions in Albania can offer an attractive sourcing option. Albania is a country that has plentiful olive trees. The number of planted trees exceeds 5 million, and is rapidly increasing, as a response to sustained demand, good prices, and government subsidies for expanding the production base (Leonetti *et al.*, 2009). Olive oil produced in Albania is not of high quality and has a large potential for quality improvement (by investing in modern technologies). A considerable number of olive processing enterprises operate at present with a daily capacity of about 400 t. They mainly use old technology based on simple pressing processes. This technology is reflected in a lowering the quality of the olive oil and high acidity of the oil. On the domestic market, local production faces strong competition from Italian and Greek imports.

5.4.3 Dairy products

The dairy sector is one of the fastest growing sectors in the agro-business sector. It provides 90% of the milk supply for Albanian consumers. This sector grew very significantly from 2005 to 2010, with milk production increasing by 15% annually. The production of high quality milk is key to establishing a profitable future for the dairy industry. The vertical integration of dairy enterprises by owning milk production farms and/or imposing quality standards to the large suppliers of milk can result in high quality milk for processing. There are 416 dairies divided between mechanized milk processing units and simple processing units in this sub-sector, though the actual number of processing units may be higher due to the existence of many small unregistered and unlicensed dairies. The latter are found in the hilly and mountainous areas, where seasonal processors (dependent on the lactation period of sheep and goats) work under production conditions with very poor hygiene. There are about 27 modern dairy processing plants in Albania operating with a capacity of 10–40 t/day, but struggling with competition from the informal market. None of these modern plants can utilize their full capacity, owing to insufficient milk supply from farmers, resulting in high cost of production. Most of these plants, 343 out of 416, are small processing plants using the traditional method of production under conditions of poor hygiene (MAFCP, 2011).

The conditions of poor hygiene in these plants in general are not in conformity with requirements of the 'Milk Law' No. 9441, nor do they meet European standards. The number of dairy plants fulfilling quality and safety production criteria is 25, versus 48 that are using traditional production techniques (MAFCP and GTZ, 2009). Furthermore, as with many of the other food industry sub-sectors, there are weak linkages between dairy enterprises and research institutions. This results in a lack of scientific information on the latest quality management philosophies and also a lack of appropriate expertise.

Apart from the modern dairy processing plants, there are many small dairies and processing units that are geographically widely distributed, most with a capacity of only 500–2000 l/day (MAFCP and GTZ, 2009). This type of dairy processing unit operates mostly in local communes to the district market. The main products of the small and medium sized dairy processing units are cheese, butter and curd. Most of them produce white cheese, and about 25% of them produce hard and semi-hard cheese and yogurt for the local market. Cheese and yogurt are the two main traditional products consumed in Albanian households. In 2010, yogurt production was 10,141 t and cheese production was 13,527 t, having slightly increased over the last 4 years (INSTAT, 2011).

The dairy processing industry is still modernizing its management structures and technologies. The VAT (value added tax) system is a major obstacle for the industry, as 20% VAT is charged on all products. Because farmers are excluded from VAT payments, the tax is levied on processed products only. This increases the retail price and reduces the competitiveness of processed food products.

In 2011, the total number of employees in the dairy processing sector is 1122 (10.3%), out of 10,804 employed in the agro-processing sector. In terms of processing industry value, the milk processing sub-sector is ranked after the flour milling and the bread production sub-sectors. Based on statistics from MAFCP (2011), investment in the milk processing sub-sector for the year 2010 was 601,000 euros, or 16% of the total investment in the agro-food industry. These figures show that the milk processing sub-sector is developing more quickly than other branches of the food industry. The most serious weaknesses in the dairy industry are in management, lack of modern technology, lack of skilled employees, lack of investment, and lack of information on food safety management systems (i.e. HACCP, ISO).

5.4.4 Alcoholic beverages production

Before the 1990s, Albania processed approximately 40,000 t of grapes and produced around 80,000 hl of wine and 30,000 hl of raki (a traditional alcoholic drink). Currently, wine production is about 110,000 hl, 37.5% higher than in the years before economic transition (MAFCP, 2009). There were about 28 wine producers on the 2011 market. Albanian grape production is primarily dominated by small, independent vineyards. Grape quality standards are generally low due to heterogeneity of cultivars and inefficient agronomic production practices. This low quality and heterogeneity has strongly affected wine and raki production. Some quality improvement trends have been recorded recently, such as improved hygiene and sanitation, storage practices, production practices, new technologies, and new management approaches, which have resulted in better wine quality. However, the quality achieved still falls short of the standards of many wineries in EU countries such as Italy, Spain, and France, which have very long-standing wine production experience, consolidated production practices and appropriate storage facilities. Approximately 70–75% of wine consumed in Albania is produced by farmers and/or small wineries. More than 80% of total Albanian wine imports come from the EU countries, with 82% of imports coming from Italy, making it the main EU supplier (MAFCP, 2010). The better quality of wine coming from Italy is the most threatening factor for the winery industry in Albania. Domestic enterprises need to significantly improve quality in terms of taste characteristics, packaging and storage facilities to be competitive in the market, despite the price being low compared with imported products. In 2010, wine and raki production employed 444 people, representing a significant increase of 50% from 2005.

Beer production before the 1990s was limited, fulfilling only domestic demand. During recent years, local beer producers have made investments in technology of production and production facilities of 30–40 million euros, with production capacity exceeding national beer consumption. The surplus beer production (approximately 20% of total domestic beer production) goes for export, mainly to neighboring countries such as Kosovo, Macedonia and Montenegro. The beer industry has used its influence (lobbying power) within the state administration to favor its position during negotiations on free trade agreements with other Balkan countries and the EU. Competition for the industry comes mostly from Greece and Serbia. These countries export 40% of the beer consumed in Albania. The total number of beer production firms in 2010 was 22, of which 14 had 1–4 employees, two had 6–10 employees, one had 11–20 employees and only five had over 20 employees. This sub-sector employed 540 people in 2010, which represented an increase of 23% compared with 2005.

5.4.5 Other processed products

Albania is rich in herbs and spices, of which it produces over 60 varieties, including cooking spices, medicinal herbs, and essential oils. About 30% of the European medicinal flora is found in Albania, of which 27 species and 15 subspecies are endemic. About 250 species are harvested for commercial/export purposes, which are used for both medicinal and culinary purposes (MAFCP, 2010). The industry employs around 100,000 gatherers or harvesters of herbs and spices. There are around 40 processors or dealers, and about ten exporting companies. Albania exports herbs and spices as either dried products or essential oils. The most important Albanian herbs and spices exported (directly or indirectly) to European countries and the USA are: sage, oregano, red juniper, black juniper, hawthorn, laurel, thyme, winter savory, wild rose, great yellow gentian, small-leafed linden, myrtle, wild apple and blackthorn. The main export countries/destinations are: the USA, Turkey, Germany, Austria, Italy, Greece, Spain, France, and Switzerland. There are also some other destinations, including Croatia, Macedonia, Hungary, Japan, Denmark, the Netherlands, and even Cook Island. Export volumes (kg) grew sevenfold in 2003, after nearly 100% annual growth in the years before. Key investment opportunities are the establishment of small on-site laboratories for quality control, processing (the application of manual or sophisticated methods to perform adequate cleaning and processing), sanitation (clean products meeting European and American Standards, preferably ASTA – American Spice Trade Association – specifications), packaging, and warehousing (clean working and storing environment). Most products exported from the Albanian agro-business sector have been medicinal plants, which experienced 16% growth from 2007 to 2008.

5.5 Conclusions

There have been fundamental changes in the agricultural production and processing structures prevailing in the Albanian agriculture sector since 1990. The breaking up of the large collectivized farms into micro farms managed by individual families is the most obvious change. This movement to private farming explains the rebound in agricultural production over the years. The quantitative increase in the volume of crop and animal production since 1990 provides a measure of the agriculture sector's development. Albanian agriculture during the first period of transition was influenced by three factors: the lack of on-farm inputs and funds to buy these inputs from abroad; the impossibility of introducing alternative mechanisms for distribution; and allocation of resources after the collapse of the centralized planning economic system.

Despite the fact that a large number of tractors and other pieces of mechanical equipment have been imported to Albania, farmers still do not have sufficient access to modern equipment. The owners of this equipment charge farmers very high prices for particular mechanized services, such as plowing, sowing, harvesting, and transporting. This has as a resulted in the slowing of mechanization, and the use of traditional methods, such as animal power and manual work, remain the only alternative for a large number of farmers. Only 74% of agricultural households use machines for plowing, about 23% use animals for plowing, and 59% use machines and manual work. The phenomenon of manual work is not linked to the lack of various agricultural mechanical means, but is related primarily to the high price of the service, the low efficiency of agricultural activity, and the very small and fragmented farm area. Additionally, in terms of agricultural inputs, the use of fertilizer has experienced a tangible reduction by 17% in 2010 compared with 2000. This reduction in the use of fertilizer is compensated for by the use of organic fertilizers, which has increased substantially (by 33%) from 2000 to 2008 (MAFCP, 2008).

There have been serious efforts from government policies to improve marketing channels and distribution by creating collection centers, especially for milk and apples (two of the main agricultural products that have suffered the most marketing problems compared with other agricultural products). Apart from that, distribution is still a problem among agricultural producers in Albania despite continued efforts, even from private entrepreneurs, to improve it. Lastly, capital allocation in agricultural production and in rural development in general is not adequate. Governmental policy such as the Agricultural and Food Sector Strategy (AFSS) has agricultural and rural development as a priority, but still there is much to be done. Taking into consideration that the rural population counts for 53% of total population in Albania, then the labor force in agriculture is underemployed and labor productivity is very low. After the 1990s, the agriculture and food processing sector have been one of the growth drivers for Albania's economy, with the production value of this sector in 2010 at 37.5% higher than in 2000. Within this, livestock production represents 55% of agricultural production.

In contrast, the agricultural and agro-processing sector must increase the level of productivity to be competitive in both domestic and foreign markets. However, the sector depends on the level of technology and knowledge used, and on the efficient use of land, labor, and other inputs. As a result, higher productivity and competitiveness are foundations for increasing production, improving market access for farmers and agro-processing businesses, increasing income, and improving the living standards of farmers and their families. Increasing the quality of agro-food products is a fundamental factor in increasing competitiveness, and will come not only from technological improvements but also from new methods and management philosophies, such as new quality management systems, ISO standards, or total quality management (TQM).

Meanwhile, despite the progress that has been identified above in the agricultural food processing sector, there are many serious aspects of food safety. Therefore, guaranteeing increased food safety for consumers will be an absolute priority. Food safety can be guaranteed through technological improvements, the application of new methods, and systems of food safety control. This can also be done by the use of integrated pest management methods, the integrated management of plant nutrition, and animal breeding, and the use of good agricultural practices and standards, which, together, will increase the importance and weight of organic products in total agricultural production.

But, the processed food market is mostly satisfied by imported products, for which food safety is assured, as domestic production is still limited and in its development stage. Except for milk and meat products, which fulfill more than 90% of the domestic production, other food and agricultural products are mostly imported from European countries (mainly Greece and Italy), and from regional countries (Macedonia and Serbia). In the agro-industry sector, the number of enterprises has increased from 1844 in 2000 to 2156 in 2010.

Also, investments in the processing industry rose 130% from 2000 to 2009, reaching 16 million euros in 2009. Most of the investments have been via private capital financing, with bank loans utilized less often. This increase in investments is accompanied by a 20% increase in people employed in the agro-industry, and a processing industry production value growth rate of 83% from 2000 to 2010.

The industrial processing of milk and meat are currently among the more advanced industries in Albania. There have been important private investments even using the most modern technology. In general though, this sector suffers from structural weaknesses, and there is still a lot of room for health and food safety conditions to improve. Prioritizing governmental support for this sector can be argued for a variety of reasons. Firstly, there is a great need to promote and guarantee food safety standards for this sector, as these are closely related to EU accession requirements. This can be achieved by direct governmental financial support for new processing technology and increasing capacities. Second, governmental support to the milk and meat industrial processing sector can urge this sector to demand quality standards for livestock products supplied from farmers, which will encourage restructuring and innovative developments both in livestock farms, and in the entire livestock production system.

Emphasis should be put on the fruit and vegetables sector as well, because the quantity of fruits and vegetables that is processed domestically is small, and foreign products prevail in the market; the development of domestic fruit and vegetable production means that domestic raw materials will always be available. Again, the development of this industry will reduce the problems of the fruit and vegetable market; it will become more complex and the supply will rapidly increase, aided by direct support to fruit and vegetable production. Moreover, industrial processing of fruit and vegetables will enhance the opportunities for agricultural employment in both urban and rural areas. Lastly, the increase of industrial processing capacities will automatically mean an increase in the commercial demand for farm products.

The Albanian agro-business sector is dynamic. Substantial investments have been made in recent years, and added value has increased at a pace of 10–15% (UNDP, 2005). Despite rapid development, however, the total production volume still does not match domestic consumption, and substantial imports of food products are necessary. The majority of the production units are very small-scale. Despite recent consolidation, the overall picture of the Albanian agro-business reveals it to be still fragmented and very small-scale compared with the country's European peers. Number wise, the sector is dominated by small companies. Agro-businesses and policymakers are aware of the need to stimulate vertical and horizontal cooperation within the agriculture sector, including consolidation. Following worldwide trends, the numbers of companies per sub-sector is declining, and the production volume per company is increasing steadily. Most processing companies operate for domestic or local markets, and work according to standards that only match the needs of these markets. Most leading companies in the Albanian agro-business sector focus on international cooperation. Some companies successfully practice the import–export business model: they import raw materials, process in Albania, and export the consumer products. This model is viable, especially in certain areas of the meat and vegetable markets, namely sausage production and frozen vegetables. The main reasons for the success for this business model are the combination of low wages, low logistics costs, and efficient operations.

References

Agraja, L. (2006) *Assessing the Comparative Advantage of Wheat Production in Albania*. CuvillierVerlag, Göttingen, Germany.

GFA Consulting Group (2010) *Food Industry Study in Southeast Europe. Final Report*. Available at: http://mazedonien.ahk.de/fileadmin/ahk_mazedonien/Dokumente/DEG-Food_industry_study-external_version-20-01-2011_02.pdf (accessed 3 September 2012).

INSTAT (1992) *Statistical Yearbook 1991*. Instituti i Statistikës (INSTAT), Tirana, Albania.

INSTAT (2011) *Statistical Yearbook 2010*. Instituti i Statistikës (INSTAT), Tirana, Albania.

Kapaj, I. (2012) Assessing quality and safety of food and beverage products in Albanian processing enterprises. PhD thesis, Universität Hohenheim, Germany. Available at: http://opus.ub.uni-hohenheim.de/volltexte/2012/680/ (accessed 3 September 2012).

Kapaj, I., Doluschitz, R. and Mane, A. (2009) Assessing quality and safety of food and beverage products. An analysis of agribusiness enterprises in Tirana District/Albania. Paper presented to: *19th IAMA [IFAMA] Annual World Forum and Symposium: Global Challenges Local Solutions, Budapest, Hungary, 20–23 June 2009*. International Food and Agribusiness Management Association. Available at: https://www.ifama.org/events/conferences/2009/cmsdocs/1053_paper.pdf (accessed 14 September 2012).

Kapaj, A., Chan-Halbrendt, C. and Deci, I. (2011) Milk market segmentation: study on the consumer preferences in urban Albania. Paper presented to: *21st IAMA [IFAMA] Annual World Forum and Symposium: The Road to 2050: Sustainability as a Business Opportunity, Frankfurt, Germany, 20–23 June 2011*. International Food and Agribusiness Management Association. Available at: https://www.ifama.org/events/conferences/2011/cmsdocs/2011SymposiumDocs/338_Symposium%20Paper.pdf (accessed 14 September 2012).

Leonetti, L., Imami, D., Zhllima, E. and Stefanllari, A. (2009) *The Olive and Olive Oil Value Chain*. Prepared by Development Solutions Associates, Tirana, Albania. Available at: http://www.eastagri.org/files/Oil-Albania.pdf (accessed 3 September 2012).

MAFCP (2008) *AFSS (Agriculture and Food Sector Strategy 2007–2013)*. Ministry of Agriculture, Food and Consumer Protection, Tirana, Albania.

MAFCP (2009) *Rural Development Strategy*. Ministry of Agriculture, Food and Consumer Protection, Tirana, Albania.

MAFCP (2010) *Statistical Yearbook, 2009*. Ministry of Agriculture, Food and Consumer Protection, Tirana, Albania.

MAFCP (2011) *Statistical Yearbook, 2010*. Ministry of Agriculture, Food and Consumer Protection, Tirana, Albania.

MAFCP and GTZ (2009) *An Assessment of the Competitiveness of the Dairy Food Chain in Albania*. Ministry of Agriculture, Food and Consumer Protection, Tirana, Albania with Deutsche Gesellschaft für Technische Zusammenarbeit (GTZ), Bonn and Eschborn, Germany.

Mane-Kapaj, A., Kapaj, I., Chan-Halbrendt, C. and Totojani, O. (2010) Assessing the comparative advantage of Albanian olive oil production. *IFAMR (International Food and Agribusiness Management Review)* 13(1), 15–26. Available at: https://www.ifama.org/publications/journal/vol13/cmsdocs/20091005_Formatted.pdf (accessed 3 September 2012).

UNDP (2005) *The Agribusiness Sector in Albania: Inward Investment Opportunities*. United Nations Development Program/European Union Trade Liberalization and Promotion Program. Report produced by independent consultants on behalf of the Albanian Foreign Investment Promotion Agency (ANIH), UNDP Albania and the European Union. Available at: http://www.trade.undp.org.al/u/d/trade/Albanian%20agribusiness%20sector.pdf (accessed 3 September 2012).

World Bank (2006) *Country Assistance Strategy for Albania for the Period FY06–FY09*. International Bank for Reconstruction and Development, International Development Association and International Financial Corporation. World Bank, Washington, DC. Available at: http://www-wds.worldbank.org/external/default/WDSContentServer/IW3P/IB/2006/01/24/000112742_20060124165609/Rendered/PDF/34329.pdf (accessed 14 September 2012).

6 Distributions and Markets in Albania

Ana Kapaj, Drini Imami and Bahri Musabelliu
Agricultural University of Tirana, Albania

6.1 Overview

This section presents an overview of the development of distribution channels, marketing, and markets for agricultural commodities in Albania. The market exchange of agricultural products is more complex and uncertain than that of other industrial products, and the distribution channels are unique. The food channeling system encompasses many activities, from harvesting, which depends on natural conditions, to processing, retailing, and consumption. An agricultural product can take one of several paths to reach the consumer. Distribution modes depend on the product, the market, the type of retail establishment, and the producer's sales skills. Some producers reach the consumer directly by selling products on farm and at farmers' markets. Others use elaborate distribution methods involving several brokers and distributors. Many producers do not have the skills or the time to market their products. For them, the use of food distributors and brokers is the only way to get their products to the markets. Most foods go through a distribution channel to reach the end consumer, whether that consumer is a shopper in a retail grocery store or a diner at a restaurant. The conventional distribution path for a packaged food product is from manufacturer to broker to distributor to retailer, but this path can vary greatly depending on the product, the target markets, and the manufacturer.

In developing countries, agricultural markets historically incite government responses to any inefficiency created by incomplete institutional and physical infrastructure, and imperfect competition. Well-functioning markets ensure that macro and micro policies change the incentives and constraints faced by micro-level decision-makers. Similarly, well-functioning markets also underpin important opportunities for farmers at the micro level for livelihood improvements that aggregate into sustainable growth. The micro-level realities of agricultural markets in much of the developing world do, however, include poor communications and transport infrastructure, limited rule of law, and restricted access to commercial finance, all of which make markets function much less effectively than they should in theory. A long list of market studies (Barrett, 1997; Platteau, 2000; Fackler and Goodwin, 2001; Fafchamps, 2004), documents considerable commodity price variability across space and seasons in developing countries, with various empirical tests of market integration suggesting significant and puzzling forgone arbitrage opportunities, significant entry and mobility barriers, and highly personalized exchange. Widespread inefficiencies result from the incomplete or unclear property rights, imperfect contract monitoring and enforcement, high transactions costs, and binding liquidity constraints that also occur in developing countries. Such market failures often motivate government intervention in markets. The challenge is greater for transitional economies, as markets are not developed and government has limited experience and financial resources with which to provide the infrastructure and services needed to improve market functions and distribution.

Albania, after five decades of centrally planned socialist economy, is transitioning to a market-oriented economy. During this transitional period, which began in the early 1990s, agriculture and those managing agribusiness systems have been facing major challenges, such as lack of good business management skills, poor decision-making by farmers and businesses, poor infrastructure and support services, and an insufficient legal framework for a well-functioning agricultural industry. The marketing and distribution sectors and government support services to farms are still adjusting to the new market-oriented market environment, resulting in suboptimal economic performances in the agriculture sector.

The Albanian market consists of about 3.9 million consumers (INSTAT, 2010). Generally speaking, the market can be divided in three segments: the 'international cities,' cities, and rural villages. Tirana, the capital and the largest city of Albania, and Durres, where the largest harbor is located, are prominent segments of the market and food trendsetters. According to the databases of the Albanian Institute of Statistics (INSTAT, Instituti i Statistikës), purchasing power in these two cities is the highest in Albania. Imported products and international food outlets can be found in these areas. However, around 1.3 million Albanians live in rural villages. The majority of these consumers either produce at least a portion of the food they consume or are closely connected to farming. According to market studies, the markets for commodities in these regions are likely to be largely informal.

As they come out of a long transition period, the agricultural product markets and distribution channels in Albania are facing significant problems in production, quality, infrastructure and access to markets. In this chapter, we will investigate the current state of the distribution channels and markets, as well as proposing that farmer cooperation is one of the best ways to improve farmer performance in the market.

6.2 Distribution Channels and Markets in the Main Agribusiness Sub-sectors

The agricultural food market is mostly satisfied by imported products, as domestic production is still limited and its development still in its infancy. Except for milk and meat products, which fulfill more than 90% of the domestic production requirement, other food and agricultural products are mostly imported from European countries (mainly Greece and Italy), and from countries in the region (Macedonia, Serbia, etc.). Given the vast range of potential agricultural product processing sectors that could have been analyzed in this chapter, priority is given here to an assessment of the following sectors of the agricultural industry: dairy products, olives and olive oil, meat, fruit and vegetables, and grapes, wine and raki. Based on official data from Albanian Ministry of Agriculture and Consumer Protection (MAFCP) and published reports, these sectors have the highest production value in Albanian agricultural production. Dairy has the highest production value at 28 million euros, followed by the olive oil sector at 15.6 million euros. The other sectors that will be discussed, meat, wine, and fruit and vegetables, have lower production values of 8 million, 6.1 million, and 2.1 million euros, respectively, but are still important to Albania's total agricultural production (UNDP, 2005). In the rest of this section, we provide an overview of the markets and distribution channels of these five groups.

6.2.1 Dairy

As already mentioned, dairy products provide the highest production value in Albanian agriculture, at 28 million euros in 2009. Milk is among the traditionally produced agricultural goods in Albania. The dairy industry, and along with it the milk collection system, are still in the course of modernizing their infrastructure and technologies. Dairy production, mainly from cows, but also from goats and sheep, has a long production tradition in Albania due to its favorable natural resources, such as plentiful forage and land for grazing. Traditional dairy products are yogurt, butter, curd, and different kinds of cheeses from cow's, sheep's and goat's milk. According to MAFCP statistics, total milk production has increased 7.1%, from 948,000 t in 2000, up to 1,116,000 t in 2009. Of this, cow's milk production accounted for 85% or 868,000 t of the total milk production in 2009 (MAFCP, 2010).

Due to the significance economic value of livestock and milk production, particularly in rural areas, the MAFCP has selected the milk sector as a priority sector for future investment. One of the government policy objectives is to improve the competitiveness of milk production in order to substitute domestic production for imports and increase export potential (Petrick *et al.*, 2003). Milk consumption in 2009 was 250 l per capita. Total milk production has gone through some slight ups and downs during the last 10 years, with production peaking in 2006 (Fig. 6.1). According to MAFCP, these slight movements in milk production trends are influenced by changes in consumer demand for dairy products. Most of the total milk production is represented by cow's milk followed by sheep's milk, and then goat's milk.

Primary production of milk is very small-scale in Albania, with large numbers of farmers each producing small quantities of milk. This may be one of the crucial challenges for the distribution system. Reaching every single farmer for milk collection is not feasible for dairy processors while, at the same time, farmers argue that it is difficult for them to reach consumers or dairy processors. In contrast with all Western European countries, there is hardly any formal vertical integration in the supply chain. Farmers bring their milk to milk collection points at their own time and convenience. Neither volume nor quality is formalized in contract agreements. Distribution to retail channels is also still in the developmental stage. In view of the fragmented structure of retail channels, the dairy industry has to deal with a large number of stakeholders in all the stages of the distribution channel: farms, wholesalers, retailers and final consumers. The results of this lack of efficient vertical integration include:

- long times to reach processing plants and markets, and consequently high pressure on quality assurance
- lack of quality control
- high distribution costs
- high pressure to pursue informal market opportunities.

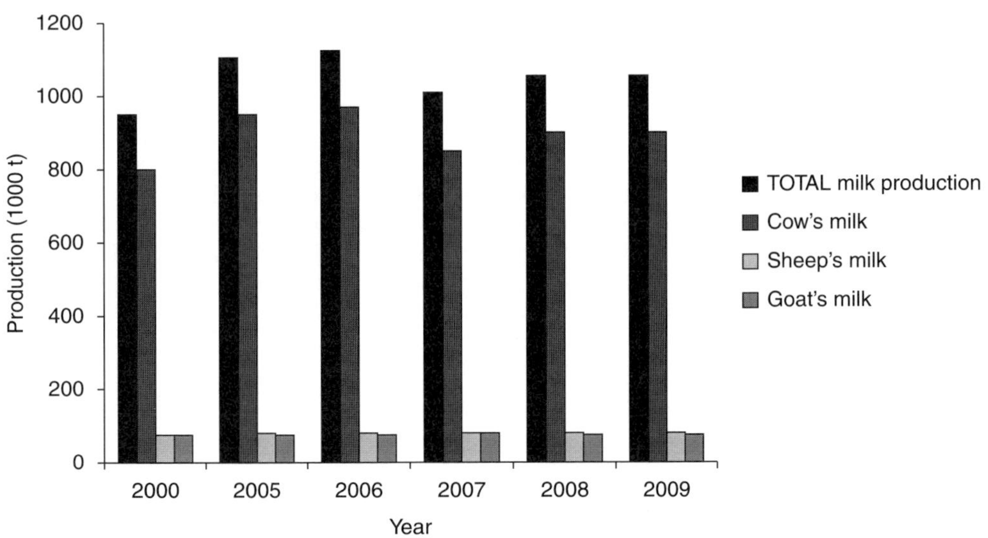

Fig. 6.1. Milk production, 2000–2009 (1000 t). Source: MAFCP (2010).

Milk production and collection systems (mainly of cow's milk) in Albania are characterized by the existence of both informal (direct selling by farmers) and formal market channels (collection and distribution by dairies). Milk production still suffers from problems of quality assurance. According the annual MAFCP survey of 2007 (MAFCP, 2006), only 23% of milk produced is going to dairies for processing (such as pasteurization), 45% is directly marketed, 19% is used for self-consumption, and 13% is used for animal feeding (calves). So a very large share of milk is consumed directly and not processed and/or controlled.

The Albanian government aims to accelerate the modernization and restructuring of agriculture and food processing, especially in the dairy and meat sectors. The Rural Development Strategy of 2007 foresees that the majority of the budget for rural development will be targeted at modernizing buildings and livestock premises in milk farm holdings, as well as enterprises in the food processing industry. The primary target group for this assistance will be the medium-sized farmers and the micro, small and medium-sized milk processing units. The government has clearly expressed its objective to foster technological adjustment of the Albanian farm and food processing sectors to meet EU standards. For that reason, from 2007, the MAFCP has supported increases in herd size and milk production as well as improvements in processing technology through the state budget and donor projects.

The dairy sector provides 90% of the milk supply for Albanian consumers. Even though there are small quantities of milk imported, domestic milk production still has the largest market share. The production statistics show that the gap between domestic and imported milk is relatively high (Table 6.1), although the quantity of imported milk quantity increased from 0.18% of domestic production in 2000 to 0.85% in 2009. Domestic milk is preferred by the Albanian market over imported milk, and this preference toward domestic dairy production is also reflected in other products, such as cheese, butter, and yogurt.

The milk sector has grown significantly in recent years. Table 6.2 shows that indicators such as investments and production values have increased from 1998 to 2008, demonstrating clearly that the sector has seen improvement. Some very significant indicators of this improvement are the sixfold higher investments (over 1 million euros) in 2008 compared with 1998 (212 thousand euros) and nearly 2.5 times higher production value in 2008 compared with 1998.

However, a problem remains in the inconsistent increases in investments and production: investments are growing faster than production. Because of the small-scale production in most of the dairy farms, the processing technology is not fully used, and this under capacity in technology use is reflected in the growth difference between investments and production.

6.2.2 Olive oil

During the privatization of farmland in 1991, some 19,000 ha of olive groves were distributed to 110,000 households, resulting in highly fragmented olive production (World Bank, 2007). In 2007, the olive sector was estimated to have produced an output value of 30 million euros (or approximately US$40 million) and olive production was spread over 118,000 farms (MAFCP, 2009). The high fragmentation of the olive production sector did not achieve economies of scale, and is thus not price competitive. Olive production has gone through ups and downs in the last 5 years (Table 6.3), with the lowest production in 2007 and the highest in 2008. This is mainly because olive groves

Table 6.1. Domestic production and imported milk quantity (as 1000 t) from 2000 to 2009. Source: Kapaj *et al.* (2011a).

	2000	2005	2006	2007	2009
Milk production	948	1076	1102	1016	1125
Imported milk	1.8	2.785	6.054	8.524	9.54

Table 6.2. Investments in, workforce and production value of the milk industry, 1998–2008. Source: MAFCP (2010).

Indicators	1998	2002	2004	2006	2008
Investments (1000 euros)	212.0	540.0	705.6	1000.6	1270.0
Number of workers	787	798	850	958	1092
Estimated production value (million euros)	12.0	20.6	26.0	27.9	28.9

Table 6.3. Olive trees, olive production and olive oil production for the period 2004 to 2008. Source: MAFCP (2010).

Items	2004	2005	2006	2007	2008
Olive trees in production (1000)	3429	3488	3603	3728	4179
Olives production (1000 t)	55.7	30.1	40.2	28.1	56.2
Olive oil production from processing plants (t)	4036	3454	4985	8979	5634

are not properly serviced. Most farmers do not use any fertilization, irrigation, pruning, disease control, or proper harvesting services (Leonetti *et al.*, 2009). Therefore, existing olive groves have a low productivity overall and olive production is not steady, but rather is characterized by oscillations.

Olive oil processors buy olives directly from farmers or through collectors/suppliers who buy olives from farmers and sell them to processors at a higher price. As the supply of olives has gone up and down over the last 5 years, processors have been competing with each other to secure them, so the role of collectors has become very important, especially in the low-production seasons.

Some processors do produce olives themselves, though. A recent survey found that almost half of olive oil processors have planted their own olive groves (Imami, 2011). Many olive oil processors have installed new Italian processing equipment from manufacturers such as Pieralisi, and Alfalavale olive oil presses. Most processors state that 50% or more of the olive oil that they process is retained by farmers for their own consumption (Imami, 2011). As the main targets of olive processing are farmers, olive processing plants are also located in the areas where olive trees are found (mainly central and southeast Albania).

There is not an exact count of olive processing facilities, but they range in number from 108 to 150 over recent years. Out of the existing known processing plants, 40 have registered brands and approved labels: 34 for olive oil production and six for table olive production (olive oil processing is generally a separate business from table olive production). MAFCP (2010) records 108 processors of olive oil, whereas the USAID (US Agency for International Development)-funded Albanian Agriculture Competitiveness (AAC) Program (Mehmeti, 2007) recorded a total number of oil mills of 150. All medium and large sized processors, and the most reputable small ones, now have their brands registered. The rest of the industry is going through a slow process of formalizing their businesses, such as branding to market their products and obtain a higher value (Leonetti *et al.*, 2009).

Another major actor in the olive oil sector is the bottlers, who are gradually becoming larger and more consolidated. New developments have also occurred in the size of the industry. We see a bimodal industry in terms of size, either quite small (wholesalers starting to process small quantities of olives as new local small oil mills are being established, etc.) or large (by Albanian standards). Bottling companies are typically supplied with olive oil from domestic olive processors in bulk, and they bottle and sell the oil in urban markets under their own brands. These companies also import olive oil from Italy or Greece, and mix it with domestically sourced olive oil for sale. They sell mainly normal and virgin olive oil in 0.5 l and 1 l plastic and glass bottles, distribute their products to (small) mini-markets and supermarkets in the Tirana and Durres areas and are the main consumers of olive oil in the urban markets (Leonetti *et al.*, 2009). The olive and olive oil sector is undergoing a process of rapid changes: numbers of new enterprises quickly increase and decline, sometimes in a period of a couple of years, while the opportunities for good profits from a growing and increasingly consumer-led demand are leading to entrepreneurial investments in the sector.

The recent development of supermarket chains is also going to generate new challenges and induce major changes for domestic producers. Opportunities for domestic olive oil brands to operate successfully with their own retail packaging will be reduced, as larger operators will find easier and more cost-effective ways to ensure a regular supply of products and more stable trade relations with farmers and importers for raw materials. Companies which are already leaders in other products sold in supermarkets will find it easier to widen their product range to include olive oil, as they already have established distribution channels. It is also expected that increasingly strong supermarket chains will impose supply contract and delivery conditions, and will take over the logistics from producers, while competition from imports will further increase in the context of trade liberalization and the penetration of international supermarket chains into the domestic market. However, informal markets and direct sales to consumers are expected to remain largely in place, despite these changes. Direct purchasing from oil mills is driven by personal relations with producers and quality. The consumer's trust in buying olive oil from a specific producer is the main reason why most of the smaller oil mills that produce high quality oil have no problem in selling their production directly to consumers.

6.2.3 Meat

The livestock industry, which includes sheep, goats, cattle, and poultry, has shown increasing growth rates in terms of production, employment, and investment. According to MAFCP data, production was two times higher in 2010 than in 2000, employee numbers have increased three times from 2000 to 2010, and investments are 1.5 times higher. Because Albania has optimal conditions for breeding sheep and goats as well as cows, the government considers the livestock sector a priority for development (MAFCP, 2007). Government policies stimulating new investments in this sector and also to support to the existing investments, are scheduled for the 2007–2013 government agenda. The livestock sector is one of the fastest growing sectors in agribusiness, and reached 15% annual growth in 2009. Investments have increased over recent years in buildings or housing for animals, slaughterhouses and technology.

Meat production is an important source of income for small Albanian agricultural holdings, especially in the impoverished rural mountainous areas in the northeast (Kapaj, 2011b). In these areas, enhancing small ruminant productivity is considered a priority by the 2007 Agriculture and Food Sector Strategy of MAFCP. Meat and byproducts from small ruminants are both used for household consumption (most Albanian farms are subsistence and semi-subsistence farms) and sold on the market. The meat processing sector comprises of about 56 companies, not taking into account a potentially large number of artisanal sausage producers, which may be over 100 in number. The total production value of the sector for 2009 is estimated to be 50 million euros. The main products of the meat processing sector are sausages and ham, and the reported employment in 2005 was 538 persons (UNDP, 2005).

There are two groups of processors: 'industrial' Albanian processors (importing most of their raw materials as frozen meat from countries such as Argentina, Brazil, and Ireland) and 'typical domestic' Albanian processors (sourcing from the domestic market), which represent 37% and 63% of the industry, respectively. The industrial processors aim to serve the domestic market and to export. As already noted, these processors largely rely on imports of raw materials and use Albanian production facilities. The companies in this group modernize their production, including development of the HACCP (hazard analysis and critical control points) and ISO (International Organization for Standardization) certification. They usually organize the sourcing of raw material with their own refrigerated trucks. In contrast, domestically sourced processors, which are the much larger group in the industry, exploit local markets, offering lower prices and domestic products, which are preferred by Albanian consumers. The main reasons for this are related to the availability of domestic supply (proper quality at an appropriate price) and the organization of the supply chain (size of production, logistics).

Most domestic meat is sold fresh and mainly through small-scale distribution channels. Small shops usually get fresh meat directly from the farmers. According to Kapaj (2011b), the best way for butchers to obtain fresh domestic meat is to get it directly from the farm, and usually they have trusted farmers from whom they get their meat. Some weak points in the meat supply chain include small-scale production, differing quality of live animals, insufficient supply, lack of slaughtering facilities (especially those complying with EU hygiene standards), quality assurance, and traceability (Kapaj, 2011b). As mentioned before, this sector is considered high priority by the Albanian government agenda. Government policies such as investments in increasing herd size and processing facilities, and the implementation of hygiene standards towards quality assurance, will be ongoing over the next few years.

6.2.4 Fruit and vegetables

Of all Albanian farms, 99% (350,000 out 353,000) have some field crops, including vegetables. In 2008, about 275,000 farms were actually producing vegetables on a scale exceeding their own consumption (AAC, 2008), but only 786 (0.3%) farms have more than 1 ha devoted to vegetable growing, and 117 farms (0.04%) more than 2 ha (MAFCP, 2009), resulting in fragmented and small-scale vegetable production. The most common crops are tomatoes, peppers, and cucumbers.

About 152,000 farms (43%) cultivate fruit trees (MAFCP, 2009), most of which are semi-subsistence farms. More than 50% of these farms have a size larger than 1 ha, mostly located in southeast Albania (MAFCP, 2009). About 61,000 farms have fruit trees (i.e. these farms can potentially produce for markets), more than 70% of which have a total area of cultivated fruit trees up to 0.2 ha (MAFCP, 2010). Only 485 farmers had a cultivated area with fruit trees of more than 1 ha in 2009. The pace of farm consolidation is still slow and the average farm size remains too low to be competitive, so the fruit and vegetables sector is still largely dominated by subsistence farming (World Bank, 2007).

Consolidators or buyers are an important category of actors in the value chain of the vegetables industry (AAC, 2008). The principal market function of the consolidators is to serve as collectors and in assembling and delivering raw vegetables from the farm to the wholesale market or to other actors such as processors or exporters. Major importers limit their activities to the import and wholesaling of imported food products, including bulk and packaged fruits and vegetables. They sell mainly to distributors in the wholesale markets, and are registered as incorporated enterprises. There are fewer than ten operators belonging to this category. Each of them makes use of 3–4 trailers a day, circulating more than 2000 million t (Mt) of vegetables a year (GTZ, 2010).

Large wholesalers/importers are generally registered as businesses residing mostly in the wholesale markets of Tirana, Lushnja, and Fier. These wholesalers sell to large distributors in the wholesale market, and the distributors then sell the product to retailers or directly to the final consumer. When they are dealing with small quantities of products, wholesalers might also sell further downstream to small retailers. Medium sized wholesalers are the main buyers at the farm level, and the source of the largest volumes arriving in wholesale markets. These medium sized consolidators generally have their own transport vehicles, most often with a capacity of 2 t (small vans) or 5 t (trucks). Small sized wholesalers deal with less than 1000 Mt of fruit and vegetables per year and have their own transport vehicles, most often a 2 t mini-van. These small sized wholesalers play a key role in supplying retailers.

Most fresh fruits and vegetables are sold in (specialized) green markets (Imami *et al.*, 2012). Green markets are situated in different parts of a large city or in certain areas of smaller towns and have modest consumer demand. Retailers and producers sell their products in such markets, and have access to a produce stand for a daily fee. The share of fresh products sold by supermarkets is still very low. For example, on average; consumers in Tirana buy 5% of the apples they consume in supermarkets (Imami *et al.*, 2012). The emergence of supermarket chains is expected to change the structure of the value chain and will strongly contribute to the consolidation of the fruit and vegetable sector. The requirements of supermarket chains (large flows, constant quality, longer terms for payments) will force consolidation among Albanian fruit and vegetable processors: they must specialize on a smaller range of products, substantially scale up their facilities, and, at the same time, increase their working capital. The establishment of the foreign owned supermarket chains Euromax and Conad has already proved an important force for the penetration of agro-food products from neighboring countries such as Serbia and Italy, respectively, and further expansion of such supermarket chains is expected to increase import competition for fresh fruits and vegetables.

There are no available statistics about the destination of various fruits and vegetables, i.e. the share that goes for table consumption and the share which goes for processing. However, it appears that only a small share of fruits and vegetables go for processing, which may be related to the relatively weak development of the processing industry. Processing plants are mainly located in the districts where the fruit and vegetable processing industry was stronger before 1990, and where raw materials are more abundantly available. All but one processor processes both fruits and vegetables (GTZ, 2010). Out of the 27 vegetable processing plants registered by MAFCP, 19 are located in Berat, Tirana, and Fier, where there is also a high concentration of fruit and vegetable production (MAFCP, 2010). The largest enterprise (Sejega) is located in the Tirana region and the second largest (Sidnej) in Berat (GTZ, 2010). In addition to registered companies, there are many other processing plants which are completely or partially operating as non-registered entities.

In almost all cases, fruit and vegetable processing is a family business, with the entrepreneurs (owners, who at the same time act as managers) having limited technical background, and directly supervising all functions in the factory (procurement, processing, marketing, administration, etc.). In the two largest companies (Sejega and Sidnej) management is better organized and based on division of responsibilities, while in the other companies the management structure is much centralized, with the owner typically controlling all activities directly. The typical fruit and vegetable processing industry is characterized by downstream integration, with direct distribution to retailers, and little or no use of wholesalers (GTZ, 2010).

To better explain the actual level of distribution channels, the results are presented below of a study on the value chain of apple orchards in Korça region of southeast Albania. This is the most important area for the apple production in Albania.

Apple value chain actors' behavior in Korça region, southeast Albania

Korça is the leading apple production region in Albania and was therefore used as the focal area for a study of the apple value chain. The main destination channel for apples is wholesalers (69.6%). The remaining apples are sold to specialized fruit and vegetable retailers and processors. Supermarkets and groceries are insignificant in terms of the apple quantity traded compared with the wholesaler market (Table 6.4). Almost all farmer-to-wholesaler transactions are done at the wholesale market. This finding confirms the important role of wholesale agriculture markets (WAMs) as a logistic platform and as a market 'matchmaking' instrument, bringing together buyers and sellers. An important segment of the distribution is apple losses (15.8%), which, according to the opinions of farmers/producers are relatively high, and where a focus on improvement is needed.

Formal (written) contracts are not common in the Albanian agricultural sector in any of the distribution channels segments, be they via wholesalers, green vendors or processors. Only one farmer confirmed that he has formal contracts with processors. None of the farmers have formal contracts with wholesalers, groceries, specialized shops or supermarkets. However, there are informal agreements in place. These are more common in relations with green vendors and processors, and most farmers state that they have an informal agreement with processors. Trade relations with the category of buyers that absorb most produce from most farms (wholesalers) in most cases is not subject to any form of either formal or informal contracts (Table 6.5).

The situation is almost the same in all of the three categories (wholesalers, green vendors and processors) of the value chain, with most actors within each category not having a contract or agreement. Processors show a slight trend towards agreements of an informal nature, but 54.5% do not use written agreements (Table 6.5). Another important aspect to analyze is the duration of the relationship between farmers and the other value chain actors (buyers). About 84% of the farmers (Table 6.6) say that their sales to the main wholesale buyers are developed on a day-to-day basis. The relationship with green vendors is similar. Farmers have slightly more stable dealings with the main buyers from different processors. About 21% of farmers supplying processors have only on-the-spot relations with them.

This lack of formal agreements is seen as a gap in the market relations between all the partners involved in apple distribution, and is also reflected in future predictions of the business, as contracts mean guaranteed markets for farmers and guaranteed supplies for buyers. Imami (2011) has found that farmers that are engaged in (formal or informal) agreements and longer term relations with buyers incur statistically significant lower losses compared with others (e.g. those who rely on on-the-spot marketing). In addition, coordination in the value chain (through contracts and longer term relations) is necessary to meet the demand for quality and safety standards, including traceability, which are commonly required by expanding supermarkets and export markets.

Table 6.4. Channels of fresh apple distribution in Korça, southeast Albania. Source: Field survey (2010).

Distribution channel	Product distribution among trade actors (%)
Wholesalers	69.6
Green vendors	3.4
Retail markets	1.7
Processors	5.8
On-farm consumption	3.7
Losses	15.8
Total	100.0

Table 6.5. Farmers' contracting/agreement relationships in apple distribution channels in Korça, southeast Albania. Source: Imami *et al.* (2012).

Type of contracting/ agreement	Wholesalers		Green vendors		Processors	
	No. farmers	%	No. farmers	%	No. farmers	%
No contract/ agreement	125	90.6	31	83.8	29	43.9
Written contract	0	0.0	0	0.0	1	1.6
Unwritten agreement	13	9.4	6	16.2	36	54.5
Total	138	100.0	37	100.0	66	100.0

Table 6.6. Time span of farmers' relations with the main buyers of the apple distribution channels in Korça, southeast Albania. Source: Imami *et al.* (2012).

Type of relationship	Wholesaler		Green vendors		Processor	
	No. farmers	Share (%)	No. farmers	Share (%)	No. farmers	Share (%)
Long-term (>1 year)	0	0.0	1	2.6	1	1.5
Midterm (2 months to 1 year)	2	1.4	0	0.0	4	6.1
Short-term (monthly)	20	14.5	9	23.7	47	71.2
On the spot (daily)	116	84.1	28	73.7	14	21.2
Total	138	100.0	38	100.0	66	100.0

6.2.5 Grapes, wine and raki

Production of grapes is an important source of income and on-farm consumption for rural households. There are about 300,000 farms (or more than 75% of the total number of Albanian farms) that are in some way engaged in the production of grapes (World Bank, 2007). Grape production is quite fragmented: 86% of farms producing grapes have less than 2 ha in production. The quantity and quality of grapes supplied have been an important constraint for the development of the wine industry, which has substantially relied on imported grapes, mostly from Macedonia during the 2000s (World Bank, 2007), although a slow process of consolidation is underway. From 1995 to 2009, the number of farms producing grapes decreased by 21%, while output more than doubled (DSA, 2011).

After privatization in 1991, wine production was interrupted or radically scaled back. The only former state winery continuing to operate on any scale was the 'Gjergj Kastrioti-Skenderbeg' winery in Durres, whose business was initially reoriented toward bottling imported wine, and then gradually started processing imported and local grapes again. Other former state wineries (such as 'Luani' in Berat, 'Rilindja' in Korce, 'Librazhd' in Librazhd, 'Sara' in Fier, and the former state winery in Permet) also continued activities on a much smaller scale.

At present, most wine and raki (an Albanian traditional brandy) is produced by farm households. A small portion of grapes for wine, particularly those of the 'Shesh' variety are also sold as table grapes, but such shares vary significantly from region to region. Most grapes are processed by farms into raki and wine (almost entirely red wine). The share of grapes processed on farm exceeds by far the share that is sold to processors (Fig. 6.2). According to MAFCP (2010) farms produced about 81,556 hl of wine and 122,522 hl of raki in 2009.

Production at home was traditionally focused on raki, which is easier to produce and traditionally preferred by rural consumers. However, due to an increase in demand for wine over the last 6 years (2005–2010), farmers have increased wine production using improvised facilities, though the product is usually of very poor quality. A few farmers have more formal relations with wineries, often as a result of support by foreign development projects. In the last few years, an increasing quantity of grapes has been sold to small wineries. There has been a drop in wine prices, particularly due to the low quality of the wine, and the increasingly difficulty of producing and selling wine in an informal market (i.e. without licenses, without paying excise taxes, etc.).

Even though grape production is an important source for raki and wine production, imported grapes are also important in the production of these beverages. During the 2000s (World Bank, 2007), some quantities of grapes were imported, mostly from Macedonia, but in the late 2000s, grape imports for wine

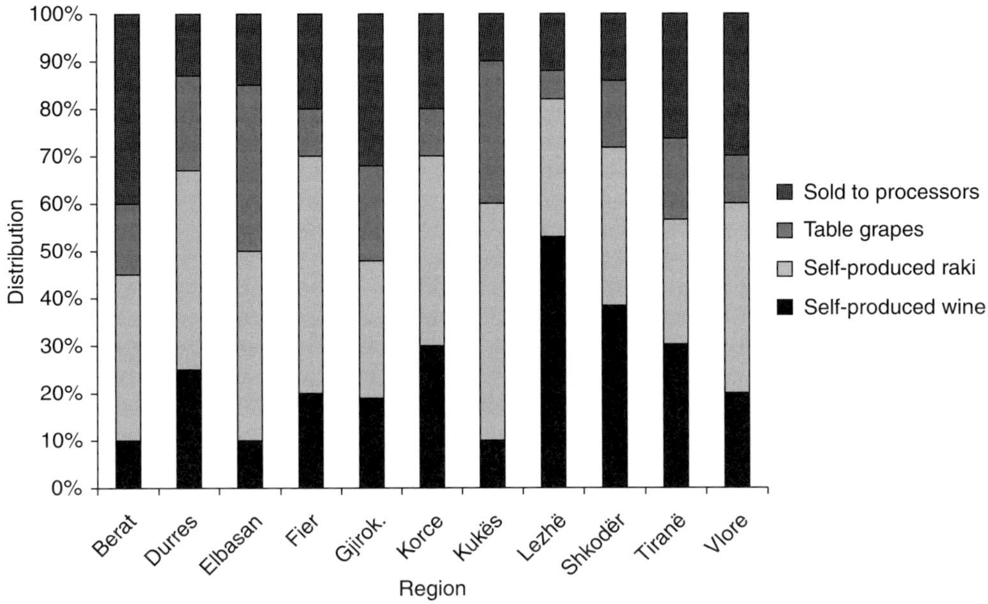

Fig. 6.2. Grape distribution and use by region in 2009. Source: MAFCP (2010).

production declined significantly, and there was an increasing connection between small and medium sized wineries and domestic grape farmers. This change is due to several factors. Firstly, increasing production of grapes and farm-made wine has pushed down prices and made on-farm processing less profitable. At the same time, increasing investments in medium-sized wineries have made commercial processors and farmers willing to scale up processing. For these wineries, using raw materials from local grapes and growers is convenient. Furthermore, the establishment of upstream, vertically integrated new medium-sized wineries, with their own newly established vineyards, ensures at least a portion of the raw material supplies that are needed. Finally, changes in the import regime have made it less profitable to import grapes from Macedonia, as was the previous custom.

As consumer income and preferences have resulted in a higher demand for quality wine, the importation of wines has increased. Imports of bulk wine are constantly decreasing, while imports of higher quality bottled wine keep growing (Fig. 6.3). The increase in demand for the importation of fine white wines provides a signal to local producers about the importance of diversifying and increasing the share of such wines produced locally.

Wines are mostly imported from Italy (about 76% of the total volume, as shown in Table 6.7), which at present is the most competitive international player on the Albanian market, followed by Montenegro. Wine is typically imported by general food importers (i.e. not specialized importers) such as 'Marketing & Distribution' and 'AGNA'. However, there are also some specialized importers of beverages, which have direct contracts with foreign producers. Imported wines are particularly competitive in the medium-to-high-end market segment, as their price/quality ratio is generally better than that of Albanian wines, especially white wines. Importers were initially focusing on cheap bottled wines, while the range of products now offered is much wider, thanks also to the increasing diffusion of specialized wine shops, wine bars and supermarket chains, which offer a much wider choice than traditional retail.

Efforts to increase the production of quality wines have so far had mixed results. Good wines produced in Albania are quite expensive. The competitiveness of Albanian quality wines

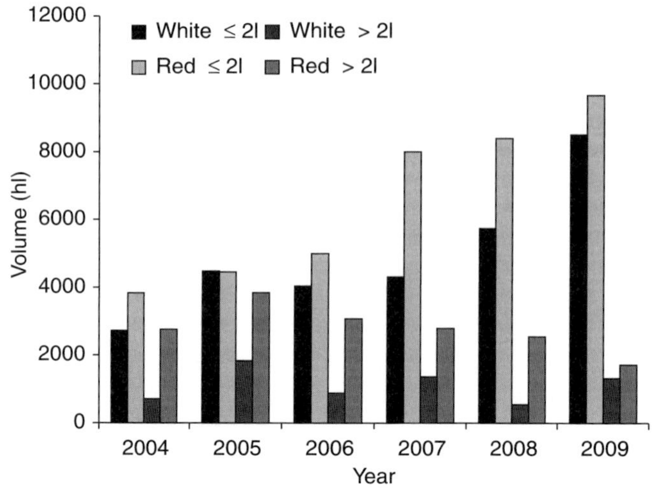

Fig. 6.3. Wine imports from the European Union, 2004–2009. Source: EUROSTAT Trade Database.

Table 6.7. Import of wine by country in 2009. Source: INSTAT (2010).

Country/region	US$	hl
World	5,770,568	25,291.1
Italy	4,324,503	19,323.3
France	515,742	1,147.8
Macedonia	232,188	1,153.5
Spain	195,176	1,091.4
Montenegro	194,717	1,499.0
Greece	99,130	594.8

is negatively affected by a global downward pressure on prices: there is a major oversupply of wine at the world level, so prices of quality wines are unrealistically low. It is now possible to buy good California wine in Europe for 1 euro/bottle and it is quite common in EU member countries to find second-tier wines with a PDO (Protected Denomination of Origin) for 1.6 euros/bottle at retail level. At these prices, few Albanian producers can realistically compete with products of the same quality (DSA, 2011). At present, the wine market is undergoing a phase of consolidation: expansion of production cannot continue at the same pace as experienced in the first decade of the 2000s, as increased domestic production has now compensated for the slowing growth of demand, and for substitute imports of cheap wine.

Several forces are contributing to shape the evolution of the competitive environment. The growth of supermarket chains will make available to consumers a larger choice of products, and will open the market for competition from imports, which are extremely competitive in the lower end of the market. Small producers will find it increasingly difficult to access the market, and will be forced to increase their product quality and specialize their trade channels (e.g. focus on supplying restaurants versus supermarkets, etc.).

6.3 Cooperatives as a Solution for Marketing and Distribution

Albanian agriculture is currently characterized by a bimodal farm size structure with a very large number of small-scale farms and a very small number of large-scale farms. More than 20 years after transition began, markets have not yet fully developed, and market-based medium-sized farms are rare (Sarris *et al.*, 1999). Many small-scale farmers produce only for personal consumption and cannot find their way to the marketplace. The transition toward a market economy seems to have passed many of these smallholders by. The small sized farms tend to produce for personal consumption, though some sell part of their products in the markets. The larger farms are market driven and only produce for selling in the market. According to a 2008 FAO (Food and Agriculture Organization) report, the destination of agricultural products was 34% for family consumption, 28% for market, 23% for processing and 15% for further production (seeds).

Research has shown that farmers have problems selling their products. Some 89% of farmers indicate a lack of development in collecting systems (UNDP, 2005). In addition, these same farmers cite a lack of storage facilities at the farm and improper grading by farmers, resulting in low and varying quality of products as further weaknesses in the supply chain. All the problems facing the small farmers in Albania that have been mentioned already, such as low quantity and quality of products, low level of technology, land fragmentation, poor infrastructure and transport to market, and lack of collection systems, have led the small farmers to low production quantities that just satisfy self-sufficiency. As self-sufficiency is a primary goal for many small farms, this production scheme has transformed Albanian farms to first producing for their own consumption. This has resulted in little attention being paid to marketing any surplus. Given the issues faced by Albanian farms, there are many potential benefits to establishing marketing and supply cooperatives. As a result of being member of a cooperative, farmers can assure markets of their products, and agro-industry enterprises can be assured of raw materials. As a consequence, farmers will consider themselves to be part of an agro-industry enterprise and, in figurative way, a part-time employee in that enterprise.

It has been argued that cooperative forms of production and marketing should be considered the most appropriate structures for Albania in overcoming the problems of fragmentation in production, the low level of technology and skills, and the unsatisfactory quality of products to qualify for wider markets (Guri, 2008). Cooperatives can make the most profit out of the increasing demand for local products, which 78% of consumers prefer to imports (Mehmeti, 2007). Organizing small farms in larger groups of appropriate structure to facilitate the trade of the agricultural products could be the solution for the problems in the marketing and distribution process. Marketing cooperatives have a role in the consolidation of individual farm activities, such as buying and selling to achieve greater economies of scale through gaining purchasing and market power. The potential benefits from cooperative efforts would be more stable prices, greater access to markets for farmers and processors, standardization of quality, and consistent supply.

A good example comes from the Shkodra (Kiri village) region in north Albania. In collaboration with Oxfam, farmers engaged in the collection of wild spices and herbs were organized into a marketing cooperative. Kiri village is high up in the mountains, a drive of 2 h from the northern Albanian town of Shkodra. The climate is very cold in winter, and the mountainous terrain can leave families isolated, particularly during the winter months. People struggle to grow enough to feed themselves on in small plots of land, and most gather wild mountain herbs to sell for cash (Oxfam, 2008). Oxfam provided the capital to start the Herb Association, paid the taxes of the business during its first year, and helped to refurbish a collection point. Oxfam also supported the association with advice on permaculture and marketing techniques, including the best way to cut, transport, and dry the herbs. It gave the Herb Association business training. Because the quality of the herbs was now higher, and they were selling in bulk, the community was able to approach new buyers and to get a better price for their products. As a result, the community now has more confidence in itself and family incomes have increased in Kiri by as much as 40%.

Although the Oxfam collaboration described above is good example of a cooperative, the fear of a renaissance of socialist cooperatives is a principal factor that has impeded the establishment of agricultural cooperatives in Albania since the start of the transition process in 1991 (MAFCP, 2007). Cooperatives in Albania before the 1990s were characterized by harsh labor conditions and low income. The memory of these organizational forms has contributed to an adverse reaction toward any form of social organization and promoted a counter-reaction towards extreme individualism. In a survey carried out in 2005 by the Center for Rural Development, a Tirana-based think tank, around 58% of the 1500 farmers interviewed considered cooperatives to be an ineffective form of economic organization, 28% were not sure about the outcome of an agricultural cooperative and only 16% expressed a positive opinion (Center for Rural Development, 2005). This reaction toward any form of social organization is rooted in the extreme collectivization of agriculture during the communist regime, the forced membership in state-controlled sociopolitical organizations, and obligatory forms of 'volunteering.' Albania was the only state of the communist block that completely replaced private property with collective property in both the agricultural and industrial sectors (Baci, 1995).

The legacy of obligatory forms of civic activities has also hindered the social initiative of the new farmers after 1991, as well as their willingness to commit their property and efforts into collective economic enterprises. This is not solely an Albanian phenomenon, but is common to other transitional economies. The more the totalitarian regime had urged obligatory membership in state-controlled organizations, and demanded volunteering work, the more people tended toward a behavior of individualism, and the less willing they were to participate in collective organizations (Howard, 2002).

Apart from the above constraints, the establishment of cooperatives has also been delayed by a lack of farmer experience in organizing cooperatives in a market economy, and the insufficient support provided by MAFCP and foreign donors. The inexperience of farmers was not confined to running a cooperative, but included, for many, also a basic knowledge of farming management techniques. In Albania, as opposed to the private craftsmen cooperatives in communist Hungary, and a private agricultural sector in ex-communist Poland, the country's isolation and extreme collectivization left farmers inexperienced in managing private farms after 1990. According to a survey by USAID (1994), 78% of farmers admitted to having insufficient experience to manage their own farms, 81% had limited knowledge of marketing, and almost all (91%) admitted that they had no idea how to organize and run a cooperative in a market economy. A study by the Center for Rural Development (2005) confirmed that little has changed since then, with 85% of farmers who still considering themselves unable to run a cooperative.

Cooperatives can benefit from a simple process of decision-making based on the 'one member one vote' principle. This is considered crucial for convincing Albanian farmers that these new forms of cooperatives are different from socialist cooperatives, as the members will now have the power to take decisions.

6.4 Conclusions and Recommendations

The structure of Albanian supply chains is an important issue for the economy of the country. The majority of Albanian production units are very small-scale. Despite recent consolidation, the overall picture of Albanian agribusiness reveals it to be still fragmented and very small-scale compared with their European peers. Small-scale primary production requires efficient and effective consolidation processes downstream in the value chain based on adding value. In most sectors this is a weak element, hampering effective production and the economic performance of food processors. Businesses and policymakers are aware of the need to stimulate vertical and horizontal cooperation, including consolidation.

The distribution channels for agricultural products are still very challenging for Albanian farmers, and are characterized by a fragmented structure. This situation is found in dairy production, especially milk production, where the dairy industry has to deal with a large number of stakeholders at all the stages of distribution. Informal markets are also a big problem for almost every agricultural product, including dairy products, olive oil, wine, vegetables and fruits.

Most companies operate for the domestic or local markets, and work according to standards that only match the needs of these markets. Some companies successfully practice the import–export business model: they import raw materials, process in Albania, and export the consumer products. This model is viable, especially in certain areas of the meat and vegetable markets, namely sausage production and frozen vegetables. The main reasons for the success for this business model are the combination of low wages, low logistics costs, and efficient operations. The business model may, however, become vulnerable as wages start to increase.

Given the actual situation of agricultural production in Albania, and the market/distribution problems, the introduction of cooperatives might be a good possibility for helping farmers' lives and activities. Cooperatives can exercise sensitive pressures in a market environment by offering high prices for products coming from members and low prices for inputs. Another function of cooperatives can be the fulfilling of demand and consumer needs. This involves all business activities between clients, members, and the cooperative concerned. The reduction of individual uncertainty by risk diversification is another important function that cooperatives can offer. They can try to lower the market power of other businesses, and by doing so, favor their members. This is realized by increasing product reliability for consumers, by associating a brand name with the product, and by offering the necessary services. The Herb Association in northern Albania's village of Kiri is a good example of just such a situation.

Another benefit of cooperatives is that they can assure quantity and quality of products, which is necessary for fulfilling market requirements. To do this, the importance of contacts must be emphasized so that cooperatives can obtain information and offer advice to encourage production and marketing practices. Finding input markets and product markets are the two most important advantages that cooperatives offer. Market availability is a big problem, especially for farmers that produce perishable products such as fruits and vegetables. The activities of cooperatives can also increase political influence on the government. As a result, legal and administrative rules can be established in favor of the business of cooperatives. These kinds of benefits can be local or regional. Furthermore, agricultural marketing and institutional researchers should conduct cost–benefit analysis on the efficiency of various solutions for improving market conditions and competitiveness for the agricultural industry. More competitive research and extension grants should be offered so as to provide quality information to enable the transformation of the country to greater cost-effectiveness, while at the same time meeting future consumer demands for taste and quality.

References

AAC (2008) *Quarterly Report, Oct.–Dec. 2008*. USAID/Albanian Agriculture Competitiveness Program, Tirana, Albania.

Baci, I. (1995) Prapambetja bujqesore [Agricultural stagnation]. *Koha Jone Journal* (Tirana, Albania), 6 January 1995.

Barrett, C.B. (1997) Food marketing liberalization and trader entry: evidence from Madagascar. *World Development* 25, 763–777.

Center for Rural Development (2005) *Agriculture towards Market Economy*. Oxfam, Tirana, Albania.

DSA (2011) *Agribusiness Sector Survey*. Report prepared for Italian Cooperation by Development Solutions Associates, Tirana, Albania.

Fackler, P.L. and Goodwin, B.K. (2001) Chapter 17: Spatial price analysis. In: Gardner, B. and Rausser, G. (eds) *Handbook of Agricultural Economics, Vol. 1B*. Elsevier, Amsterdam, Netherlands, pp. 972–1024.

Fafchamps, M. (2004) *Market Institutions in Sub-Saharan Africa*. MIT Press, Cambridge, Massachusetts.

GTZ (2010) *Analysis of Fruits and Vegetables Sector*. GTZ (Deutsche Gesellschaft für Technische Zusammenarbeit) Office, Tirana, Albania.

Guri, G. (2008) Agricultural cooperatives: an imperative for the development of social economy? Lessons from rural Albania. Paper presented to: *ICA Research Conference, The Role of Co-operatives in Sustaining Development and Fostering Social Responsibility, Riva del Garda, Trento, Italy, 16–18 October 2008*. Available at: http://www.acdivoca.org/acdivoca/CoopLib.nsf/d40b394e0533f5b285256d96004f1ad4/d37fbd097d28ed7d852575e0005671f1/$FILE/AGRICULTURAL%20COOPERATIVES%20AN%20IMPERATIVE%20FOR%20THE%20DEVELOPMENT%20OF%20SOCIAL%20ECONOMY%20LESSONS%20FROM%20RURAL%20ALBANIA.pdf (accessed 4 September 2012).

Howard, M.M. (2002) The weakness of post-communist civil society. *Journal of Democracy* 13(1), 157–169.

Imami, D. (2011) Analysis of agrifood value chain actors' choices in Albania. PhD thesis, University of Bologna, Italy.

Imami, D., Skreli, E., Cela, A. and Zhllima, E. (2012) *Analyzing Consumer Preferences for Apple in Albania*. AUT Working Paper, Agricultural University of Tirana, Albania.

INSTAT (2010) *Statistical Yearbook 2009*. Instituti i Statistikës (INSTAT), Tirana, Albania

Kapaj, A., Chan-Halbrendt, C., Deci, E. and Kapaj, I. (2011a) Milk consumers' preferences in urban Albania. *Advances in Management and Applied Economics* 1(3), 197–215.

Kapaj, A., Kapaj, I., Deci, E., Jaupi, A. and Pagria, I. (2011b) Identifying some key indicators benchmarks (competitive benchmark[s]) helping meat processors in Albania improving their management decisions. *Valahian Journal of Economic Studies* 2(16)(Issue 1), 49–54. Available at: http://www.vjes.eu/images/2011/VJES%20-%20Volume%202/Vol.2(16)_Issue1_2011_Ana%20Kapaj.pdf(accessed 4 September 2012).

Leonetti, L., Imami, D., Zhllima, E. and Stefanllari, A. (2009) *The Olive and Olive Oil Value Chain in Albania*. Prepared by Development Solutions Associates, Tirana, Albania. Available at: http://www.eastagri.org/files/Oil-Albania.pdf (accessed 3 September 2012).

MAFCP (2007) *Statistical Yearbook 2006*. Ministry of Agriculture, Food and Consumer Protection, Tirana, Albania.

MAFCP (2009) *Gjendja Aktuale e Ullishtarisë dhe Perspektiva e Zhvillimit te saj (Actual Situation of the Olive Sector and Development Perspectives)*. Ministry of Agriculture, Food and Consumer Protection, Tirana, Albania.

MAFCP (2010) *Statistical Yearbook 2009*. Ministry of Agriculture Food and Consumer Protection, Tirana, Albania.

Mehmeti, I. (2007) *Olives Value Chain Assessment*. USAID/Albania Agriculture Competitiveness Program, Development Alternatives, Inc (DAI), Tirana, Albania.

Oxfam GB (2008) *Albania: Development of Agriculture Cooperatives. From Poverty to Power: Case Study*. Oxfam GB Office, Tirana, Albania. Available at: http://oxfamilibrary.openrepository.com/oxfam/bitstream/10546/111954/1/fp2p-cs-albania-development-agricultural-cooperatives-140608-en.pdf (accessed 14 September 2012).

Petrick, M., Schreiber, C. and Weingarten, P. (2003) Competitiveness of milk and wine production and processing in Albania. Presented to: *IAMA (Leibniz Institute of Agricultural Development in Central and Eastern Europe) Forum, October 2003, Halle, Germany*.

Platteau, J.-P. (2000) *Institutions, Social Norms, and Economic Development*. Harwood Academic Publishers, London.

Sarris, A., Doucha, T. and Mathijs, E. (1999) Agricultural restructuring in Central and Eastern Europe: implications for competitiveness and rural development. *European Review of Agricultural Economics* 26, 305–329.

UNDP (2005) *The Agribusiness Sector in Albania: Inward Investment Opportunities*. United Nations Development Program/European Union Trade Liberalization and Promotion Program. Report produced by independent consultants on behalf of the Albanian Foreign Investment Promotion Agency (ANIH), UNDP Albania and the European Union. Available at: http://www.trade.undp.org.al/u/d/trade/Albanian%20agribusiness%20sector.pdf (accessed 3 September 2012).

USAID (1994) *Albanian Agriculture in the Market Economy*. US Agency for International Development. Tirana. Albania

World Bank (2007) *Albania: Strategic Policies for a More Competitive Agriculture Sector*. Report No. AAA18-AL, World Bank Sustainable Development Sector Unit, Europe and Central Asia, Tirana, Albania. Available at: http://siteresources.worldbank.org/INTALBANIA/Resources/Strategic_Policies_for_more_competitive_Agriculture_sector.pdf (accessed 3 September 2012).

7 Quality Assurance and Food Safety in Albania

Luciano Leonetti,[1] Ilir Kapaj[2] and Florian Xhafa[3]
[1]*Development Solution Associates, Tirana, Albania;* [2]*Agricultural University of Tirana, Albania;* [3]*Tirana Business University, Albania*

7.1 Introduction

In the last decade, safety and quality became the important issue in agribusiness development. Consumers, in general, have become more conscious of quality, and organizations are now judged more on their overall quality performance than on their financial performance alone (Schiefer, 2002). This evolution is part of a more general reorientation of the concept of value chain and quality, which changed from being production oriented to customer oriented. Also, integrative approaches, system thinking, the focus on advanced technologies, and belief in human capacities have had a considerable impact on current quality management (Schiefer, 2002). Safety and quality has become a vital distinctive feature for competition in the world market of food products. To obtain a good quality end product, quality is more and more managed along the whole food chain, from the supplier of raw materials to the consumption. Consumers' demand, competitive pressure, and legal provisions force actors in agribusiness and the food industry to use quality management as a strategic issue in innovation and production (Luning *et al.*, 2002).

Different types of quality management and food safety systems exists. Such as HACCP (hazard analysis and critical control points), the ISO (International Organization for Standardization) series, and BRC (British Retail Consortium Standards). HACCP is a system used by the food industry to ensure that all food consumed is safe to eat. It is a systematic approach to hazard identification, assessment of risk, and its control. The HACCP system involves a structured approach for the control of food safety from farm to fork. ISO has a membership of 158 national standards institutes from countries large and small, industrialized and developing, in all regions of the world. The organization develops voluntary technical standards that add value to all types of business operations, and these standards contribute to the dissemination of technology and good business practice. They support the development, manufacturing, and supply of more efficient, safer and cleaner products and services. The standards also make trade between countries easier and fairer. Each country adopts and applies a legal framework that regulates these aspects of products and services in order to develop and implement them, and in relation to its international trade needs and ambitions. The basic regulations are detailed in *Codex Alimentarius*, a collection of international food standards, guidelines and codes of practice that contribute to the safety, quality and fairness of international food trade. The *Codex* has been developed under the FAO (Food and

Agriculture Organization)/WHO (World Health Organization) Food Standards Programme. However, much legislation requires more stringent rules, such as the application of the HACCP principles and system, or EUREPGAP (common standards for farm management practices; name derived from Euro-Retailer Produce Working Group/Good Agricultural Practices) and GLOBALG.A.P. (which is involved in setting voluntary standards for certification of agricultural products around the globe) practices. These sets of rules are intended to guarantee the safety of products traded, and the relevant certification is a tool to speed up and facilitate trade, especially international trade.

The application of quality management standards, such as ISO 9001: 2000/2008 (Quality Management System) and ISO 22000 (Food Safety Management Systems—which are requirements for any organization in the food chain) are also intended to guarantee the quality and safety of food. As a general rule, legislation can impose compliance with certain systems (e.g. HACCP), or even the adoption of certain standards, but cannot request that an enterprise be certified for applying such standards. Bearing a safety and/or quality certification is a major tool in international competition, so that ever increasing numbers of enterprises seek to obtain such certification, even if it is not compulsory by law. The importance of being certified with compliance of ISO 9001:2008 and ISO 22000 standards is confirmed by the figures of the certificates released. At the global level, more than 1.1 million certificates have been released for the ISO 9001: 2008 standard, while the growth in the number of certificates released for the ISO 22000 standard in recent years has been the fastest among all ISO certificates (International Standards Organization, 2011).

However, complying with quality and safety systems mandated by laws requires large and increasing investments and recurrent costs. Impact analyses must be made when introducing new legislation imposing more stringent quality assurance schemes. At the enterprise level, accurate cost–benefit analyses are necessary to assess which voluntary quality management scheme(s) are appropriate and functional to enhance the competitiveness of the enterprise, taking into consideration the actual and prospective activity and market conditions. The same criterion should be applied before deciding whether to apply for certification.

This chapter describes the legal framework and the situation of quality assurance and food safety in Albania. The account includes an analysis of the evolution of the system during the transition period, and a description of the present legal and institutional framework. The results are also included of several studies on the application of food quality and safety systems in Albanian Food and Beverage Enterprises (FBE), and of the system of services provided by public and private subjects to implement quality and food safety systems at an enterprise level.

7.2 The Food Safety System in Albania

The situation and evolution of food safety and quality assurance in Albania is the result of two processes proceeding at different speeds. One of these processes is continuous pressure from the European Union (EU) and other international organizations to adopt a system of norms and technical requirements, and to establish institutions and services; these steps are necessary to achieve several goals. The first of these is to reduce the risks for Albanian consumers and producers for cross-border diffusion of animal diseases through trade. The second is to allow the gradual integration of Albania into international markets. The third is to harmonize the Albanian legal framework with the EU system of quality assurance and food safety, a key criterion for EU accession. This process has led to the adoption of primary legislation which is largely in line with the legal provisions of the most advanced countries. At the same time as this process is proceeding, another process is the slow and largely incomplete development of secondary legislation and institution building in Albania. This is creating a wide gap between legal provisions and the actual situation in terms of food quality and safety law enforcement agencies and institutions. The progress that has been achieved is largely the result of a market-driven spontaneous improvement of standards, especially in the retail sector and in some processing industries.

In 2007, a World Bank document highlighted the needs for improvement in different aspects of food safety. Based on this document, and despite significant progress, key constraints persist in the national food safety control system in Albania in legislation, institutional capacity, infrastructure, control and enforcement, and private investment (World Bank, 2007). Compared with that assessment, the main progress has been achieved in harmonizing the Albanian legal framework to that of the EU, while progress in law enforcement, infrastructures and controls been more limited. Private investments are increasing, but these are spurred more by competition among value chain operators than by the much limited available incentives or the existing weak controls.

7.3 The Evolution of the Legal Framework

It is important to observe the legal framework that is related to the implementation of the food and safety standards in Albania. Changes in the legal framework are guided by the need to update approaches and methods to improve food safety, moving from mandatory compliance with standards towards preventive-oriented regulation and risk-based control systems, and from test-based to risk-based inspection services. The introduction of partial reforms started after the dismantling of the planned economy in 1991, but the first efforts to reform the system of public health related to food production and consumption, food safety, and consumer protection were made after the accession of Albania to the World Trade Organization (WTO) in 2000, which made it necessary to change the norms regulating the quality and safety characteristics of food products.

Accession to the WTO introduced a distinction between requirements (obligatory, defined by regulations) and standards (voluntary). Due to this provision, a long transitional period started, in which Minimal Technical Safety Requirements (MTR-Safety) and Minimal Technical Quality Requirements (MTR-Quality), fixed by secondary legislation must gradually be substituted by the standards fixed by the General Department of Standardization as the source of the norms indicating the characteristics of food products. To prevent legal loopholes potentially affecting consumer protection, these norms foresaw the obligatory indication in the label of the standard applied. This provision is not fully coherent as it cannot be mandatory to apply a standard that by definition is voluntary. However, over the last 11 years, this has been the tool to ensure the compliance of food products with the technical requirements defined by the (only) existing norms. At the same time, the characteristics of some products sold without a label, as with most fruit and vegetables, are only partially defined by other, more general norms set for consumer protection, such as the law on chemical residuals.

In December 2001, the European Commission (EC) presented its vision for the reform of food legislation through the *White Paper on Food Safety*. This is a landmark document that showed the path for the evolution of food safety legislation in all EU member countries (International Business Publications, 2011). The cornerstone of EU food legislation is Regulation 178/2002, also known as the 'General Food Law'. A number of other norms and regulations followed, covering a variety of subjects, such as animal products, consumers' health protection, and the responsibility of producers for the safety of products. Fulfilling this food legislation became a prerequisite condition for countries wanting to join the EU (as in the case of Albania) or simply willing to trade with EU member countries.

The first important step in the direction of primary legislation harmonized to the reoriented EU set of norms (the so-called *acquis communautaire*) was the approval in 2004 of the veterinary law (Law 'On Veterinary Service and Inspectorate' No. 9308/94), which for the first time shifted the focus of rules and regulation on the identification, movement and control of live animals and animal products from productive issues to public health issues, in a system that controlled 'from farm to fork'.

In 2006, the Albanian Ministry of Agriculture, Food and Forestry was restructured, and became the Ministry of Agriculture, Food and Consumer Protection (MAFCP). As part of the restructuring, a General Directorate for Consumer Protection was established, and this became one of the four General Directorates of the Ministry. This agency was responsible for all of the upstream segments of the food chains, ranging from animal health and plant protection, to the licensing and control of food processing plants. Furthermore, it has oversight over the release of labels and all aspects related to food safety, with the exception of controls on the use of fertilizers and the relevant impacts on soil pollution.

Roles and responsibilities in the control of food in retail outlets and in hotels, restaurants, and catering services remained largely undefined, even if in the cities the control was (and is) carried out by the city council veterinary service. In many cases, this service gave an extensive interpretation of its role, initially limited to controlling the condition and provenance of food of animal origin. The most visible function of city council veterinary inspectors consists of supervising the slaughtering activities performed within the boundaries of the city, and controlling the meat sold in the city meat shops to make sure that it all bears a veterinary seal and/or a certificate indicating the provenance, or a label.

Over the last 20 years, the strictness with which rules and regulations have been applied in outlet controls has been quite limited, owing to lax application of existing laws, complicated rules for effective law enforcement (inspectors should seek support from police to close a shop not complying with the rules), and structural limits of the supply chain. Due to the lack of appropriate slaughtering facilities, the actual application of rules would have resulted in a comprehensive interruption of fresh meat supplies to urban areas. Still, according to the World Bank (2007), 'Only three slaughterhouses in Albania are considered suitable for meat production, but utilization is extremely low, implying that extensive production takes place in uncertified premises.' At the end of 2007, activities for the establishment of a National Food Authority (NFA) started, with the support of EU and UN development projects. The objective was to create an organization that, supported from some other cooperating structures (such as the National Reference Laboratory), would reach a full capacity in risk assessment, crisis management and risk communication.

In March 2008, Law No. 9863/08 'On Food' was approved, ruling on the overall matter of food safety and consumer protection, and outlining the functions of the NFA as in charge of performing all the food safety controls along the food chain 'from farm to fork'. Among other issues, this norm introduced, for the first time, the mandatory application of some elements of the HACCP system not seen in the previous version of the law, which had been in force since 1995 (Law No. 7491/95). This new law started a process of transition toward a centralization of all control functions, to be gradually transferred to the NFA, even if some functions remained with the veterinary inspectors of the ministry and the city council veterinary services. According to the agreements with the EU, this sharing of functions was intended to remain only in a transitional period; but in fact, with the amendments introduced in 2011, the ministry's veterinary and plant protection services will continue to play a role in controls at the primary production level. This process is still ongoing.

Also in 2008, the law No. 9902/08 'On Consumer Protection' was approved, replacing the previous norm (Law 'On Consumer Protection' No. 9135/03). This law follows a horizontal approach to the overall protection of consumers and aims to avoid misleading them. Food products are explicitly excluded, as ruled by Law 9863/08 'On Food', although the articles included in this law include some key rules that condition the norms on labeling. In 2011, the promulgation of a new veterinary law (Law No. 10465/11) introduced major changes in the responsibilities of ministry veterinary inspectors (i.e. those from the Directorate of Animal Health and Plant Protection working in central and regional offices), limiting the scope of their intervention in favor of the NFA inspectors.

The key legislation for the whole sector remains Law 9863/08 'On Food', which is in a phase of substantial reshaping, assigning much more powers to the NFA, and defining the responsibilities of the different inspection services (those from Ministry Directorate of Animal Health and Plant Protection, those from the NFA, and those from the city councils). Also, the law will make mandatory the application of the full HACCP system in all food processing activities, although the relevant certification will remain fully voluntary. At present, only the application of GHP (Good Hygiene Practices) is mandatory.

7.4 The Food Safety Situation and its Evolution

Food safety in Albania is improving, but the pace of improvements is slow when considering the progress achieved in other development indicators and income growth. Many food chains, particularly meat and dairy products chains, are risky, which shows the slow movement of this sector compared with other sectors of the Albanian economy, such as construction and services. However, even in the worst periods experienced after the beginning of transition, no major food-borne diseases epidemics were recorded, even if such diseases remain an important public health issue. In most cases, the abundant use of preservative agents and the prudent and conservative consumption habits of customers have contributed to reducing short-term risks.

During the planned economy, food safety levels were strictly regulated, as was the overall economy of the country. Some food chains related to animal production (i.e. fresh meat and dairy) were more controlled than now, even if epidemiologic data show some limits in the effectiveness of such controls. Compared with the present situation, the old system was much less efficient, but more reliable in terms of food safety controls. This is due to the fact that it was based on a few large enterprises that were fully controlled by the state, and required limited and strictly controlled movements of animals.

Immediately after the end of planned economy (beginning in the 1990s), the supply chains of the large processing facilities established during the old economy collapsed. Some of these processing facilities were soon privatized and continued to work, but using only a small part of their full capacity. The new owners of those processing plants who had managed to survive had, in most cases, a more limited capacity for investing in the renovation of obsolete premises and equipment. As a result, the whole process of modernization proceeded at a very slow pace, with negative consequences in terms of competitiveness, quality, and safety of the products. In general, during the first 10 years of the transition period (i.e. 1991–2000), investments in agribusiness have been limited and mostly devoted to establishing family micro businesses, which, with a few notable exceptions, were mostly in meat processing.

At the beginning of the transition, legislation loopholes and lapsed enforcement of existing laws prevented a complete collapse of the food chains and allowed local food producers to remain in business. From 2000 onward, the economy as a whole started a more sustained process of development. The same factors that in previous years had allowed domestic agri-food processors to survive (the large number of informal businesses, minimal dimensions, ability to source all necessary raw materials without logistic problems, very little investment, and ability to keep the prices low) started to hamper and delay the consolidation and development of agri-food business. Food processing companies willing and able to grow as formalized small and medium-sized businesses (SME) had to face unfair competition from a huge informal food processing sector. This sector was composed of improvised processing micro facilities, not bound by any rules and able to sell at very low prices.

Furthermore, these small producers had (and still have) little knowledge about food quality and safety standards and requirements, and are not in the situation to apply food safety and quality management schemes (World Bank, 2007). The scarce knowledge of food processors about quality assessment (QA) schemes, as assessed by the 2007 World Bank study, was confirmed in 2010 by the SME Training Needs Assessment (TNA) performed by the EU. The limited progress achieved in the three years in this field shows the importance of a substantial awareness campaign addressed to agri-food SME.

Roadside sales of fresh and home-processed products, and uncontrolled door-to-door direct sales of food products in Tirana (milk and lamb meat) and other main cities have also contributed to the increased risk of food-borne diseases and the reduced capacity of regular businesses to remain in the market without accepting compromises in the quality and safety of products. Under these conditions, making major investments in food safety would be not sustainable. The few exceptions were represented by the companies that managed to start some export activities (first fish, and later eggs), or those that grew enough to find it possible and profitable to invest in their credibility in terms of safety and quality. These conditions were first achieved by the leading enterprises in meat processing, and after that in the milling industry. These processing units established modern processing plants and efficient marketing logistics; some of them even managed to get the plant certified for export in the demanding EU common market.

Food-borne diseases are still an important public health problem in Albania. Some infections, such as hepatitis, salmonella, typhoid fever, and toxic infections slowly decreased after 1990. At the same time, gastrointestinal infections, after decreasing between 1990 and 2000, surged after 2000; after reaching a peak in 2003 and starting to decrease, they still remaining at levels higher than in 1990 (data from the Ministry of Health, sector of Hygiene and Epidemiology). Concerns for public health are mostly related to the meat and milk food chains. In the case of meat, the main issue is related to the fresh meat supply chains. The main issues with this are the lack of controls on animal movements, and in the widespread practice of slaughtering animals in inappropriate premises, without appropriate veterinary control. In contrast, the processed meat value chain is one of the most controlled in Albania. In the (unprocessed) milk and dairy chains there are several critical points creating threats for human health, ranging from microbiological contaminations at different stages of collection and distribution, to threats due to animal diseases, the most serious one being bovine brucellosis. The spread of this disease increased throughout the 1990s and until 2002. In some districts, such as Saranda and Gjirokaster, incidence exceeded 10% of the small ruminant population, but after this date, incidence at the national level was stabilized at a level between 0.7% and 0.8% of the total small livestock number (World Bank, 2007). In parallel with the spread of the disease, the number of cases of brucellosis among humans increased from a few tens of cases in the first part of the 1990s to over 3000 in 2004.

Even though increasing revenues, market segmentation, and the consequent higher attention that consumers pay to food safety and quality are speeding up the process for change and adaptation to EU standards, the process is proceeding at a different pace in the various components of the food safety system. The harmonization of the primary legislation to the EU set of norms (the *acquis communautaire*) is well advanced, even if some recently introduced legislation will most probably need amendments and integrations. Adoption of secondary legislation is lagging behind. To be applied, the norms of food safety need the specifications of MTR-Safety and MTR-Quality in each category of products. Other MTR are also required for hygiene rules to be applied, and for definition of the conditions that a product must satisfy to be legally traded. So far, only eight MTR-Safety and no MTR-Quality have been approved.

The application of the food safety legal framework depends on the functionality and effectiveness of the organizations responsible for all the food safety controls along the food chain. The main role is played by the newly established NFA, but important functions are also attributed to the Veterinary and Plant Protection Directorate of the MAFCP, to the veterinary services of the main municipalities, and to the National Reference Laboratory located in the Veterinary and Food Safety Institute (ISUV, Instituti i Sigurisë Ushqimore dhe Veterinarisë). In this domain, the system is still in the phase of transition, though a complete overhaul was introduced in summer 2011, in parallel with the drafting of the new 'Food Law', and with the introduction of the new law on veterinary services. However, functions and responsibilities are not yet clearly defined, and not fully tested in their actual functionality. A role is also played by the Epidemiology Department of the Ministry of Health, which monitors the evolution of food-borne diseases. The public network of food laboratories at regional level, and the National Reference Laboratory, are part of this system of public services. The services to food processing enterprises are also improving, and the number of service providers is increasing, but the quality and reliability of such services is variable. A certain number of consultants and advisors, coming from developed countries familiar with quality standards, are available to support enterprises in introducing the EUREPGAP and GLOBALG.A.P. practices, establishing and maintaining the HACCP system, and complying with the relevant ISO standards. But the quality of services is uneven because the market is still too small and new to have made a selection of the best consultancy provider in the field of quality management.

ISO and HACCP certifications are given by Albanian companies and Albanian branches of international companies, although there is no internationally recognized regulatory body supervising the activity of these companies. The Directorate for Accreditation of the Ministry of Economy should perform this function, but it is not a member of the European Accreditation body. Private veterinarian practitioners (PVP) are also playing a specific function in the food control chain, being contracted from the State Veterinary Service (SVS) for vaccinations against quarantine diseases, and being responsible for communicating to the government authority on possible diseases outbreaks.

Awareness of food chain actors (producers, food processors, traders, consumers) of the importance and relevance of food safety and quality assurance schemes is still low. Among producers and processors, it is not clear what kind of quality assurance scheme is important (or mandatory), and quality assurance schemes are often seen as a marketing tool (Leonetti *et al.*, 2010). Among consumers, quality is largely identified with origin, and customary retailers are more trusted than quality schemes guaranteed by public authorities or certifications (Imami *et al.*, 2010).

7.5 The Present Legal Framework for Food Safety and Consumer Protection

When the new 'Food Law' enters into force, the Albanian legal framework will be based on a set of primary legislation that can be considered well advanced on the path of harmonization to EU *acquis*. At present (January 2012), the law is in a final draft stage. With the introduction of this new law, the main primary norms of the legal framework will be:

1. the updated law 'On Food'
2. the law 'On Veterinary Services' (Law No. 10464/11)
3. the law 'On Consumer Protection' (Law No. 9205/03), now in need of updating.

In addition, there are several laws regulating different aspects of food safety, such as Law 9362/05 'On the Plant Protection Service' (as amended by Law 9980/08 and Law 10137/09), Law 8531/99 'On Chemical Fertilisers Control Services' (as amended by Law 10137/09), and Law 9199/04 'On Production, Processing, Certification and Marketing of "Bio" Products' (as amended by Law 10137/09). For the main groups of products, there are specific laws as well, such as Law No. 8944/02 'On the Production, Origin and Trade of Olive Oil', and Law No. 9441/05 'On Production, Collection, Processing and Marketing of Milk and Milk-based Products'. The process of building up an organic and coherent system of secondary legislation is still far from being completed, as the bylaws in force have been largely made on the basis of previous primary legislation, now either no more in force, or partially substituted by new norms. The most important parts of secondary legislation are:

1. DCM No. 1344/08 'On the Approval of Rules on Labelling of Food Products'
2. DCM No. 209/08 'On Maximum Levels of Pesticides in Food Products of Plant Origin'.

However, the most important and specific bylaws for quality assurance are the MTR-Safety bylaws defining the requirements in collecting and processing food products, and the methods to be used in processing and analyzing samples of these products, and the MTR-Quality bylaws defining the characteristics of the product required for grading (size, defects, maximum share of products in a lot that can be not compliant with the grade, etc.). So far, six MTR-Safety bylaws are in force:

1. General Hygiene Conditions for Food Establishments and Food Business Operators
2. The Implementation of GHP (Good Hygiene Practices)/GMP (Good Management Practices) pre-requisite Programs and HACCP-based Procedures in Food Establishments
3. Specific Hygiene Requirements for Meat, Meat Products, Meat Preparations, Minced Meat and Mechanically Separated Meat
4. Specific Hygiene Requirements for Milk Production Establishments, Raw Milk and Dairy Products
5. Live Bivalve Mollusks, Fish and Fishery Products
6. Specific Hygiene Requirements for, and Official Controls of, Other Products of Animal Origin, which includes requirements for a wide range of products, including, among others, eggs, honey, snails, frog legs, treated stomachs, bladder and intestines.

The priority given to these first six MTR bylaws appears to be guided by three main factors. The first of these is the necessity to give an overall coverage to safety control systems in processing plants (MTR 'General Hygiene' and 'HACCP'). The second is the need to regulate safety issues in the two food chains considered as the most critical ones by all previous sector studies (meat and milk/dairy also influenced the priority MTR). The third factor is the need to provide safety regulations for the other animal products that Albania is already exporting, and wants to export in larger quantities.

Safety of products in other major agribusiness branches, such as olives and olive oil, wine, fresh and processed fruit and vegetables, and medicinal and aromatic plants (MAPs), is not yet regulated in the form of comprehensive MTR-Safety, but older norms regulate some safety and quality aspects. In addition, the (voluntary) standards of the General Directorate of Standardization (GDS) are still in force. In some cases (as for all packaged products bearing a label), it is obligatory to indicate on the label the standard used, which, in practice, is making somehow compulsory the otherwise voluntary standards. The incompleteness of MTR-Safety and MTR-Quality represents a serious handicap in effectively applying the legislation on food safety, in improving the system of food quality and safety control (i.e. the control of compliance with MTR), and in developing the marketing practices commonly applied in the market (e.g. a recognized grading system of fresh produce is a key issue for market segmentation). Now that the primary legal framework has been largely updated, completing the MTR-Safety and introducing the MTR-Quality should represent a priority in finalizing the Albanian quality assurance legal framework, and harmonizing it with EU legislation.

7.6 Institutions and Public Services for the Application of the Food Safety System

Implementation of food safety systems by the public services is important for the Albanian food processing industry. The reference institution in Albania for quality assurance is the NFA, which was established in 2009, but started to be really operative only in the summer of 2011, with the transfer to the NFA, within the MAFCP, of staff dealing with food inspections. The NFA is tasked with carrying out all of the inspections necessary to verify the conformity of food producing enterprises (farms, processing plants, wholesale markets, etc.), with the regulations, and its tasks include the assessment of new food processing enterprises. The limits of these tasks are not yet clearly defined. The border between the functions of MAFCP and NFA inspectors is still unclear, and is one of the reasons delaying the updating of the Law 'On Food'. Initially, it was proposed that NFA inspectors completely replace all other inspection services dealing with the food chain, including the phytosanitary and veterinary inspectors of the MAFCP, and municipal inspectors dealing with local slaughterhouses and retail outlets. So the amended Law 'On Food' has not yet been finalized, and more changes are still possible.

Referring to the new guidelines, the MAFCP has the function to ensure and control all animal and plant diseases, including those that are transmissible or that can generate food-borne diseases. The SVS is contracting private veterinarians to perform the vaccination campaigns paid for by the state, to ensure that animals affected by quarantine diseases are culled and disposed of (and farmers compensated). The SVS is also responsible for the alert system related to outbreaks of animal diseases.

The role of municipal council veterinary inspectors, as another actor for the applications of food safety systems, is still particularly unclear. In the past, their main function was to control the livestock markets, verify that slaughtering activities performed in the boundaries of the municipality were carried out according to the regulations, and also verify that all the meat carcasses retailed had been controlled (as demonstrated by a certificate and a veterinary seal). These inspectors also had the task of policing retail outlets for valid licenses and of making sure that the public health and environmental norms were being respected. In practice, because most slaughtering activities have been informally performed since 1991, the main work of the inspectors has consisted of controls in shops. Over time, the function of these inspectors in some cities has been widened to cover the safety of all food products. With the establishment of the NFA, their role has been reduced to only checking for the validity of licenses and ascertaining that other municipal regulations are respected, with no power on the control of food, which has been transferred to NFA inspectors.

However, the various inspection bodies end up sharing tasks, as the shift from a regulation-focused approach (general regulations) toward an enforced self-regulatory approach (specific regulations for each inspected body) requires significant changes to the inspection procedures that have been applied so far, primarily relying on hazard-based inspections of processes and premises (applying HACCP principles), rather than the sampling and laboratory testing of products (World Bank, 2007). Considering the broad field of responsibilities to be covered, the total number of inspection personnel, about 700 in 2007 (World Bank, 2007), will need to increase, although at present it is actually decreasing. The current decrease is due to municipal inspectors being moved to other tasks. The number of veterinarians foreseen in the organization charts of the NFA and MAFCP in 2011 (214 units) is lower than that working for the MAFCP alone (243 in 2006) before the reform (Leonetti *et al.*, 2011).

Apart from above mentioned institutions, another important one is ISUV, which, as already noted, acts as the National Reference Laboratory. Among other functions, the ISUV carries out the analyses of food products for the NFA, tests water quality for aquaculture production units, ensures the monitoring of animal diseases, and participates in the European avian flu and rabies surveillance and alert system. The ISUV also keeps the national livestock register. Animal identification is a key tool for traceability and quality assurance in animal products. Initially intended as a tool for production management (i.e. to record inputs and outputs from breeding activity), with time, the main function of the register has changed to become one of the cornerstones for the traceability and safety of food of animal origin.

7.7 The System of Private Services to Enterprises

In the food sector, there is a growing demand by enterprises for HACCP and ISO 9001: 2000/2008 (Quality Management System) certifications. The majority of the large food processing companies are GMP and/or HACCP certified, with the exception of the fruit and vegetable processing sector, where only a few companies have such certification. The first ISO 9001 certification in Albania was released in 2000. At present, the certified companies are mostly those that deal with milk processing (six in 2010, almost 10%) (Schröder *et al.*, 2010), meat processing (seven in 2010, 14%), (Mautner Markhof *et al.*, 2010) animal feed production, fish (six in 2012), flour production (all the three leading companies), and mineral water. So far, there have been few Albanian companies certified ISO 22000 (Food Safety Management Systems – requirements for any organization in the food chain), though again, those that are so certified come mostly from the milk processing sector. Certification for ISO 14000 (Environmental Management) would be also relevant for Albanian agribusiness, but so far no enterprises have been so certified. Demand for EUREPGAP and GLOBALG.A.P. certifications is extremely limited and inconsistent. A few (less than ten) EUREPGAP certifications have been released, mostly with support from development aid projects. Certification for organic production is on the rise, and has been supported for many years by grants and assistance from development projects, mostly funded by the Swiss government.

The system of private services to enterprises is provided by three types of service providers. The first of these are private consultants and consulting enterprises, which advise enterprises on practices to be introduced, and investments to be made, to achieve compliance with standard quality and safety systems and legal requirements. The second type of providers are certification organizations that release quality management and food safety certifications (HACCP, ISO 9001, ISO 22000, etc.); organizations dealing with organic production are also included in this category. Third, food laboratories also provide services to private customers. The level of expertise and up-to-date information provided by private consulting businesses are very important for the food industry. The quality management and food safety systems aforementioned are new approaches. As a result, the role of these consulting businesses is getting more important. The following section discusses in more detail the profile of these three groups of service providers.

7.7.1 Private consultants and consulting service

In most cases, consulting services in Albania are provided by individual consultants in private practice. Consulting on food safety, quality management and quality assurance is usually not their main business, as the market is new and there is not enough (exclusive) work for such a consultancy. For the same reason, the consultancy market is also not really consolidated, so the quality of consulting is quite variable, and there has not yet been a real selection based on quality of services. In some cases, these enterprises are also agribusiness entrepreneurs, who are partners or have been partners in food processing industries. Indeed, some of the most reputed consultants have this background. Another group of consultants consists of university professors, as well as employees of international certification companies.

Almost all active consultants have received training from development aid projects or worked abroad. The most active programs for HACCP advisory services were financed by GTZ (Deutsche Gesellschaft für Technische Zusammenarbeit, now GIZ, Deutsche Gesellschaft für Internationale Zusammenarbeit), while the Swiss Cooperation (Swiss Agency for Development and Cooperation, SDC) focused its activity on organic certification, and USAID (US Agency for International Development) on EUREPGAP. However, all donors and some UN agencies (such as FAO, UNDP (UN Development Programme), and the World Bank) have financed some training on food safety and quality assurance schemes. The need to improve the quantity and quality of consultants assisting in the implementation of good practices on farms (EUREPGAP and GLOBALG.A.P.) and in processing plants (GHP, HACCP, GMP, ISO) has been repeatedly identified as a key issue (World Bank, 2007; Leonetti *et al.*, 2010).

7.7.2 Certification organizations

Certification organizations in Albania are branches of foreign specialized companies, but a few Albanian companies are active as well. The local branches of foreign companies maintain the market leadership, with the exception of the BIO (organic) certification, where the Albanian ALBINSPECT (certification body) is by far the market leader. ALCEBO (certification body), established in 2008 with support of the Dutch government, has been the first Albanian organization offering GLOBALG.A.P., HACCP and ISO certification.

7.7.3 Food laboratories providing services to entrepreneurs

Most of the laboratories working for food safety and quality controls are owned by producers' associations (such as the Albanian Dairy and Meat Association (ADAMA), a breeders' association), or certification bodies (such as ALCEBO), or are public laboratories that also provide services to private subjects for a fee (such as ISUV, or the food laboratory of the Agricultural University of Tirana). An increasing, but so far still limited, number of food processing enterprises have their own food laboratories. Livestock markets and fruit and vegetables wholesale markets should have a laboratory (and an NFA inspection post) to comply with EU regulations, but at this time they do not.

7.8 The Food Safety Situation from the Perspective of Consumers

It is also important to understand consumer perception of food safety. Albanian consumers associate quality and safety with traditional farm products, and have more trust in personal relations with a known supplier, when the product is directly purchased from producers, than in any quality assurance scheme (Leonetti and Kristo, 2005, 2010; Leonetti *et al.*, 2009; Imami *et al.*, 2010). This is particularly true if the scheme is guaranteed by the state (Imami *et al.*, 2010). Albanian consumers also prefer domestic production (Chan-Halbrendt *et al.*, 2010a; Leonetti and Kristo, 2010), when it is considered as such (Leonetti *et al.*, 2009), which is often not the case, as demonstrated by the fact that imported products always dominate in the high-end segment of the market (Leonetti *et al.*, 2009).

The generalized lack of trust in contractual agreements or quality schemes guaranteed by an authority or organization is reflected in all relationships along the value chain (World Bank, 2007; Imami *et al.*, 2010; Leonetti *et al.*, 2010). In the past, personal relations have been the only accepted way of confidence building. The exception is the few cases when an Albanian enterprise or a product of a specific origin has managed to build up a reputation for quality and reliability (Leonetti and Kristo, 2005, 2010; World Bank, 2007, 2010).

When there is no reputed brand or personal relationship with the vendor, high prices are often considered one of the few reliable proxies for quality (Chan-Halbrendt *et al.*, 2010b). In this respect, it is possible to use as a case study a recent consumer survey (Imami *et al.*, 2010), which showed that consumers would overwhelmingly trust a respected butcher's recommendation for lamb without a veterinary stamp more than veterinary-stamped lamb from an unknown butcher. In the same survey, 56% of respondents answered that the main source of guarantee came from knowing the butcher, and 24% in knowing the origin of the animal, while only 12% answered that the veterinary stamp (i.e. the basic quality assurance required by the law) was the most important guarantee. Finally, about two thirds of consumers answered that they would switch to veterinary certified meat (stamp) only, if it was 100 or 200 lek (ALL) cheaper (i.e. 20% cheaper, considering current prices) than meat recommended by the butcher.

7.9 Awareness and Use of Food Safety and Quality Assurance Schemes in the Food and Beverage Industry

The need for increasing the awareness and knowledge of food producers, and in particular, of food processors about safety and quality assurance schemes, and their importance and use in domestic and international markets, has increased. A large number of enterprises in the food industry suffer from a lack of information in the field of quality management, and they need theoretical and practical training in this field. The need and urgency to fix this situation, particularly in the light of the fast-changing Albanian legal framework for food safety, was first reported by UNDP (2005) and the World Bank (2007). On this issue, the World Bank (2007) document contained a clear-cut statement: 'There is a lack of awareness and understanding of the HACCP principles in the country, both among enforcers and in the private sector.' Three recent studies, which are profiled below, show that in the last years the situation has improved, but not as much as is necessary.

7.9.1 Quality assurance awareness and implementation survey in Tirana district

The first and most specific research on quality assurance focused on the level of implementation of quality standards and quality management systems in the food and beverage industry in the Tirana region of Albania, and on the problems that this industry is facing (Kapaj *et al.*, 2009). The aim of the research was to identify the most appropriate incentive policies to support the management of enterprises in successfully implementing these quality management systems (Civici, 2003; World Bank, 2007). The research was based on a stratified random sample of 35 food processing industries, based on a list provided by MAFCP. The strata (groups) were based on the number of employees of the enterprise, as reported in MAFCP records (i.e. as declared by the enterprises). It is important to emphasize that, as the list was supplied by MAFCP, only formally established enterprises were surveyed.

Two main questions were submitted to the enterprises. In the first, those surveyed were asked to self-assess their level of knowledge of EU legislation on the quality and safety of the products. In the second, they were asked whether they apply quality management standards. Other important questions relevant to willingness to receive training on food safety and quality assurance schemes, and sources of information on customer preferences, were also included in the questionnaire. As expected, there was a direct correlation between the size of the enterprises and their knowledge of EU legislation. All large enterprises have full knowledge, or know most of the EU legislation, while knowledge is lower among medium sized enterprises, and much lower among small enterprises. Altogether, 48% of enterprises interviewed responded that they know the EU legislation on quality and safety very well, and 26% that they mostly know this legislation (Fig. 7.1).

Answers to the second key question demonstrated the direct correlation between the application of quality management standards, and the market size that the enterprise serves (local, regional or national markets). Specifically, 25% of the enterprises that apply quality management standards trade in local markets, whereas 75% of the enterprises trade at national level. A different reading of these percentages provides a more evident measure of the gap between enterprises facing competition on the national market, and those limiting their actions to the local market. That is, 88.9% of the enterprises that do not apply these standards are operating in local markets, while only 11.1% are operating in the national market. Both correlations between size of the enterprises and the knowledge of EU legislation, and between market outreach and the application of quality management systems, are statistically significant ($P < 0.05$).

Enterprises were also asked whether they wanted to invest in training their employees in quality and safety. The results showed that around 85% of the enterprises included in the survey expressed their willingness to invest in training while 12% did not. Of those surveyed, 74% of the enterprises mentioned a lack of financial means as the main constraint to investments in technological renovation and training in the field of quality and safety. As a consequence, the extent of implementation of quality and safety management systems in the food industry in Albania is not at a satisfactory level, although the willingness of enterprises to implement such systems mostly exists.

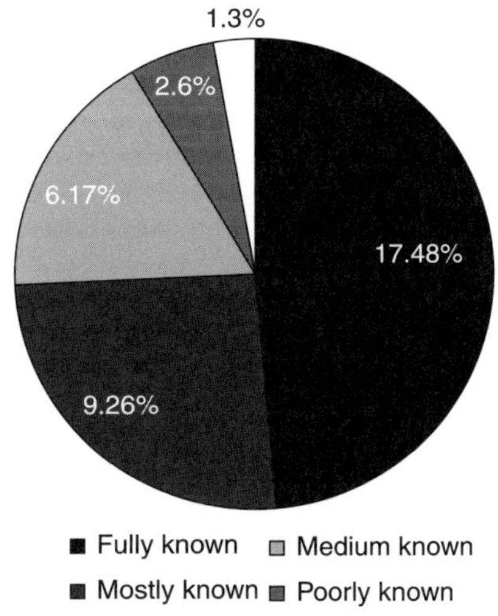

Fig. 7.1. Level of knowledge of European Union legislation on food and beverage quality and safety: respondents' self-assessment. Source: Kapaj *et al.* (2009).

These findings confirm the need for a 'stick and carrot' approach (i.e. law enforcement in parallel with incentives for complying with standards) for speeding up the process of adaptation to and compliance with EU safety and quality standards. The need for such an approach has been previously assessed (World Bank, 2007), and is also the policy approach followed by the Albanian government and the EU. These entities have sought to accomplish this task via the establishment of the NFA for enforcement and guidance to enterprises, and access to the IPARD (Instrument for Pre-Accession Assistance for Rural Development) financial facility, which provides financial support to enterprises willing to comply with EU safety and quality standards.

In order to explore the understanding of enterprises for consumer preferences, they were asked for ways of gathering this information. The purpose of this question was to ascertain the quality and the extent of information that enterprises use to evaluate consumers' preferences toward their products. Because one of the main pillars of new quality management approaches such as ISO, TQM (total quality management), etc., is the consumer, the ability of enterprises to gather the necessary information in terms of quantity and quality of consumer preferences becomes very important. Following this further, the quality of information depends on its source.

The results of this survey show the paramount importance of supply chains in Albania. That is, enterprises obtain quality and safety information though feedback received by the shops they own, or via independent retailers, who are in most cases directly supplied by the processors themselves, without the intermediation of wholesalers, or through cash and carry. The majority (56.5%) of enterprises use company surveys for gathering information on consumer preferences for quality and safety. To understand better how companies obtain this data, more detailed information was gathered through non-structured questions in interviews. These showed that the most important method used by enterprises to obtain information on consumer preferences for quality and safety is the 'company survey' (Table 7.1). Enterprises that practice decision-making by participation (employee involvement) account for 21.7% of the total, whereas 11% use statistical information from government institutions. The rest use nothing, or other sources. No respondents responded that they use 'scientific studies from universities' as a source of information (Table 7.1).

This survey highlights the lack of understanding of EU legislation by food and beverage enterprises in Tirana district. Only 48% of the enterprises interviewed have full knowledge about this legislation while 43% have a basic knowledge regarding EU legislation on quality and safety. Furthermore, 8% do not know anything about EU legislation. Compared with data from 2005, the number of enterprises that have full knowledge about EU legislation on quality and safety has increased by 17% (Deçolli, 2005). This is a result of continued efforts by food enterprises in Albania to improve the quality and safety of their products, as well as the success of various activities by state institutions in Albania. These activities provide the necessary and appropriate information to food enterprises on quality and

Table 7.1. Sources of information on consumer preferences. Source: Kapaj *et al.* (2009).

	Responses[a]		
Source of information	No.	%	% of cases
Company surveys	26	56.5	74.3
Decision-making by participation	10	21.7	28.6
Statistical information	5	10.9	14.3
Nothing	1	2.2	2.9
Other sources	4	8.7	11.4
Total	46	100.0	131.4

[a]Multiple answers were allowed.

safety. Increased competition in the domestic food market due to free trade agreements with neighboring countries, and because of the reduction of most customs duties, have confronted Albania's food industry with the need to improve product quality.

An important factor is that in Tirana district, consumers are more educated about quality and safety, and their demand for quality food is continuously increased. Even so, only 23% of the 35 enterprises surveyed had established quality management systems, while 77% did not apply these standards. Nevertheless, the direct relation that was demonstrated between market outreach and the adoption of quality assurance schemes does show that Albanian companies are responding to competition pressure and evolving consumer demand. At the same time, financial constraints represent a limit to the pace of adaptation. Large companies seem to be experiencing sufficient competitive pressure to improve their quality management schemes and to be able to find, with some difficulty, the resources to respond to the challenge. At the same time, smaller enterprises dealing with local demand have less awareness, less competitive pressure and also fewer resources. Efforts in enforcing the existing legislation and addressing incentives for updating should be focused on this group of enterprises, while informal or semi-informal enterprises should be pushed and supported to formalize or close.

7.9.2 The impact of implementation of HACCP, awareness raising and capacity building in the milk industry in Albania

A second survey, performed in 2007–2009, was focused on measuring the impact of raising awareness of HACCP and capacity building in Albania's milk industry. A total of 620 samples from containers, plastic packaging, the hands of personnel, and raw milk were collected over a 2-year period in a medium sized milk processing factory located in middle Albania. Samples were analyzed for total coliforms, *Escherichia coli*, *Salmonella*, somatic cells (screening test), and pH, before and after intervention. Contingent to the study, an awareness raising campaign and training of the factory's employees were carried out. Furthermore, before and after a training session, a questionnaire was used to evaluate employees' knowledge of the safe handling of milk and its products. The pH and *Salmonella* parameters were found to be within the allowed levels in all samples, both before and after the training. However, high bacteriological contamination was detected (*E. coli* in 25% of samples, and total coliforms in 80% of samples) before staff training and the implementation of HACCP (Molla *et al.*, 2012). The bacteriological contamination originated from the low level of hygiene and poor sanitation conditions in the factory, and the lack of implementation of an HACCP system. Interventions through the training of personnel and the gradual implementation of HACCP improved the situation. After this was done, contamination by *E. coli* was reduced to 20%.

7.9.3 SME training need assessment

The TNA Project (TNA, 2010) was carried out in Albania in 2010 within the framework of the EU-financed project 'Supporting SMEs to Become More Competitive in the EU Market.' The TNA exercise was carried out on a sample of 250 registered (i.e. working in the formal sector) SMEs of all sectors, including agri-food processing (but not primary production). The survey showed that 44% of respondents did not have enough information on quality management systems, and that most of the others did not base their answers on knowledge of the law (Table 7.2). This law imposes

Table 7.2. Importance of quality assurance. Source: TNA (2010).

Answer	%	Answer	%
We don't have enough information about it	41	HACCP (hazard analysis and critical control points)	9
ISO 9000	26	Other relevant quality assessment (QA)	16
ISO 14000	3	No response	4

the application of some of the HACCP system and, in the future, will require the full application of HACCP system.

In many cases, certification is perceived by the food processor as a marketing tool (Leonetti *et al.*, 2010), without a real understanding of the quality assurance scheme that would be most functional for the business (domestic market, export, behavior of competitors, etc.). One of the TNA key findings was that 'Albanian SMEs are in great need of systematising their Quality Management Systems (QMS), to structure and document their business operations, so as to assure an effective transfer of working methods, techniques, control functions, and other operations and technology functions from the owner/manager to other managers and staff. As a second step in the QMS process, Albanian enterprises need to start thinking about achieving internationally recognised QMS standards, such as ISO and HACCP. For sectors like food-processing obtaining HACCP is no longer a novelty but an essential requirement so as to sell in to the EU market.'

7.10 Incentives and Support Schemes for the Introduction of Quality Assurance Schemes

At present, some state subsidy schemes are available to promote investments in quality assurance and quality products. As part of the gradual adaptation of agricultural policies to the EU CAP (Common Agricultural Policy), the support to agri-food enterprises has been gradually shifted from input subsidies to a system based on competitive grants given to companies willing to make new investments. The present system of state subsidies is mostly addressed toward increasing the primary production base. A small part of subsidy allocations is devoted to promote some quality products (contribution to the production of extra-virgin olive oil), quality assurance schemes (organic productions), and measures to improve food safety (a subsidy for livestock identification).

Some support schemes are providing facilities to contribute to the costs of consulting for the introduction of quality assurance schemes. The main providers of these are financed by ALBINVEST (a state organization to promote investments in Albania) and by EBRD (European Bank for Reconstruction and Development)-BAS (Business Advisory Services), funded with external aid funds. Several development aid projects are also providing different kinds of support for quality assurance programs, such as training for food safety and quality (the EU SME project), and competitive grants for quality certification (World Bank), but all these support activities will be phased out in the next 1–2 years.

The system of state subsidies will be further adapted, with the objective of gradually introducing the support system already in force in EU member countries. Based on structural funds (funds dedicated to each sector), the new system will be co-financed by the EU and its member countries. During the pre-accession period, Albania will gradually access the resources made available by IPARD, co-funded (as the structural funds will be) from the EU and the Albanian government.

IPARD will provide a 50% contribution (in the form of grants) to investments in primary production and food processing, initially in three sectors (milk and dairy; meat and meat products; and fresh and processed fruit and vegetables). In the future, these grants will be extended to all the sectors considered strategic under the Cross-Cutting Rural Development Strategy. These grants are explicitly aimed at improving the competitiveness of commercial farms and food processors through compliance with EU quality and safety standards. Scaling up production is included in the eligible investment objectives (provided that there is proven market space), but the primary objective is to introduce EU quality and safety standards. To be eligible, applicants must comply with Albanian norms on quality and safety at the moment of the application, and the supported investments must comply with EU norms when investment is completed. It is expected that IPARD will be made available to Albania starting in 2013 or 2014. A provisional scheme, called 'IPARD-like' will be introduced in 2012.

7.11 Perspectives on the Improvement of Food Safety and Evolution of the System of Incentives and Deterrents

The overall analysis of the system of quality assurance in Albanian agribusiness leads to the conclusion that progress is being made, but slowly. Scarce awareness of and lack of trust in the respect of rules are the two key elements preventing a much faster diffusion of quality assurance schemes. 'Bringing agri-food establishments up to EU standards will be a major challenge for Albania, particularly for meat and dairy' (World Bank, 2007). This consideration has been confirmed and specified in later studies (Imami *et al.*, 2010; Leonetti *et al.*, 2011). While some progress has been achieved, especially in the harmonization of the general legal framework, and in the compliance of larger companies with some basic quality assurance schemes (GHP, GMP) and with HACCP, the gap between the progress of legislation and the actual implementation of norms is widening, or at least not decreasing.

However, in the next few years, the legislation, now almost harmonized, will become more stable, and secondary legislation will be gradually completed, and possibly more strongly enforced. In parallel with this, more incentives will be made available through IPARD for the enterprises willing to adopt quality assurance schemes. Five main issues should to be primarily addressed to improve food quality and safety in Albania, and to introduce quality assurance schemes in a larger number of enterprises.

The first issue to be addressed is a major one, that of reducing the size of the informal economy. The large number of informal or semi-informal businesses providing no quality assurance is a major obstacle for the enterprises willing to compete according to the rules. According to the World Bank Enterprise survey of 2007 'Practices of the informal sector' this is the second ranked most important business environment constraint in Albania (after regular supplies of electricity), with 17.6% of enterprises mentioning it as the most important constraint, well above corruption (11%) (World Bank Enterprise, 2012). This constraint is also considered more important than in other countries of the region (where it counts for less than 12% according to the World Bank Enterprise (2012).

A second issue to be dealt with is the need to increase awareness of the importance of food safety and quality assurance schemes throughout the whole food chain. According to TNA (2010), 106 entrepreneurs out of 241 do not know what kind of quality assurance scheme is important for their activities; and according to Kapaj *et al.* (2011), the majority (52%) of sampled agri-food processing SME have a knowledge of EU food safety legislation ranging from incomplete to nil. Also, the majority of consumers are convinced that the food they consume is very safe and trust their customary retailer much more than any public quality assurance scheme (Imami *et al.*, 2010).

The next issue that must be tackled is the introduction of an effective and accountable system of 'sticks' and 'carrots' (World Bank, 2007). Laws must be much more vigorously enforced, in parallel with the introduction of support schemes for enterprises willing to invest in introducing quality assurance schemes. The institutional and legal framework for both 'sticks' (the NFA) and 'carrots' (IPARD) is available. Consistent action has to be taken in a transparent and fair way. As part of this effort, secondary legislation for MTR-Safety and MTR-Quality should be introduced as soon as possible, and the human and technical resources of the agencies in charge of performing controls (NFA) and awarding support (IPARD) should be strengthened and improved in quality.

The fourth issue is that services to enterprises must be improved as well. The shortage of local expertise available to assist the food industry in implementing EU food safety requirements is an issue that was identified some years ago (World Bank, 2007). However, efforts to overcome this constraint have not yet given sufficient results (Leonetti *et al.*, 2009). The regulatory functions of certifying companies are also in need of improvement, as Albanian certification is often not recognized abroad.

Lastly, continuing and increasing the training and awareness effort for food safety and quality assurance schemes will play a major role in any industrial and agricultural development policy in Albania. Enterprises must more seriously address the issue of investing in training their employees, not only expressing a willingness to invest, but also taking concrete actions. All the managers of agri-food enterprises must raise the level of basic knowledge related to the quality assurance schemes that increasingly play a key role for competitiveness in national and international markets. They must gradually place quality as the main development focus for their enterprises. Training should be addressed not only to managers, but to other employees as well, especially those in the sector of production and quality control, who must acquire more thorough and extensive knowledge in the field of quality and safety.

With the proceeding of the pre-accession process, the introduction of quality assurance schemes in Albanian agribusiness becomes no more a matter of increased competitiveness, but a matter of survival. Increasing pressure will come from more intense competition from imported products, as well as from the EC. As a prerequisite for accession to EU, the EC will expect Albania to proceed on the parallel paths of enforcing the harmonized legislation on food safety and quality, and increasing access to the financial facilities provided to encourage investments in this field.

References

Chan-Halbrendt, C., Imami, D., Zhang, Q., Zhllima, E. and Leonetti, L. (2010a) Latent class analysis of consumer preferences for wine in Tirana. Paper presented to: *20th IAMA [IFAMA] Annual World Forum and Symposium: Navigating the Global Food System in a New Era, 19–22 June, Boston, USA*. International Food and Agribusiness Management Association. Available at: https://www.ifama.org/events/conferences/2010/cmsdocs/134_paper.pdf (accessed 5 September 2012).

Chan-Halbrendt, C., Zhllima, E., Sisior, G., Imami, D. and Leonetti, L. (2010b) Consumer preference for olive oil in Tirana, Albania. Paper presented to: *20th IAMA [IFAMA] Annual World Forum and Symposium: Navigating the Global Food System in a New Era, 19–22 June Boston, USA*. International Food and Agribusiness Management Association. Available at: https://www.ifama.org/events/conferences/2010/cmsdocs/135_paper3.pdf (accessed 5 September 2012).

Civici, A. (2003) *The Situation and Competitiveness Level of the Agro-food Sector in Albania*. MAFCP Report, Ministry of Agriculture, Food and Consumer Protection, Tirana, Albania.

Deçolli, P. (2005) *Albanian Enterprises of Food and Beverages Processing; Situation, Analyses and Recommendations in the Products Quality and Safety Field*. SEED/ANE (South East Europe Enterprise Development/Agiensiae Nxitjes së Eksporteve), Tirana, Albania.

Imami, D., Leonetti, L. and Zhllima, E. (2010) Analyzing consumer preferences for lamb meat in Albania. In: *Improving the Performance of Livestock Sector in Albania*. DSA Project Report, Development Solution Associates Tirana, Albania.

International Business Publications (2011) *Albania Business Law Handbook*. International Business Publications, 5th edn, 2008, updated to 2011.Washington, DC, USA.

International Standards Organization (2011) *The ISO Survey of Certifications 2010*. International Standards Organization, Geneva, Switzerland.

Kapaj, I., Doluschitz, R. and Mane, A. (2009) Assessing quality and safety of food and beverage products. An analysis of agribusiness enterprises in Tirana District/Albania. Paper presented to: *19th IAMA [IFAMA] Annual World Forum and Symposium: Global Challenges Local Solutions, Budapest, Hungary, 20–23 June 2009*. International Food and Agribusiness Management Association. Available at: https://www.ifama.org/events/conferences/2009/cmsdocs/1053_paper.pdf (accessed 14 September 2012).

Kapaj, A., Kapaj, I., Deci, E., Jaupi, A. and Pagria, I. (2011) Identifying some key indicators benchmarks (competitive benchmark[s]) helping meat processors in Albania improving their management decisions. *Valahian Journal of Economic Studies* 2(16)(issue 1), 49–54. Available at: http://www.vjes.eu/images/2011/VJES%20-%20Volume%202/Vol.2(16)_Issue1_2011_Ana%20Kapaj.pdf (accessed 4 September 2012).

Leonetti, L. and Kristo, I. (2005) *The Food Chain Structure of Small Ruminants Meat and Dairy Products in Albania*. Report from UNDP (United Nations Development Programme) Project SMS, UNDP, Tirana, Albania.

Leonetti, L. and Kristo, I. (2010) *The Value Chain of Small Ruminants Meat Supplying the Tirana–Durres Urban Area*. Report from Project "Improving the Performance of Livestock Sector in Albania", UNDP (United Nations Development Programme), Tirana, Albania.

Leonetti, L., Imami, D., Zhllima, E. and Stefanllari, A. (2009) The olive and olive oil value chain in Albania. In: *Formulation of a Technical Assistance Project in Support of Albania's Olive Oil Sector*. DSA Project Report, Development Solution Associates, Tirana, Albania.

Leonetti, L., Imami, D., Kongoli, R., Nani, A., Rapushi, P. and Xhafa, F. (2010) *Fruit and Vegetables Sector Study*. IPARD (Instrument for Pre-Accession Assistance for Rural Development) Report from Project "Capacity Building for Implementing the Rural Development Strategy", Tirana, Albania.

Leonetti, L., Zhllima, E. and Imami, D. (2011) *Private Veterinary Practitioners Survey*. PAZA-Project: Improving Consumer Protection Against Zoonotic Diseases, DSA Project Report, Development Solution Associates, Tirana, Albania.

Luning, L.A., Marcelis, W.J. and Jorgen, W.M.F. (2002) *Food Quality Management: a Techno Managerial Approach*. Wageningen Academic Publishers, Wageningen, Netherlands.

Mautner Markhof, M., Musabelliu, B. and Zusi, A. (2010) *Meat Sector Study*. Meat IPARD (Instrument for Pre-Accession Assistance for Rural Development), Report from Project "Capacity Building for Implementing the Rural Development Strategy", Tirana, Albania.

Molla, L., Bizhga, V., Tafaj, L. and Petrela, E. (2012) Evaluation of the impact of implementation of HACCP, of awareness raising and of capacity building in the milk industry in Albania (forthcoming). Presented to: *Proceedings of Fifth International Scientific Conference BALWOIS 2012, Conference on Water, Climate and Environment, Ohrid, Macedonia, May 28th to June 2nd, 2012.*

Schiefer, G. (2002) *Quality Management in Agriculture and Food: Management Principles System Requirements and Development Directions*. University of Bonn-ILB Press, Bonn, Germany.

Schröder, E., Uruçi, M. and Meço, M. (2010) *Milk Sector Study*. Milk IPARD (Instrument for Pre-Accession Assistance for Rural Development), Report from Project "Capacity Building for Implementing the Rural Development Strategy", Tirana, Albania.

TNA (2010) *Supporting SMEs to Become More Competitive in the EU Market*. Training Need Analysis for SMEs, Report from Project "EU SME Project, Albania", Tirana, Albania.

UNDP (United Nations Development Programme) (2005) *UNDP Marketing and Distribution Chain Study. Detailed Reports and Annexes*. Report from Project "Support for Trade Regulation and Trade Promotion", Tirana, Albania.

World Bank (2007) *Albania: Strategic Policies for a More Competitive Agriculture Sector*. Report No. AAA18-AL, World Bank Sustainable Development Sector Unit, Europe and Central Asia, Tirana, Albania. Available at: http://siteresources.worldbank.org/INTALBANIA/Resources/Strategic_Policies_for_more_competitive_Agriculture_sector.pdf (accessed 3 September 2012).

World Bank Enterprise (2012) Enterprise Surveys. Available at: www.enterprisesurveys.org (accessed 5 September).

8 Consumer Preferences for Selected Agricultural Products in Albania

Drini Imami,[1] Edvin Zhllima[1] and Catherine Chan-Halbrendt[2]
[1]*Agricultural University of Tirana, Albania;*
[2]*University of Hawai'i at Mānoa, USA*

8.1 Introduction

Since the beginning of the transition from a centrally planned to a market oriented economy in the early 1990s, economic growth, fast urbanization, regional trade liberalization, and gradual integration into the European Union (EU), have contributed to a rapid evolution of consumption patterns and lifestyle in Albania. The changing structure of income that is emerging from an increasing per capita gross domestic product (GDP), and redistribution among classes of the population and among regions (with the urban population benefiting more from the growth than rural population), has given rise to a larger urban middle income class of consumers. The food demand from these emerging urban middle income consumers, combined with the gradual consolidation of the retail sector, and the recent establishment and expansion of the first supermarket chains, has strong implications for the agri-food industry.

In the past, Albania had a strictly planned command economy in which consumer choices were not determined by direct decision-making at the farm level, but by production oriented central targets, a similar situation to that in various countries of Eastern Europe (Albania belongs geographically to southern Europe) (Lerman, 2000). The cooperatives and state farms were isolated from consumer preferences, and production was based on regional orientations and top-down decision-making according the state plan. Citizens were insulated by imported food flows, especially in the last decade of the regime when the government was oriented toward self-sufficient production. Agricultural privatization and the liberalization of the trade regime in the 1990s opened the way for a direct linkage between the consumers and farmers' crop production choices, as well as between consumers and the traders importing agri-food products.

After the privatization of agriculture, the production system started to practice the principles of private and free market practices, but the subsistence oriented and fragmented farm structure, as well as the fragile and technologically obsolete processing technology, left little mechanism for orienting production toward Albanian consumer preferences. Imports boomed in the first transition years owing to the limitations of domestic production, and the improvement of the Albanian dietary requirements arising from the increase in the average income level, and the westernization of the products entering into the market.

The need to match increasing demand, the requirement for quality by consumers within the farming and processing industry, and the objective to improve the trade balance (which was heavily negative for the majority of agri-food products in the first decade of transition), obliged the government to design strategies and to intervene with farm support after 2000. A fuel subsidy scheme was implemented during the period 2000–2006, and since 2007 the Ministry of Agriculture, Food and Consumer Protection (MAFCP) has introduced production subsidy schemes for the support of various agriculture sub-sectors, including orchards, vineyards, olive groves with the aim of increasing farm size, efficiency, and competitiveness. However, such schemes, which have been subject to continuous revision, have not been based on studies of market behavior and, particularly, on consumer analysis. The MAFCP has established a Crosscutting Strategy for Consumer Protection and Market Surveillance for the period 2007–2013. This is focused on empowering (in terms of information and transparency), as well as protecting (in terms of economic interests), the safety, quality, selection and diversity of food, and access to food, for Albanian consumers. But at both the policymaking level and the enterprise level, no serious steps are being taken toward understanding Albanian consumer preferences.

The average per capita income and expenditure in Albania has been increasing, and the poverty level has declined from 25% to 12% due to the positive influence of economic growth. This increasing consumer wealth has been not accompanied by the adoption of a more consumer oriented strategy by either Albanian businesses or policymakers. As this chapter reveals, there have been few attempts by researchers and other stakeholders to study consumer preferences in the country. Understanding consumer preferences is important, not only for agri-food entrepreneurs, but also for government policymakers so that they can tune their policies and allocate their resources more efficiently and in line with market demand. The objective of this chapter is to describe consumer preferences for some of the main agricultural products in Albania, relying on various data sources, previous studies conducted by authors of the chapter and an extensive literature review.

In the next Section (8.2), an analysis is made of the main trends in consumption, and other key issues, such as supply. Data are presented on consumer patterns and dynamics over the past few decades for the main food categories, along with comparisons with the rest of the world. Section 8.3 summarizes consumer preferences and purchasing behavior for the main agri-food products by summarizing recent consumer preference studies on processed agriculture products (olive oil and wine), meat (small ruminant meat), fruit (apples), and milk. The final Section (8.4) presents recommendations to agro-industry stakeholders and policymakers.

8.2 Main Consumption Trends and Key Issues

In this section we analyze in detail consumption trends for the main food items in Albania, drawing comparisons with regional and world trends. Comparisons with south Europe are meaningful, as Albania is situated in southern Europe and has similar climatic conditions to its southern European neighbors. As reported by FAOSTAT (Food and Agriculture Organization Statistics database), south Europe includes: Albania, Andorra, Bosnia and Herzegovina, Croatia, Gibraltar, Greece, Italy, Macedonia, Malta, Montenegro, Portugal, San Marino, Serbia, Slovenia, and Spain. Although geographically, Albania does not belong to Eastern Europe, comparisons are made with this group of countries because Albania, being a transition economy, has a similar background. As reported by FAOSTAT, Eastern Europe includes: Belarus, Bulgaria, the Czech Republic, Hungary, Poland, the Republic of Moldova, Romania, the Russian Federation, Slovakia, and Ukraine.

The statistics that are analyzed are from FAOSTAT, and due to lack of data on consumption, supply per capita is used as a proxy for demand based figures: supply per capita = (production + imports − exports + changes in stocks (decrease or increase))/population. In the case of cereals and, to some extent, meat, there should not be big gaps between supply (per capita) and consumption (per capita) as there are minimal losses and changes in stock, so the terms 'consumption (per capita)' and 'supply (per capita)' can be justifiably interchanged. In contrast, for fruits and vegetables, which are more perishable, there is a (significant) gap between supply and consumption; in this case, supply trends can only be considered as an indicator of per capita consumption trends.

During the communist era (beginning after World War II and collapsing in the early 1990s), food production and imports were not determined by consumer preferences, but rather by the state. The distribution of food to consumers was strictly planned, and implemented by the state-owned distribution and retail system. In the latter part of the communist era, the Albanian diet was dominated by cereals, whose consumption exceeded 200 kg/year per capita. The government policy for self-sufficiency in food production in the 1970s caused a big shift from quality to quantity, and from meat and dairy products to grains. During the 1980s, Albania experienced a production crisis. Grain production started to fall on a per capita basis, and animal production failed to provide an adequate supply of proteins (Erebara, 2011). In the second part of the 1980s, most food items were scarce on the grocery store shelves, and some items (such as meat) were rationed to families. Each family was entitled to purchase no more than 0.5 kg of butter, ten eggs, 1 kg meat or 1 kg sausages a week (Erebara, 2011).

After the transition to a market economy, which began in the early 1990s, Albanian consumers shifted their consumption from mainly cereals-based foods to more meat, fruit, and vegetables (Berisha and Mara, 2005). This major change was a result of substantially higher income per capita, on one hand, and market liberalization on the other hand (World Bank, 2003, 2007). As part of the liberalization of trade, and combined with expanded retail and storage capacities, and increases in income and the standard of living, the consumption of meat, fruit and vegetables has increased significantly in recent years. By the late 2000s per capita consumption of fruit, vegetables and meat was two to three times higher than during the pre-transition period (Fig. 8.1). In the period 1961–2007, the per capita supply of cereals dropped, while the supply of vegetables, fruit and meat more than doubled.

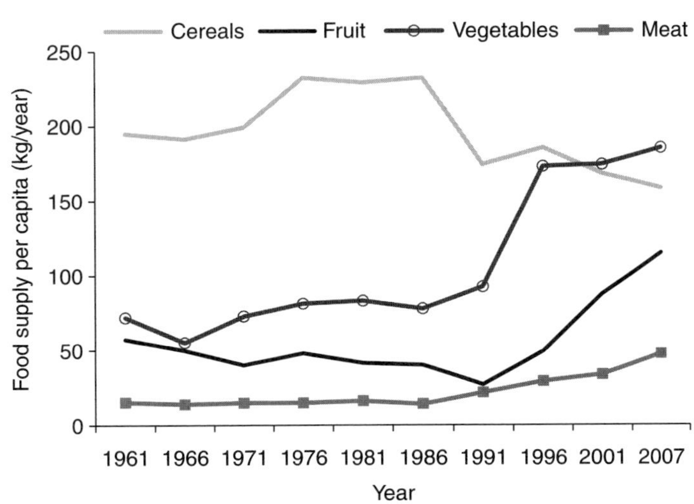

Fig. 8.1. Per capita trends of main consumer basket food supply in Albania (kg/year), 1961–2007. Source: FAOSTAT.

Meat per capita supply has almost tripled since the pre-transition period in Albania, and exceeds the growth rate of the rest of the world during the same period. Specifically, Albanian per capita meat supply was 19% higher than the world average in 2007, though it is still 38% lower than the European average, 48% lower than in south Europe, and 23% lower than in Eastern Europe (Fig. 8.2). This gap indicates that, as Albanian income is expected to further increase in the future, meat consumption per capita is likely to increase to resemble that of its richer neighbors.

Consumption of cereals, which was high in Albania and Eastern Europe during the communism era, declined substantially after the transition (Fig. 8.3). This decline has been caused by a shift to higher consumption of meat, fruit and vegetables. However, despite the sharp decline, Albanians still consume, on average, more cereals per capita than the world average, as well as exceeding the European average. Thus, the expectation is that as incomes further increase, more consumers will tend to consume fewer cereals in favor of the other more expensive food products.

Per capita fruit supply has also increased substantially over the past few decades in Albania. The increase in consumption of fruit is fuelled by improvement in post-harvest and storage facilities and increased domestic production, in addition to increases in income. By 2007, Albanian per capita fruit supply was considerably higher than in the world overall, as well as higher than the European average, and close to the south European average (Fig. 8.4).

Similar to the fruit situation, per capita vegetable supply has increased substantially over the past few decades. In 2007, per capita vegetable supply was significantly higher than that of the world overall, the European average, and the south European average (Fig. 8.5). The increase in the consumption of vegetables in Albania, as with fruit, has been caused by increased income, combined with increased domestic production (the domestic product was more accessible for

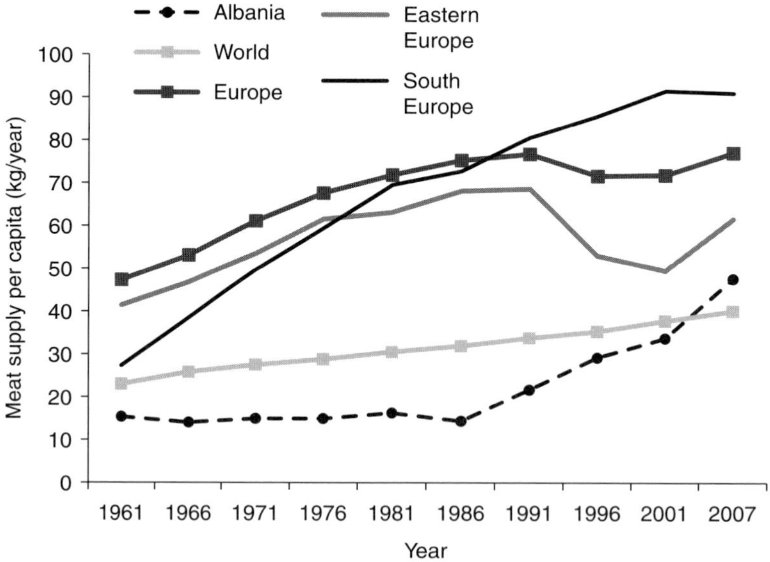

Fig. 8.2. Per capita trend of average meat supply in Albania, Europe and the world (kg/year), 1961–2007. Source: FAOSTAT.

local consumers at lower prices). While in the case of fruit an additional factor positively affecting consumption was enhanced storage capacities (enabling storage of fruits for months), for vegetables this is less important, but the development of the greenhouses has enabled the availability of domestic vegetable production/supply throughout the year.

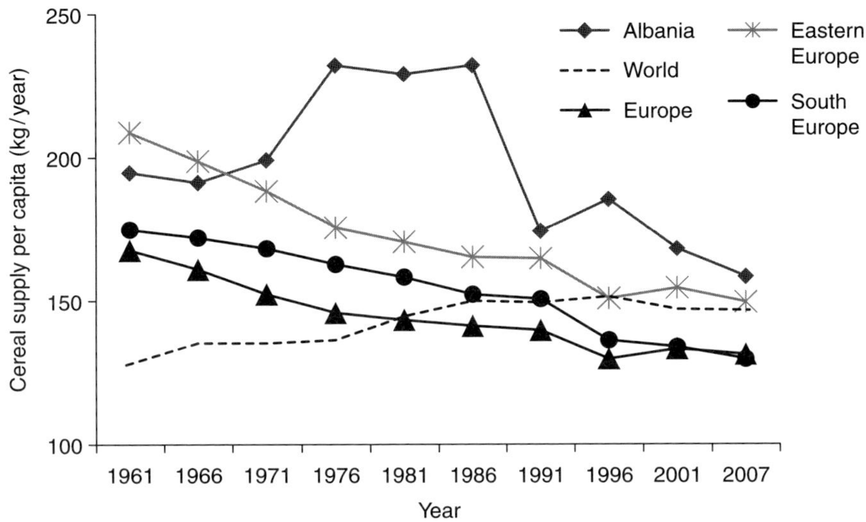

Fig. 8.3. Per capita trend of average cereals supply in Albania, Europe and the world (kg/year), 1961–2007. Source: FAOSTAT.

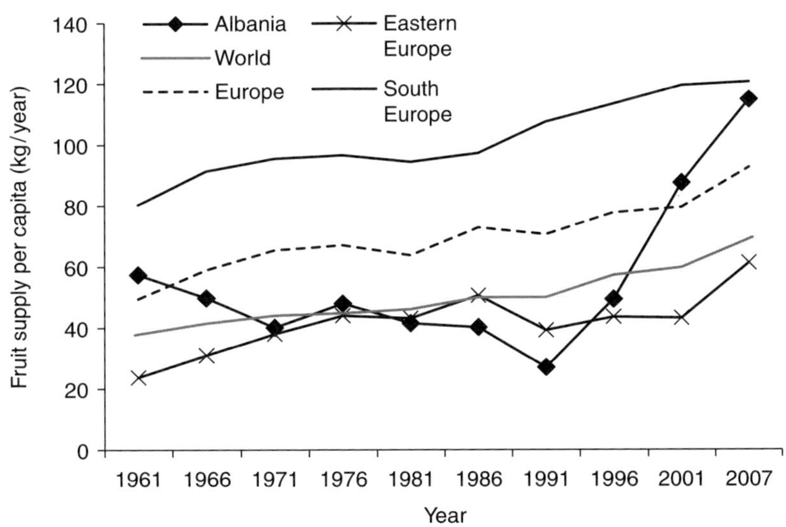

Fig. 8.4. Per capita trend of average fruit supply in Albania, Europe and the world (kg/year), 1961–2007. Source: FAOSTAT.

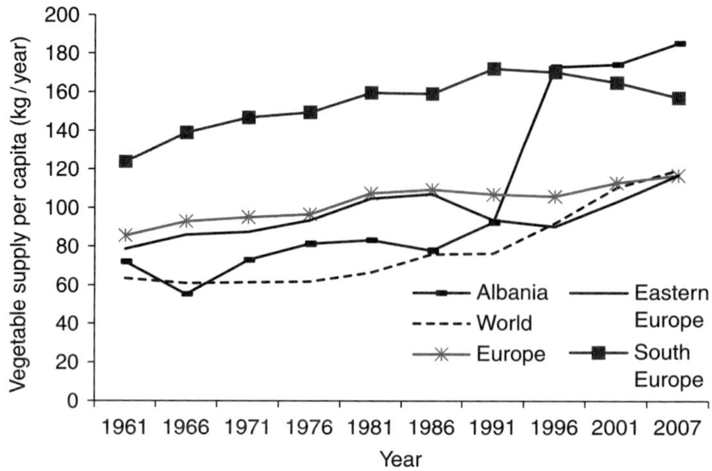

Fig. 8.5. Per capita trend of average vegetable supply in Albania, Europe and the world (kg/year), 1961–2007. Source: FAOSTAT.

8.3 Consumer Preferences and Purchasing Behavior for the Main Agri-food Products

The section will summarize recent consumer preference studies for olive oil, wine, small ruminant meat, apples, and milk. Furthermore, conclusions will be drawn as to the product attributes and the factors driving shifts and patterns in consumers' preferences.

8.3.1 Consumer preferences for olive oil

Olive oil consumption in Albania has strongly declined over the past few decades. One reason is the decrease in domestic production, resulting from heavy damage to olive plantations during the transition (INSTAT, 1998). Albanian average per capita consumption remains higher than the world average, but is significantly lower than the European average, and merely a fraction of that in south Europe—which includes the leading olive oil producers and consumer countries such as Greece, Spain and Italy (Fig 8.6).

In addition to changes in domestic production and income patterns, another factor which may explain the current and future market potential for olive oil consumption is consumer preferences. The following discussion is based on an analysis of consumer studies in olive oil conducted in June 2009, which asked 200 olive oil consumers in Tirana (Albania's capital and biggest urban market) their product attribute preferences. The objective of the survey was to assess consumer preferences for the main product attributes, group consumers by preferences and socio-demographic factors, and assess their willingness to pay for certain improvements in product attributes. A conjoint choice experiment was used to design the survey, and a latent class approach was used to analyze the data. The study identified the five most important attributes for survey respondents to consider when given different product profiles to choose from. The attributes used, and their components, were determined by focus group analysis and an extensive literature review. They were: type (extra virgin, virgin and normal), origin (import versus domestic), place of purchase (shop versus supermarket), taste (pungent versus bitter) and price. The socio-demographic characteristics (gender, education, income, age, and knowledge of different types of olive oil) of the consumers surveyed were also recorded. The results demonstrated that there were six distinctive classes (1–6) of consumer preferences.

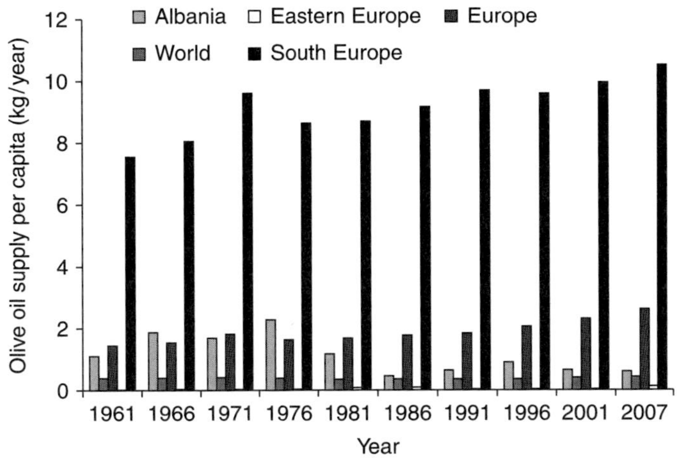

Fig. 8.6. Per capita trend of olive oil supply in Albania, Europe and the world (kg/year), 1961–2007. Source: FAOSTAT.

The parameters of the model constructed and importance of the various product attributes are reported in Table 8.1.

Of the six classes of consumer preferences that were found, Class 1 respondents preferred extra virgin domestic olive oil and were willing to pay higher prices, but were from lower income households. This was contrary to expectations, as, in general, it is expected that price and product demand have an inverse relationship for food items. The respondents from this class considered price as the most important of the five product attributes, although it may be that Albanians see olive oil as healthy and, as such, are willing to pay more for it. In addition, the average consumer in Tirana was found not to be familiar with the quality of olive oils. Currently, Albania has weak food quality and safety assurance enforcement, and consumers have to rely on their uninformed judgments and experience to gauge quality. As a result, Albanian consumers behave just as any other consumers in the presence of asymmetric or vague information on the product, and use price as an indicator of quality (Cicia *et al.*, 2002; Sandalidou *et al.*, 2002; Anania and Nistico, 2004).

Class 2 respondents considered origin of the olive oil to be the most important attribute. Classes 1 and 2 combined accounted for almost 75% of the respondents who preferred domestic olive oil. However, even with high preferences for domestic olive oil and higher prices, the reality is that, with over 1000 t of imports, some of the consumers in this group might still be purchasing imported olive oils. As mentioned above, without prior knowledge of or familiarity with the quality standards of these products, price is often used as an indicator of quality. As olive oil is a staple part of the Albanian diet, people, inasmuch as possible, tend to purchase good quality product to consume. As a result, Albanians might choose to purchase imported olive oil as they interpret the higher prices as indicating a good quality olive oil.

In Class 3, which comprised older females who generally prefer imported olive oil and higher prices, the imported products were again likely to be seen as being of higher quality. Imports are perceived as being produced in a more secure, safe, and controlled environment (Civici *et al.*, 2004). As the females and retired males who usually take care of everyday food purchases tend to be more risk averse, they prioritize food safety when selecting their everyday food purchases. In this study, the respondents' preference toward imported products appears to be a result of their perception of the relationship between imports and safety. The group exhibits no significant preferences based on type of olive oil—they might not have knowledge of the different olive oil types. This lack of product knowledge can lead, once more, to a reliance

Table 8.1. Estimated parameters (positive, negative, or relative importance) of a consumer preference olive oil model, based on data from a survey in Tirana, Albania. Source: Chan-Halbrendt *et al.* (2010).

Attributes	Class 1	Class 2	Class 3	Class 4	Class 5	Class 6
Type						
Extra virgin	+[a]					+[a]
Virgin						+[a]
Normal	–[a]					–[a]
Origin						
Imported	–[a]	–[a]	+[a]			
Domestic	+[a]	+[a]	–[a]			
Place of purchase						
Shop		–[b]				
Supermarket		+[b]				
Taste						
Pungent		+[b]	–	–[a]	+[a]	
Bitter		–[b]		+[a]	–[a]	
Price	+[a]	+[a]	+[a]	+[a]	+[a]	
Socio-demographic factors						
Gender						
Male			–[a]			+[b]
Female			+[a]			–[b]
Education						
Elementary						
High School			–[b]			
University						
Income	–[a]		+[a]			
Age			+[b]			
Knowledge of differences between types of olive oil						
Yes			–[a]	+[a]		
No			+[a]	–[a]		
Relative importance of attributes by class (%)						
Class size	47.61	26.44	8.34	7.12	5.46	5.03
Type	16.19	3.53	10.27	3.24	12.47	**87.26**
Origin	14.32	**49.88**	**47.39**	5.90	6.45	1.81
Place of purchase	0.34	5.29	0.42	7.02	4.64	3.43
Taste	1.26	6.60	0.59	**50.09**	**52.20**	1.21
Price	**67.88**	34.7	41.33	33.75	24.24	6.29

[a]Significant at the 0.05 level; [b]Significant at the 0.10 level.

on price as a marketing cue for product quality (Zeithaml, 1988, quoted in Veale and Quester, 2009; Gabor and Granger, 1966; Kupiec and Revell, 2001; Verdú-Jover *et al.*, 2004; Volckner and Hofmann, 2007), which results in a preference for higher priced olive oils. Other studies show that origin of product is also taken as a proxy for quality (Cicia *et al.*, 2002; Scarpa and Del Giudice, 2004). High income families, such as those found in Class 3, are more likely to be able to afford and be willing to pay the higher price demanded for olive oils imported from well-known producers, such as Spain, Italy and Greece.

Classes 4 and 5 showed statistically significant preferences only for taste and price, with taste ranked as the most important attribute. Together, these two classes make up about 13% of the total number of respondents. Although Class 5 did not show any significant socio-demographics, Class 4 showed significant knowledge in the differences between the types of olive oil, which allowed them to form a specific olive oil preference. This is useful for producers to label their products for marketing, as this can be exploited to enhance product recognition. Some classes had a specific taste preference for bitter (Class 4) and pungent (Class 5) oils, and both were willing to pay higher prices for their preference. Though this population segment is small, it creates the potential for small niche markets that cater to specific consumers. As these consumers are willing to pay higher prices for their preferred product, these niche markets can produce high quality olive oil directed towards such consumers.

Class 6 had significant preference for their preferred types of olive oil, and considered this attribute to be the most important. This class was also the only class that showed a negative, though not statistically significant, price preference. Type of olive oil, i.e. extra virgin and virgin, seemed to entirely determine their purchasing decisions.

With regard to each class rating on which product attribute is most important, it can be seem from Table 8.1 that Class 1, with the most respondents, rated price as most important. Classes 2 and 3 rated origin as the most important. Classes 4 and 5 rated taste as most important, and Class 6 rated type as most important. In general then, marketers who cater to the mass should focus their selling strategies on price and origin. However, there are niche markets for special taste and type of olive oil.

8.3.2 Consumer preferences for wine

Consumption of wine has increased significantly since transition owing to increased income and market liberalization, although average per capita consumption is still far below the world average. It also lags behind the European average, particularly the south European average. Per capita wine supply is less than 20% of that in south Europe (Fig. 8.7).

To assess consumer wine preferences, a conjoint choice experiment technique and latent class analysis were used. Two hundred random respondents were interviewed and completed a survey in Tirana during July and August 2009. The product attributes used, and their components, were determined by focus group analysis and a literature review. The attributes presented to the respondents to gauge their wine preferences were: type (white, red, rosé), origin (import, Italian, domestic), taste (dry, sweet, neutral), aroma (light versus full) and price. Socio-demographic

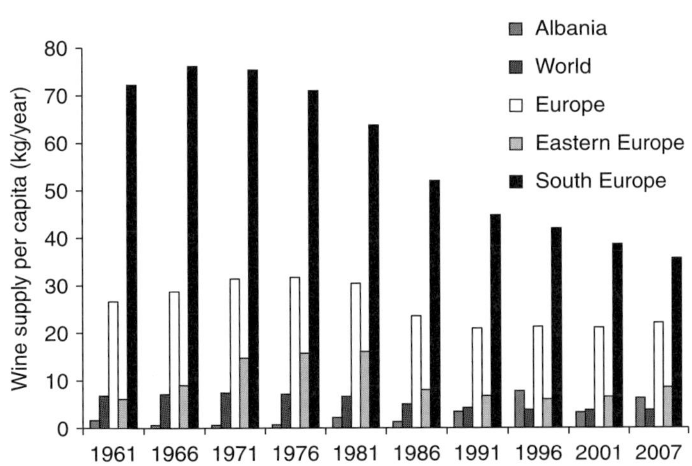

Fig. 8.7. Per capita trend of wine supply in Albania, Europe and the world (kg/year), 1961–2007. Source: FAOSTAT

variables analyzed included household size, gender, age, education, and income. The model parameter estimates are shown in Table 8.2, which does not include all the socio-demographic variables. Using latent class analysis, a six-class model was found to be the best in explaining the consumer preferences.

Respondents in Class 1 (38.29% of the sample population) prefer red wine to white wine, prefer Italian wine, and do not like imports. In addition, they have a strong preference for dry wines over sweet wines, and prefer moderate prices to higher prices. These consumers can be characterized as consumers who prefer moderately priced sweet, red wine imported from Italy.

Respondents in Class 2 (30.50% of the sample population) prefer rosé/light red wine over white wine, and Italian imports at low prices. Class 2 respondents can be characterized as consumers with a preference for low-priced Italian rosé/light red wines.

Respondents in Class 3 (14.02% of the sample population) significantly preferred red and rosé/light wines to white wine, but did not

Table 8.2. Estimated parameters (positive, negative, or relative importance) of consumer preference model for wine, based on data from a survey in Tirana, Albania. Source: Zhllima *et al.* (2013).

	Class 1	Class 2	Class 3	Class 4	Class 5	Class 6
Attributes						
Type						
White	$-^{b}$	$-^{a}$	$-^{b}$		$+^{b}$	
Red	$+^{a}$		$+^{b}$		$-^{b}$	
Rosé/light red		$+^{a}$	$+^{b}$		$-^{b}$	
Origin						
Imported				$-^{b}$		
Italian	$+^{b}$	$+^{b}$		$-^{b}$		
Domestic	$-^{b}$			$+^{b}$	$-^{a}$	$+^{a}$
Price (lek/0.75 l)						
300		$+^{b}$	$+^{a}$	$+^{b}$	$+^{a}$	$-^{b}$
600	$+^{b}$	$+^{b}$				$-^{b}$
900		$-^{b}$			$-^{b}$	
1200	$-^{b}$	$-^{b}$	$-^{a}$	$-^{b}$		$+^{b}$
Taste						
Dry	$-^{b}$					
Sweet	$+^{a}$					
Neutral						
Aroma						
Light						$+^{a}$
Full						$-^{a}$
Respondents' demographics						
Household size					$-^{b}$	
Gender						
Female					$-^{a}$	$+^{a}$
Male					$+^{a}$	$-^{a}$
Age	$-^{b}$			$+^{b}$		
Relative importance of wine attributes by class (%)						
Class size	38.29	30.50	14.02	5.96	5.83	5.39
Type	15.15	7.14	**68.00**	2.89	**54.28**	7.24
Origin	**47.47**	10.29	9.33	**53.58**	13.16	12.69
Price	22.22	**77.71**	16.00	39.12	21.71	**64.21**
Taste	13.13	4.29	6.33	1.22	6.25	7.33
Aroma	2.02	0.57	0.33	3.20	4.61	8.52

[a]Significant at 0.05 level; [b]Significant at 0.01 level.

like higher prices. Thus, they are a group of consumers who prefer low priced rosé/light red and red wine.

Respondents in Class 4 (5.96% of the population) significantly preferred domestic wine and, as a result, did not like Italian and other imported wines. Like Class 3, they highly preferred lower prices to higher ones. Thus they are a class of wine consumers who prefer local and inexpensive wine.

Respondents in Class 5 (5.83% of the population) significantly preferred white wines, and did not like red or rosé/light red wines. They did not like domestic wines, and preferred the lowest prices. They can be considered a group of white wine consumers who like cheap and all imported wines.

Class 6 comprised only 5.39% of the sample population. This is the only class in the study population which significantly preferred a light to a full-bodied aroma, in addition to the other attributes. The group also preferred higher prices, and strongly preferred domestic wine. They can be characterized as consumers who prefer domestic wines with light aromas and are willing to pay a high price for these local, high quality wines.

In terms of relative importance, the most important attributes are different by class. Classes 1 and 4 rated origin are most important; Classes 3 and 5 rated type as most important, and Classes 2 and 6 rated price as most important. Taste and aroma were not rated among the most important attributes by any classes. The implication for marketers is that they should focus on selling strategies that include either type, origin or price.

The results of this study can provide valuable market information to wine producers, researchers, extension specialists, and farmers. With information on six different classes of consumers, the wine industry should develop different strategies to produce different types of wine for specific groups of consumers in Tirana. Generally speaking, a large portion of the population likes less expensive Italian red or rosé/light red wine. In socioeconomic terms, younger consumers tend to like sweet, red, Italian, medium priced wines, while older consumers prefer domestic, lower priced wines. Single males and males in smaller households tend to like less expensive white wines, while females prefer domestic high quality wines, and are more sensitive to aroma.

The results of this research confirm (or are in line with) other research findings on beverage and food products. There are clear marketing implications in the apparent contradiction of two of the relatively small, but not insignificant, consumer groups (Classes 4 and 6), in that they both choose domestic wine but one group prefers lower prices, and the other higher prices. The preferences of the latter group contradict economic price theory, which states that the higher the price, the lower the demand—although previous research on foodstuffs in transitional developing economies has shown that a preference for higher priced goods indicates a preference for quality products (Alhabeeb, 2007). Therefore, in this study, it is reasonable to find that one group prefers more expensive wine, as in the olive oil study (Section 8.3.1), because a willingness to pay higher prices is indicative of another product attribute that associates price with the level of food safety. This tendency is stronger among women (Mehta, 2003), converging with this study's findings that female buyers (Class 6), prefer the more (or most) expensive domestic wines, compared with Class 4 consumers who like cheaper domestic wine and tend to be older.

Given the results of this study, the domestic agro-industry could easily target two types of consumer groups using different product attribute–price strategies. First, Class 4 respondents, older consumers who prefer cheap domestic wine without any clear preference for type and aroma, and Class 3 consumers who prefer cheap red wine without preference for origin and aroma, can be targeted jointly by the domestic wine industry through more aggressive marketing of cheaper domestic red wine. Second, domestic producers can target Class 6 consumers, who have preference for expensive, light, high quality domestic wine. However, all three of these classes together comprise a small proportion of the consumers (25%). For the remaining consumers, who currently prefer less expensive Italian imports, of mostly red wines, the marketing strategy should be to improve the quality and the quality perception of domestic wines, and to price them competitively. In the long run, this will be important to the government strategy for substituting imported wines with those that are domestically produced. This strategy will require standardization of local wineries, the provision of support to increase size and efficiency—at both the farm and the processing level, and perhaps a change in wine cultivars and an improvement in farming practices.

8.3.3 Consumer preferences for meat: the case of small ruminant meat

Production and consumption of lamb and goat kid meat is an important component of the traditional diet in Albania, as pasturing was, and still is, an important activity in mountainous areas. During the 1960s and 1970s, the farming of small ruminants (as well as cattle farming and other branches of agriculture) was subject to collectivization under the planned economy, and this resulted in a decrease in production. Therefore, consumption of meat in general, and consumption of small ruminant meat specifically, decreased remarkably after this period, because of collectivization, and while Albania was fairly isolated from the world economy.

In the period 1978–1982, the government implemented its last major agricultural reform, that of livestock collectivization. But the government was not able to collect all the animals, and halted the process after tens of thousands of cows were culled by farmers, who refused to deliver them without payment (Erebara, 2011). As a result, between 1979 and 1984, the mass slaughter caused first an increase in meat production (from 62,000 t in 1980 to 74,000 t in 1981), and then a reduction in the period 1983–1984; the number of cattle fell from 606,000 in 1980, to 573,000 in 1984 (MAFCP, 1981, 1986, 1992 see Statistical Yearbooks 1980, 1985, 1991).

The privatization of the agriculture sector as part of the transition into a market economy, combined with an increase in income, resulted in an increase of production and consumption of lamb and goat kid meat during the 1990s and 2000s. The estimated yearly per capita consumption in urban areas is 3.5–3.8 kg of lamb and mutton, and 0.6 kg of kid goat meat, while consumption in rural areas is much higher, with an estimated consumption of more than 7 kg of lamb and mutton, and 3 kg of kid goat meat (Leonetti and Kristo, 2010). In all, Albanians consume, on average, much more mutton and goat meat than either the world or the European average. In fact, Albanians consume almost twice the quantity of these meats than do south European consumers (Fig. 8.8).

In order to investigate consumer preference for small ruminant (lamb) meat in Albania, a conjoint choice experiment study with latent class analysis was conducted with 378 randomly

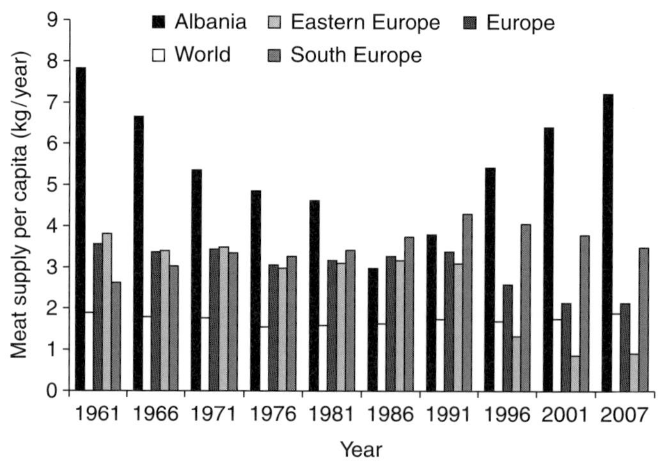

Fig. 8.8. Per capita trend of lamb/mutton and goat meat supply in Albania, Europe and the world (kg/year), 1961–2007. Source: FAOSTAT.

selected residents of Tirana, Vlora, Saranda, and Durres during the period of July–August 2010. These four towns are situated in central and southwest Albania, and, together, make up more than 65% of Albania's population; they have a high concentration of purchasing power. Geographical coverage, interview locations and numbers interviewed in this study were based on focus group findings, as for the olive oil and wine preference studies. The attributes chosen for analysis were price, origin (domestic highland, domestic lowland, imported), weight (7, 10, 13 or 16 kg), and type of certification (certificated from an unknown butcher or not certificated from a known butcher). The preferences of the four classes of consumer groups defined by the analysis are shown in Table 8.3.

Most estimated parameters of the model were significant at the 0.01 level, with the exception of some of the prices and weights (13 and 16 kg). All price variables were negative, as expected. All of the four classes preferred meat from domestic highland origins, followed by domestic plain/lowland origins, and, lastly, imports. There was variability in the preferences for carcass weight and certification by class.

Respondents in Class 1 strongly preferred lamb meat from domestic highlands, but not domestic plain/lowland or imported lamb meat. For butchered weight, respondents were more likely to buy 10–13 kg meat and less likely to buy heavier meat. In addition, respondents in Class 1 were more likely to buy meat from a known butcher, even if it did not have veterinary stamps. The price coefficient was negative (as expected) and significant.

Respondents in Class 2 significantly preferred domestic highland and domestic plain/lowland meat, but were less likely to buy imported lamb meat. Like Class 1 respondents, they were more likely to buy medium sized meat and less likely to buy higher weight meat. The certification preference also paralleled Class 1 in that respondents were likely to buy from known butchers without certification. Also for this class, the price coefficient was negative (as expected) and significant.

Table 8.3. Estimated parameters (positive, negative, or relative importance) of consumer lamb meat preference model, based on data from a survey in four towns in central/southwest Albania. Source: Imami *et al.* (2011).

	Class 1	Class 2	Class 3	Class 4
Attribute				
Price	–[b]	–[b]		
Origin				
Domestic highland	+[b]	+[b]	+[b]	+[b]
Domestic lowland	–[b]	+[b]	+[b]	+[b]
Imported	–[b]	–[b]	–[b]	–[b]
Weight (kg)				
7			+[b]	–[a]
10	+[b]	+[b]	+[b]	+[b]
13	+[b]	+[b]	–[b]	
16	–[b]	–[b]	–[b]	
Certification				
Veterinarian certificate from unknown butcher	–[b]	–[b]	–[b]	+[b]
No veterinarian certificate but from trusted butcher	+[b]	+[b]	+[b]	–[b]
Relative importance of attributes by class (%)				
Class size	37	30	19	15
Price	19	33	3	2
Origin	63	29	48	52
Weight	11	16	43	7
Certification	6	22	7	39

[a]Significant at 0.05 level, [b]Significant at 0.01 level.

Respondents in Class 3 were more likely to buy domestic meat, but strongly preferred highland meat to that from the plain/lowland; their lowest origin preference was for imported meat. Class 3 was likely to buy the lowest weight meat, and was more likely to buy meat from known butchers without government quality and safety certification. Price was not significant in this class.

Respondents in Class 4 were most likely to buy domestic highland meat, followed by domestic plain/lowland meat, and they were least likely to buy imported meat. Respondents were more likely to buy medium weight meat and less likely to buy the lower weight meat. Attitudes toward certification differed for this class, in that Class 4 respondents were more likely to buy meat with approved government certification. Price was not significant in Class 4 either.

The results shown and discussed above indicate the significant attributes and the relative preference directions of the attributes for each of the four classes defined. Relative importance analysis can show, in addition, which attribute among all the significant attributes is the most important for each of the classes. This analysis indicated that origin was the most important attribute for most of the classes, but particularly so for Classes 1, 3 and 4, in which the importance was close to or over 50%. For Class 2, price and origin were equally important in purchasing decisions. For Class 3, origin and weight were equally important, while the group placed very little importance on price and certification. For Class 4, origin was very important, followed by certification; however, price and weight are not important to this class.

Another aspect analyzed in this study was willingness to pay. This is represented by implicit prices, which show the maximum amount of money that consumers are willing to pay in exchange for a different attribute level for the good concerned. As previously discussed, origin is very important for all classes. For Classes 1 and 2, domestic highland meat is the preferred meat. Respondents in Class 1 and 2 are willing to pay an additional 882 and 13 lek (ALL)/kg, respectively, for domestic highland meat as opposed to domestic plain/lowland meat, and an additional 101 and 276 ALL/kg for domestic plain/lowland meat as opposed to imported meat (Table 8.4). These two classes would switch to veterinary certified (stamped) meat only if it were about 100 or 200 ALL/kg less expensive than meat without formal certification, but recommended by the butcher.

To conclude, within the given prices and weight constraints provided by the survey, origin is by far the most important factor for consumers when choosing to buy lamb meat (similar to the results of studies by Grunert, 1997, and Bernabe'o and Tendero, 2005). The attribute preferences are followed by weight and certification. However, the importance of weight, as well as that of price and certification, seem to vary significantly across the four consumer classes identified.

All identified consumer groups prefer small (10 kg) carcass weight, while only Class 3 prefers a lower (7 kg) carcass weight. The three largest classes (Classes 1–3) show a statistically significant dislike for the 16 kg carcass weight compared with lighter weights. These findings are partially in line with a similar study on consumer preferences for lamb meat in the Mediterranean region; Bernabe'o and Tendero (2005) found that Spanish consumers generally preferred suckling lamb (butchered between 25 and 30 days, and with a carcass weight of less than 7 kg) compared with 'ternasco' lamb (butchered between 60 and 90 days, with a carcass weight of 10–14 kg). These estimates may be translated into a range of cost–benefit options for breeders, taking into consideration feeding, breeding schedule, and location.

As long as proper labeling is in place to avoid cheating by butchers or retailers, it appears that competition from imported lamb meat is not a major issue for the Albanian lamb meat industry. Domestic highland lamb meat is

Table 8.4. Willingness of consumers in four towns in central/southwest Albania to pay for changes of origin of lamb meat, as lek (ALL)/kg. Source: Imami *et al.* (2011).

Origin	Class 1	Class 2
Change from imported to domestic plain/lowland meat	101	276
Change from domestic plain/lowland to domestic highland meat	883	13

strongly preferred to lowland meat, which can be used to advantage if labeling or other marketing tools are available to inform and convince consumers of the meat origin. Therefore, enforcement of origin identification should be a priority for the government and other stakeholders, while producers should consider introducing and promoting their own brands (i.e. producer associations' brands).

There is obvious distrust in the public food safety enforcement system, as most consumer groups do not choose veterinarian stamped meat as a primary source of quality and safety information. This has strong implications for policymakers. Strengthening the implementation capacities, and improving the image, of the public veterinarian and food safety regulation system should be considered a priority in the context of ongoing institutional and legal reforms. Conversely, there is a relatively strong trust among most consumer groups in local butchers. It should be noted that trusted local butchers may be perceived to have the capacity not only to guarantee safety, but also to guarantee other aspects, such as quality and origin. This exceeds the scope of the current veterinarian certification provision, in which safety is the primary concern. Therefore, it may be concluded that all marketing, promotion and communication strategies of different stakeholders should largely rely on butchers to reach consumers.

8.3.4 Consumer preferences for fruit: the case of apple

Immediately after the collective farm system collapsed, around 60% of Albanian orchards and vineyards were abandoned or uprooted, resulting in lower production and consumption of fruit. Apples and other fruits started to be replanted in the 1990s as agricultural land was privatized by agricultural households (World Bank, 2007). The consumption of apples has grown significantly over the past two decades as domestic production has increased, and trade liberalization has enabled higher imports, and incomes have increased. Albanian per capita supply is twice the world average, and is comparable to the European average (Fig. 8.9).

In order to understand consumer apple preferences, a study using a conjoint choice experiment and latent class analysis was conducted in Tirana during summer 2010. Some 250 respondents were interviewed in person. The attributes analyzed included apple variety (Starking, Golden, Granny Smith, red yellow

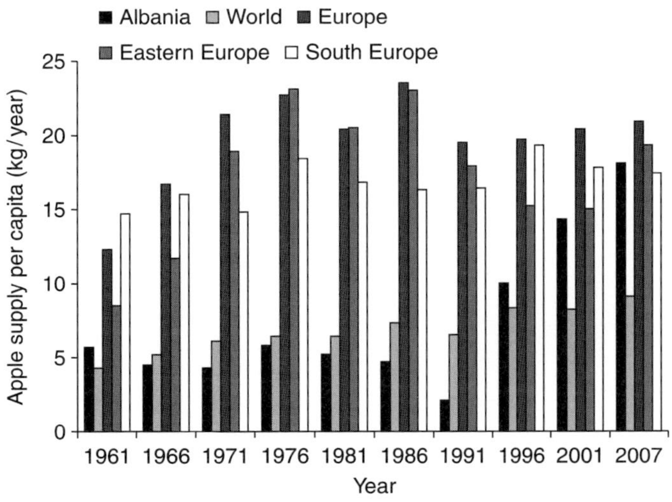

Fig. 8.9. Per capita trend of apple supply in Albania, Europe and the world (kg/year), 1961–2007. Source: FAOSTAT.

striped), origin (domestic, imported), size (big, small), and price. The preferences of the four classes of consumer groups defined by the analysis are shown in Table 8.5.

Class 1 represents almost half of the respondents (44.7%), while Class 2 represents 30.5%. These are the two largest classes, and together make up more than three quarters of the total number of consumers. Classes 3 and 4 (which are also the smallest) represent 14.3% and 10.6%, respectively, of consumers. The results of the study results reveal that origin, variety, and size are quite important attributes. Variety/color is so important that Class 1 can be called the 'green class', Class 3 the 'yellow' class, and Class 4 the 'red' class. Class 1, the largest, is willing to pay more than 17 ALL/ kg for green versus yellow apples. Apple color had previously been found to be important by Jerko and Kovačić (2007).

There is a clear preference for domestic versus imported apples in two out of the four classes (Classes 1 and 2). For the Class 1 consumers, there is not only a clear preference for domestic versus imported apple, but a willingness to pay a premium of ALL 13.6/kg for domestic apples compared with imported apples. The preference for domestic apples is supported by other studies conducted in Albania (USAID-AAC, 2009), and in the Balkans region (Jerko and Kovačić, 2007), which have also found the same preference by Croatian consumers for domestic apples. The distinction between domestic and imported apples does not seem to be very objective, however. More than one third of consumers (35.5%) believe that they are able to distinguish domestic from imported apples on their own, and 41.5% make the decision on origin by asking the trader.

The study also found that fruit size is important in three out of four classes (Classes 1–3). It needs to be emphasized that the preference for large sized apples (Classes 1 and 2 consumers) and for small sized apples (Class 3 consumers) implies consumers' willingness to pay for that attribute. Manalo (1990), using a conjoint choice analysis and Carew and Smith (2004), using a hedonic pricing model, also found a willingness to pay for large apples. Apple size is associated with either apple quality or production method. From the survey results, larger size, when preferred, implies better quality, phrased as 'better appearance', or 'better taste'; and smaller size, when preferred, implies 'more naturally' grown produce, or more convenience in consumption.

Table 8.5. Parameter estimates of consumer apple preference model, based on data from a survey in Tirana, Albania. Source: Skreli *et al.* (2012).

	Class 1	Class 2	Class 3	Class 4
Variety				
Red (Starking)				+[a]
Yellow (Golden)	–[a]		+[a]	
Green (Granny Smith)	+[a]		–[a]	–[a]
Red yellow striped	–[a]		–[a]	–[a]
Origin				
Domestic	+[a]	+[a]		
Imported	–[a]	–[a]		
Size				
Big (8 cm)	+[a]	+[a]	–[a]	
Small (5 cm)	–[a]	–[a]	+[a]	
Price		–[a]	–[a]	
Relative importance of attributes by class (%)				
Class size	44.7	30.5	14.3	10.6
Variety	44.5	3.6	52.9	83.9
Origin	35.2	10.2	4.1	5.3
Size	12	6.1	22.6	1.8
Price	8.3	80.1	20.4	8.9

[a]Significant at 0.01 level.

The implications of the study results may be considered from both marketing and policy viewpoints. In terms of marketing, it is important that actors throughout the apple value chain (farmers, collectors/traders), think not in terms of producing and selling simply 'apples', but producing and selling 'green apples', 'yellow apples', and 'red apples'. This is important in terms of targeting different consumer segments. Additionally, it is important in terms of government policy. The government may consider supporting domestic production of apple varieties that are more preferred by consumers—the study shows that green apples are preferred to other varieties, and that consumers are willing to pay additional money for them. When it comes to origin, a quite important warning can be imposed though: there is difference between a moral/subjective preference for domestic apples, and a lack of objective differentiation between domestic and imported apples. Therefore, designing a certification program for 'domestically produced apples' is likely to better channel customer support to domestically produced apples.

8.3.5 Consumer preferences for milk

Dairy activities have a long tradition in Albania owing to the favorable climate for dairy production. Cattle production is dominant in the plains, while in the hills and mountains, sheep and goat production are more suitable. Traditional dairy products are yogurt, butter, curd, and different kinds of cheeses from cow's, sheep's and goat's milk. Albanian milk supply per capita (without specifying destination or use) is significantly higher compared with that in Europe and in the world: it is almost five times higher than the world average, and 2.5 times higher than the European average (Fig. 8.10).

A conjoint choice experiment study with latent class analysis was conducted to determine differences in consumers' milk preferences in Albania's three main urban districts: Tirana, Fieri, and Berati (Kapaj *et al.*, 2011). Some 250 consumers (143 from Tirana, 61 from Fieri, and 46 from Berati), who regularly buy milk from large and small retailers, were surveyed from December 2010 to January 2011. Four attributes were presented to respondents to choose their preferred milk product: fat content (0, 1.5, or 3.2%), type and origin (UHT—ultra high temperature—import, UHT domestic, fresh pasteurized), place of purchase (local shop versus supermarket), and price. Results from the latent class analysis show that a four class model is the best fit. Table 8.6 gives the estimated parameters and significance for the four classes.

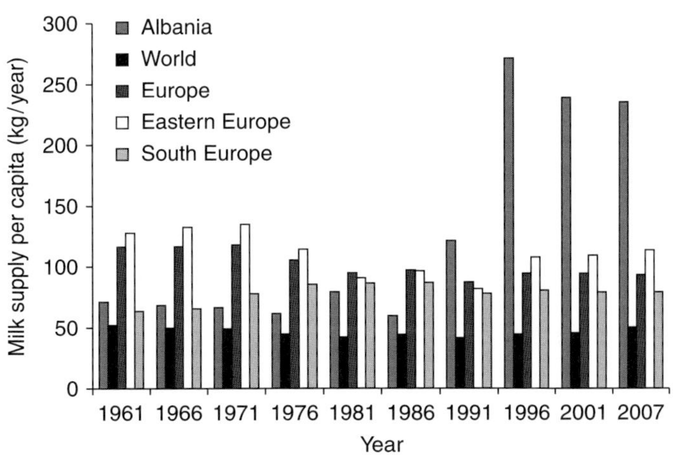

Fig. 8.10. Per capita trend of milk supply in Albania, Europe and the world (kg/year), 1961–2007. Source: FAOSTAT.

Table 8.6. Estimated signs of parameters for milk preference model, based on data from a survey of the three main urban districts in Albania (Tirana, Fieri, and Berati). Source: Kapaj *et al.* (2011).

	Class 1	Class 2	Class 3	Class 4
Attributes				
Fat content				
0%	+[a]	+[a]	–[a]	–[a]
1.5%		+[a]	+[a]	
3.2%	–[a]	–[a]	–[a]	+[a]
Type and origin				
UHT imported	–[a]		+[a]	–[a]
UHT domestic	+[a]			–[a]
Fresh pasteurized milk	–[a]			+[a]
Price			+[a]	–[a]
Place of purchase				
Supermarket	–[a]	–[a]		+[a]
Local shop	+[a]	+[a]		–[a]
Relative importance of attributes by class (%)				
Class size	9.2	28.1	42.6	20.1
Fat level	25.14	87.89	70.76	22.27
Type and origin	63.76	3.67	11.64	56.98
Place of purchase	7.97	3.92	5.41	12.83
Price	3.12	4.51	12.19	7.91

[a]Significant at 0.05 level or lower.

Class 1 has 9% of consumers who prefer 0% fat content milk, and dislike high fat level milk. They also prefer domestic UHT milk rather than imported and fresh milk, and they prefer to buy the milk in trusted local shops; the type of the milk is the most important attribute.

Class 2, which has 28.1% of the respondents, prefers no or lower fat content milk (0% and 1.5%), and prefer to purchase their milk in trusted local shops. For this class, the price and the type of milk are not significant; the fat level is the most important attribute.

Class 3, with 42.6% of the respondents, is the largest group, and has also chosen the fat level as the most important attribute. This class includes consumers that prefer low fat content (1.5%), and UHT imported milk. For this class, the place of purchase is not significant, though they prefer to pay a higher price for their milk. As found with the other commodities, this could be attributed to Albanian consumers' perception of higher price relating to quality.

Class 4 (20.1% respondents), prefers milk with a high fat level, freshly pasteurized and purchased in the supermarket. This class, again, chose fat content as the most important attribute.

The consumers of milk in the three main urban districts of Albania are being oriented to low fat milk of the UHT type. The majority of them have a strong preference toward imported milk, which shows the large market space that the domestic industry has to compete for in the market. Domestic producers still have to reward the preference of this segment of consumers for a higher price of milk by providing better quality, reducing the amount of out-of-date fresh pasteurized milk, and maintaining an aggressive distribution and monitoring strategy so that they cover trusted shops, and increase their dominance on supermarket shelves.

8.4 Conclusions

Albania is a country that has experienced a very centralized and autocratic socialist regime, in which food production (and imports) was determined by top-down policy directives and oriented to fulfill basic nutritional requirements, rather than toward consumer demand and preferences. The distribution of domestic production and the selection of food imports was based on a strictly planned state-owned distribution and retail system, which, in the last decade of its life, suffered population growth, diminishing returns from the agriculture sector and an autocratic international trade regime. Many food staple products were scarce, especially on the urban grocery store shelves, whereas others (such as meat and milk) were rationed to families.

In recent years, as the market has opened up, and trade has liberalized, combined with private farming and more modern retail and storage capacities, there has been a marked increase in supply of most agri-food products. This, in turn, has given rise to a significant increase in the consumption of meat, fruit and vegetables, owing to increases in income, and demand for a variety of foods instead of the more staple foods, such as cereals, that were available during the communist era. By the late 2000s per capita consumption of fruit, vegetables and meat was two to three times higher than during the pre-transition period. In parallel, the consumption of cereals per capita has declined in recent years.

As Albania's economy grows, its market develops, and the urban population increases, consumption trends and quantity are changing, so it is crucial for producers and suppliers to pay attention to consumption trends and food product preferences. From the consumer preference studies described in this chapter, there is definitely a preference for domestically produced product. However, there are concerns about the perceived quality and safety of various Albanian agri-food products. There is obvious distrust in the government food safety enforcement system, while privately enforced quality and safety systems are almost non-existent. Often, consumers will use higher price as a gauge of quality, or prefer imports because of their more transparent food safety labeling and standards. This has strong implications for policymakers. Strengthening the implementation capacities, and improving the image, of the government food safety regulation system should be considered a priority in the context of ongoing institutional and legal reforms, such that the preference for domestic produce and products increases to match consumer preferences.

The information provided by the various studies included in this chapter is useful for marketing experts, policymakers, producer associations, and retailers. There are obviously differences among consumer groups as to their preferences for a particular product. Farmers can use this information to decide which would be the most profitable and preferred attributes of the products that they can produce, and farmer associations or production groups may want to consider introducing their own brands to distinguish their products from imports. Retailers and intermediaries (wholesalers) can brand/label their products to promote preferred origin and other preferred attributes.

Finally, the study findings are of interest to policymakers and donor programs active in the agri-food sector; these play a crucial role in shaping sector production in the context of the market trends. However, the market opportunity for import substitution can only occur if there is a good distribution infrastructure to the major markets, and easily accessible market intelligence information available to farmers and processors.

References

Alhabeeb, M.J. (2007) Consumers' cues for product quality: an application of the signaling theory. In: *Allied Academies International Conference, Jacksonville 2007. Proceedings of the Academy of Marketing Studies* 12(1) 1–5. Available at: www.sbaer.uca.edu/research/allied/2007/Marketing%20 Studies/1.pdf (accessed 6 September 2012).

Anania, G. and Nisticò, R. (2004) Public regulation as a substitute for trust in quality food markets: what if the trust substitute cannot be fully trusted? *Journal of Institutional and Theoretical Economics (JITE)* 160, 681–701.

Berisha, A. and Mara, V. (2005) Role of diets in Balkan countries, as part of Mediterranean diets, in people's health. *Albanian Journal of Agricultural Sciences (Revista Shqiptare e Shkencave Bujqësore)* 1, 34–39.

Bernabe'u, R. and Tendero, A. (2005) Preference structure for lamb meat consumers. A Spanish case study. *Meat Science* 71, 464–470.

Carew, R. and Smith, E.G. (2004) The value of apple characteristics to wholesalers in western Canada: a hedonic approach. *Canadian Journal of Plant Science* 84, 829–835.

Chan-Halbrendt, C., Zhllima, E., Sisior, G., Imami, D. and Leonetti, L. (2010) Consumer preference for olive oil in Tirana, Albania. Paper presented to: *20th IAMA [IFAMA] Annual World Forum and Symposium: Navigating the Global Food System in a New Era, 19–22 June 2010, Boston, USA.* International Food and Agribusiness Management Association. Available at: https://www.ifama.org/events/conferences/2010/cmsdocs/135_paper3.pdf (accessed 5 September 2012).

Cicia, G., Giudice, T.D. and Scarpa, R. (2002) Consumers' perception of quality in organic food: a random utility model under preference heterogeneity and choice correlation from rank orderings. *British Food Journal* 104(3–5), 200–213.

Civici, A., Guri, F. and Zhllima, E. (2004) Wine production and consumer behavior in Albania. *Albanian Bulletin of Agricultural Sciences* 3(3), 34–41.

Erebara, G. (2011) The demographic component in Albanian food and economic crisis of 1980s. Master Programe in Economic History, School of Economics and Management, Lund University. Available at http://www.lunduniversity.lu.se/o.o.i.s?id=24965&postid=1976170 (accessed February 2012.

Gabor, A. and Granger, C.W.J. (1966) Price as an indicator of quality: report on an inquiry. *Economica* 33, 43–70. August), 107–117.

Grunert, K.G. (1997) What is in a steak? A cross-cultural study on the quality perception of beef. *Food Quality and Preference* 8(3), 157–174.

Imami, D., Chan-Halbrendt, C., Zhang, Q. and Zhllima, E. (2011) Conjoint analysis of consumer preferences for lamb meat in central and southwest urban Albania. Paper presented to: *21st IAMA [IFAMA] Annual World Forum and Symposium: The Road to 2050: Sustainability as a Business Opportunity, Frankfurt, Germany, 20–23 June 2011*. International Food and Agribusiness Management Association. Available at: https://www.ifama.org/events/conferences/2011/cmsdocs/2011SymposiumDocs/367_Symposium%20Paper.pdf (accessed 6 September 2012).

INSTAT (1998) *Census of Agriculture Economic Households in Albania*. Instituti i Statistikës (INSTAT), Tirana, Albania.

Jerko, M. and Kovačić, D. (2008) The importance of apple attributes: a comparison of self-explicated and conjoint analysis results. Paper prepared for presentation at: *12th EAAE Congress, 'People, Food and Environments: Global Trends and European Strategies', Ghent (Belgium), 26–29 August 2008*. European Association of Agricultural Economists. Available at: http://ageconsearch.umn.edu/bitstream/44401/2/098.pdf (accessed 6 September 2012).

Kapaj, A., Chan Halbrendt, C. and Deci, E. (2011) Milk market segmentation: study on the consumers preferences in urban Albania, Paper presented to: *21st IAMA [IFAMA] Annual World Forum and Symposium: The Road to 2050: Sustainability as a Business Opportunity, Frankfurt, Germany, 20–23 June 2011*. International Food and Agribusiness Management Association. Available at: https://www.ifama.org/events/conferences/2011/cmsdocs/2011SymposiumDocs/338_Symposium%20Paper.pdf (accessed 6 September 2012).

Kupiec, B. and Revell, B. (2001) Measuring consumer quality judgments. *British Food Journal* 103(1), 7–22.

Leonetti, L. and Kristo, I. (2010) *The Value Chain of Small Ruminants Meat Supplying the Tirana–Durres Urban Area*. Report from Project "Improving the Performance of Livestock Sector in Albania", UNDP (United Nations Development Programme), Tirana, Albania.

Lerman, Z. (2000) Agriculture in transition economies: from common heritage to divergence. *Agricultural Economics* 1481, 1–20. Available at: http://departments.agri.huji.ac.il/economics/lerman-article1.pdf (accessed 6 September 2012).

Manalo, A. (1990) Assessing the importance of apple attributes: an agricultural application of conjoint analysis. *Northeastern Journal of Agricultural and Resource Economics* 19, 118–124.

MAFCP (1981) *Statistical Yearbook 1980*. Ministry of Agriculture, Food and Consumer Protection, Tirana, Albania.

MAFCP (1986) *Statistical Yearbook 1985*. Ministry of Agriculture, Food and Consumer Protection, Tirana, Albania.

MAFCP (1992) *Statistical Yearbook 1991*. Ministry of Agriculture, Food and Consumer Protection, Tirana, Albania.

Mehta, M.D. (2003) Public perceptions of food safety: assessing the risks posed by genetic modification, irradiation, microbiological contamination and high fat/high calorie foods. *Pierce Law Review* 1, 69–84.

Sandalidou, E., Siskos, Y. and Baourakis, G. (2002) Customers' perspectives on the quality of organic olive oil in Greece: a satisfaction evaluation approach. *British Food Journal* 104(3–5), 391–406.

Scarpa, R. and Del Giudice, T. (2004) Market segmentation via mixed logit: extra-virgin olive oil in urban Italy. *Journal of Agricultural and Food Industrial Organization* 2(7), DOI: 10.2202/1542-0485.1080.

Skreli, E., Imami, D., Dishnica, T., Jaupi, A., Cela, A. and Fortuzi, A. (2011) Consumers' preferences for apple fruits in Tirana market using a conjoint analysis. Paper presented to: *21st IAMA [IFAMA] Annual World Forum and Symposium: The Road to 2050: Sustainability as a Business Opportunity, Frankfurt, Germany, 20–23 June 2011*. International Food and Agribusiness Management Association. Available at: https://www.ifama.org/events/conferences/2011/cmsdocs/2011SymposiumDocs/378_Symposium%20Paper.pdf (accessed 6 September 2012).

USAID-AAC (2009) *Value Chain Analysis of Apple*. Technical Report prepared by Development Alternatives, Inc., Tirana, Albania for US Agency for International Development/Albanian Agriculture Competitiveness Program (AAC), Tirana, Albania.

Veale, R. and Quester, P. (2009) Do consumer expectations match experience? Predicting the influence of price and country of origin on perceptions of product quality. *International Business Review* 18, 134–144.

Verdú-Jover, A.J., Lloréns Montes, F.J. and Fuentes-Fuentes, M. del M. (2004) Measuring perceptions of quality in food products: the case of red wine. *Food Quality and Preference* 15, 453–469.

Volckner, F. and Hofmann, J. (2007) The price–perceived quality relationship: a meta-analytic review and assessment of its determinants. *Marketing Letters* 18, 191–196.

World Bank (2003) *Rural Development Strategy*. World Bank Office, Tirana, Albania.

World Bank (2007) *Albania: Strategic Policies for a More Competitive Agriculture Sector*. Report No. AAA18-AL, World Bank Sustainable Development Sector Unit, Europe and Central Asia, Tirana, Albania. Available at: http://siteresources.worldbank.org/INTALBANIA/Resources/Strategic_Policies_for_more_competitive_Agriculture_sector.pdf (accessed 3 September 2012).

Zeithaml, V.A. (1988) Consumer perceptions of price, quality, and value: a means–end model and synthesis of evidence. *Journal of Marketing* 52, 2–22.

Zhllima, E., Chan-Halbrendt, C., Quango, Z., Imami, D., Long, R., Leonetti, L. and Canavari, M. (2013) Latent class analysis of consumer preferences for wine in Tirana, Albania. *Journal of International Food and Agribusiness Marketing* (in press). Manuscript available at: https://dl.dropbox.com/u/15850633/Publications/2013_JIFAM_WineTiranaLCA_acceptedmanuscript.pdf (accessed 14 September 2012).

9 Trade in Albania

Engjell Skreli[1] and Alex McCalla[2]
[1]*Agricultural University of Tirana, Albania;*
[2]*University of California, Davis, USA*

9.1 Introduction

Albania adopted a very liberal trade policy at the beginning of its transition toward a market economy in the early 1990s. Membership in the World Trade Organization (WTO) in the year 2000 formalized its already consolidated liberal trade regime. Free trade agreements with the European Union (EU), and with Balkan and other countries complete Albania's liberal trade policy profile with a preferential trade regime.

However, despite its very liberal trade regime with Europe, Albania's trade performance in terms of exports, imports and trade flows suggests that the country still remains a rather closed economy. In the period 2008–2010, Albania's trade averaged US$3239 per capita. This is lower than for other transitioning countries. This is ten times less than for Slovenia, 40 times less than for Hong Kong and 50 times less than for Singapore. The trade openness index (the ratio of the value of trade flows to gross domestic product, or GDP) in 2010 was only 50%, which is obviously lower than for other transitioning countries, such as Croatia (52%), Romania (69%), the Former Yugoslav Republic of Macedonia (FYROM) (96%), Bulgaria (97%), and Slovenia (125%) (WTO, 2012b).

With an export to import ratio of 1:3, the country has a large trade deficit, which reached 21% of GDP in 2010 (IMF, 2011), and is higher than that of other countries in transition (WTO, 2012b). Agricultural trade, in particular, had an import ratio of 1:11 in 2010, which contributed to the deepening of Albania's trade deficit.

There have been efforts to describe Albania's national trade (ACIT, 2009, 2010), and more in-depth studies to understand the capacity of gravity models (see Section 9.3.3) to explain regional trade patterns in the Western Balkans (Christie, 2001; Xhepa and Agolli, 2005). Other studies have addressed the impact of preferential trade regimes for western Balkan countries (Mançellari, 2005; Pere, 2009). Despite these efforts, no study has been undertaken to analyze Albania's agricultural trade in a comprehensive way. Hence, the main objective of this chapter is to understand the factors giving rise to Albania's agricultural trade by exploring the explanatory power of various international trade theory models. This is useful for selecting the best policy options for aiming to improve the contribution of agriculture to reducing Albania's trade deficit. In this context, the following questions have been addressed in the five main sections of the chapter:

- What is Albania's performance in terms of national and agricultural trade?
- What are the main traded agricultural products and the main trading partners?

- Which trade theory best helps to explain Albania's agricultural trading patterns?
- What kind of trade policy has Albania adopted?
- What strategic options does the government face in terms of trade policy?

Overall, the chapter provides a rather comprehensive review on Albania's agricultural trade over the last 15 years (1996–2010), although the authors recognize that, due to lack of data, some of the questions raised are only partially answered. That having been said, the questions raised, and the information missing, may motivate other researchers to enrich the current 'picture'.

9.2 Albania's Trade Performance During Transition to a Market Economy

9.2.1 Albania's national trade dynamics

During the communist regime, Albania was an economically closed country. The policy of economic self-sufficiency was applied to every sector, including industry, agriculture, construction, and services. Trade was therefore limited to 'the indispensable' to support self-sufficiency.

Trade data for the period 1996 to 2010 reveal that Albania remained a relatively small and 'closed' country in terms of trade performance for the whole period of transition toward a market economy. Despite an increase in trade flows (exports plus imports), which in 2010 amounted to slightly more than 6 billion US dollars—a sixfold increase over 1996—overall trade flows remain modest (Fig. 9.1). Albania had an openness index of 50% in 2010 (WTO, 2012a), which further indicates that, despite its liberal trade regime, Albania remains a rather 'closed' country in terms of trade flows with the rest of the world.

Albania's export base is low, and country is very dependent on imports. The country's imports amounted to US$4.6 billion in 2010 (Fig. 9.1), which represents a fivefold increase over the 1996 figure. The increase in imports has been driven by the country's still poor infrastructure, and the need to invest in the fast growing sectors, namely construction and services. Consequently, major import items include electrical power, machinery and equipment, and construction raw materials, namely cement and iron. Food imports are also quite important (ACIT, 2010). However, despite impressive export growth in relative terms (exports grew more than sevenfold between 1996 and 2010), Albania's import to export ratio of 3:1 in 2010 (UN COMTRADE, 2012) remains very unfavourable. The main exports include

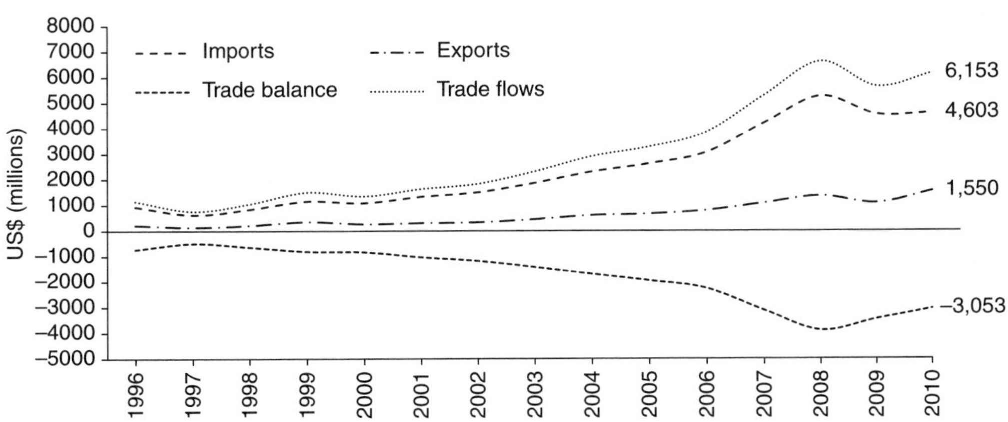

Fig. 9.1. Albania's national trade performance, 1996–2010. Source: UN COMTRADE (2012), and authors' own work.

minerals, electric power, clothes and garments produced by low skilled labor, medicinal and aromatic plants (MAPs), and fish.

9.2.2 Albania's agricultural trade dynamics

Agricultural trade represents a small part of national trade flows. In 2010, agriculture contributed to only 15% of national trade flows (Table 9.1), which is only half of the contribution made in 1996. In fact, this is lower than agriculture's 18 % contribution to GDP in 2009 (Bank of Albania, 2011), indicating lower and decreasing agricultural trade performance compared with the rest of the Albanian economy.

Agricultural trade data supports the import-dependent trade model discussed above. Import dependency is even stronger for the agricultural sector than for the whole economy; the import to export ratio for agriculture in 2010 was 9.4 (Table 9.1), which is three times higher than for the whole economy. The agricultural imports for 2010 amounted to US$837 million (Fig. 9.2), which is more than triple that in 1996. The bulk of imported products are represented by cereals, tobacco, spirits, meat, and fruit and vegetables. Imports include both final products and products to be processed by the nascent Albanian agro-processing industry.

Table 9.1. Albania's agricultural trade; selected indicators, 1996 to 2010. Sources: United Nations Commodity Trade Statistics Database (UN COMTRADE), and authors' own calculations.

	Year														
Trade item	1996	1997	1998	1999	2000	2001	2002	2003	2004	2005	2006	2007	2008	2009	2010
Agricultural trade to national trade (%)	31	26	25	23	20	18	18	18	17	16	16	15	15	15	15
Import to export ratio	9	6.2	7.0	10.7	8.7	9.3	18.1	10.0	9.2	8.5	8.7	8.8	10.0	10.0	9.4

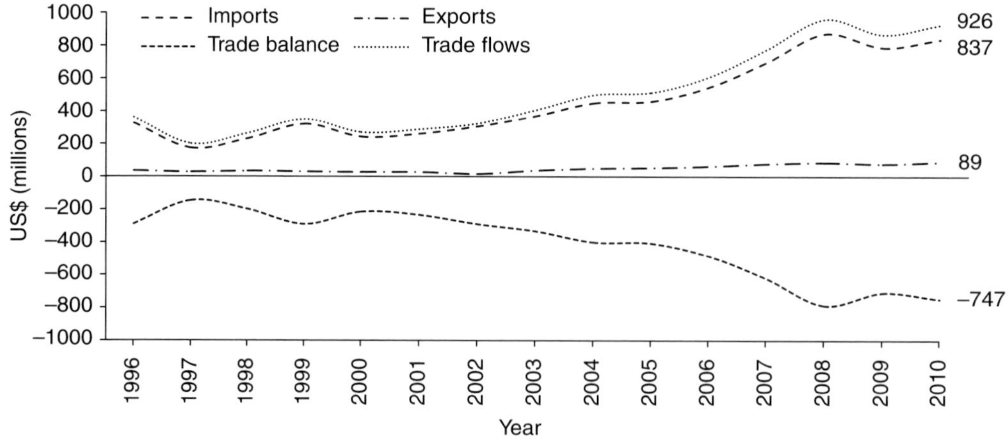

Fig. 9.2. Albania's agricultural trade performance, 1996–2010. Source: UN COMTRADE (2012), and authors' own work.

The low export base of agricultural products, combined with sizable imports contributes to a deteriorating agricultural trade balance. Although increasing, exports of Albanian agricultural products are very low. In 2010, the country exported only US$89 million in agricultural and food products (Fig. 9.2). An extremely high import to export ratio (9.4 in 2010), with virtually no substantial change since 1996, highlights a structural problem in Albanian agriculture. The elements of that problem are: tiny farm size; virtually nonexistent value chain governance; and poor quality, safety and marketing institutions and infrastructures. The main export items are natural resources, namely MAPs, fish, and modest quantities of fruit and vegetables.

9.3 Geographic Distribution of Agricultural Trade

9.3.1 Albania's agricultural trade flows by main trade partners

Traditional partners (Italy, Greece, Germany, Turkey, and Bulgaria), but also newcomers (Poland, Brazil, Russia, and Croatia) are now among Albania's main trading partners (Fig. 9.3). Italy is now Albania's first trading partner, outranking Greece, which was the leading trade partner until 2005 but has now fallen to second place. Serbia is currently Albania's third largest trading partner, though it is important to clarify that trade with Kosovo is included in Serbian data, as UN COMTRADE statistics do not present trade data for these countries separately. Traditional trading partners, such as Germany, Turkey, and Bulgaria, remain important, though some new and quite interesting partners begin to appear as important Albanian trading partners, including Poland, Brazil, Russia, and Croatia.

The emergence of new trading partners is quite interesting in terms of the insight they give into theoretical models to explain agricultural trade. Poland is a rather big agricultural producer in the EU, Brazil and Russia are world agricultural superpowers with large land areas available, and Croatia is a country close to Albania. It therefore becomes obvious that several models, including factor endowment, distance, size of economy, and other factors, may be important in explaining Albania's agricultural trade. Understanding the factors determining the current geographical distribution of trade and their trends is important in terms of trade policy focus, such as: promoting trade with special attention to the most important trading partners, including reshaping trade diplomacy; removing barriers to trade with the

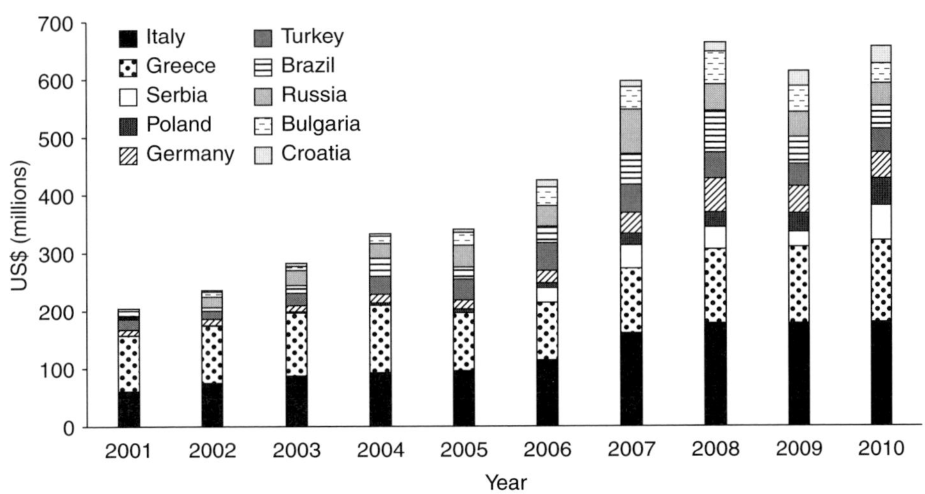

Fig. 9.3. Albania's geographic distribution of agricultural trade, 2001–2010. Source: International Trade Center (ITC) calculations, based on UN COMTRADE statistics.

main trading partners; facilitating free movement of business people; and the like.

9.3.2 Agricultural exports and imports by main trading partners

Albania's main trading partners are EU countries. As already noted, the evidence shows that Italy is the largest Albania's trading partner for both exports and imports, followed by Greece (Table 9.2). Beyond this already generally known information, one can observe that Italy and Greece make up around 60% of exports and around 30% of imports. Additionally, Albania has important trade relations with other EU countries, including Poland, Germany, Austria, France, and Bulgaria. While Germany and Austria are Albania's traditional partners in agricultural trade, Poland is a newcomer.

In 2010, Albania traded more with Balkan countries. Trade data show that, starting from 2006, Albania has increased trade with Serbia, which is now the third most important country from which Albania imports, and the fourth most important country to which Albania exports. Additionally, Croatia and FYROM are now among Albania's ten top trading partners.

Distant countries are also becoming important trading partners for Albania. Brazil and Russia are major countries from which Albania imports large quantities of agricultural products. They are powerful newcomers; Brazil and Russia are ranked as in the top ten world exporting countries of agricultural products (ITC, 2012) and Albania's imports from these countries follow an exponential trend. Imports from Brazil have increased 11-fold between 2001 and 2010, and imports from Russia have increased 17-fold over the same period. The imports from more distant countries are quite

Table 9.2. Albania's agricultural exports and imports for last 10 years with its main trading partners. Values in US$ (millions). Source: ITC, and authors' own calculations.

	Year									
Country	2001	2002	2003	2004	2005	2006	2007	2008	2009	2010
Exports										
Total	**27.5**	**17.0**	**37.1**	**49.3**	**53.9**	**62.8**	**78.5**	**86.8**	**78.9**	**89.2**
Italy	8.2	3.2	15.9	18.5	20.3	26.3	30.8	35.7	35.8	38.6
Greece	7.8	6.2	7.0	6.8	8.7	8.7	13.0	9.9	9.0	13.5
Germany	3.4	4.6	4.5	5.7	6.0	8.2	9.7	10.7	7.5	8.1
Serbia	0.0	0.0	0.0	0.0	0.0	7.7	11.6	6.7	1.7	6.7
USA	1.5	1.3	2.0	3.0	3.7	2.7	3.0	4.0	5.0	4.4
Austria	0.3	0.3	0.3	0.3	0.2	0.5	0.9	6.8	5.3	4.1
FYROM[a]	0.2	0.5	0.5	0.9	0.8	1.2	3.8	2.6	2.5	2.3
France	1.6	1.6	1.4	1.5	1.8	2.3	1.3	2.2	1.6	1.9
Spain	0.2	0.4	0.3	0.5	0.5	0.4	0.5	0.5	1.9	1.8
Turkey	0.7	0.7	0.9	1.3	1.2	1.1	1.4	2.9	1.7	1.8
Imports										
Total	**260.7**	**306.6**	**370.2**	**448.9**	**460.1**	**546.4**	**698.3**	**873.2**	**788.3**	**836.5**
Italy	52.4	71.6	71.2	74.1	75.6	87.8	129.5	141.6	141.4	140.4
Greece	88.3	92.9	102.3	110.3	91.6	91.4	99.1	118.7	123.4	128.3
Serbia	0.0	0.0	0.0	0.0	0.0	17.3	28.8	31.3	24.2	53.6
Poland	0.3	0.8	2.3	4.1	6.8	8.4	19.6	24.4	31.6	45.9
Brazil	3.5	5.9	13.3	31.5	21.2	28.5	53.9	71.8	46.6	39.9
Russia	2.3	18.4	25.7	25.2	37.4	35.9	76.3	45.5	43.1	38.8
Turkey	17.5	14.0	20.5	29.5	34.5	46.7	47.2	42.7	37.1	38.2
Germany	6.5	6.2	6.3	9.1	9.6	13.1	26.2	48.1	39.3	38.0
Bulgaria	9.3	8.5	7.0	12.6	22.0	31.9	38.7	56.2	44.2	32.6
Croatia	3.9	2.6	5.4	3.9	5.4	9.9	10.2	14.8	25.5	28.3

[a]FYROM, Former Yugoslav Republic of Macedonia.

likely due to Albanian agriculture's shift to labor intensive products and away from land intensive products; Albania imports wheat from Russia and meat from Brazil, both land intensive products. Another important fact is reduced distance in terms of lower transport and communication costs.

9.3.3 Does a gravity model explain Albania's agricultural trade flows?

Gravity models are mathematical models based on an analogy with Newton's gravitational law: every object in the universe attracts every other object with a force that depends on product masses and (inversely) on the distance between them. In the case of trade, the gravity model states that the volume of trade can be estimated as an increasing function of economic size, or the national incomes of trading partners, and a decreasing function of the distance between them (Balistreri and Hillberry, 2006).

Distance as a factor explaining Albania's geographic distribution of trade

Distance—embracing physical, cultural and language, political and administrative distance—is a significant factor explaining agricultural trade. Albania's top three trading partners—Italy, Greece, and Serbia—make up 41% of the country's trade. Italy, Greece, Serbia, and FYROM account for more than two thirds of exports, and Italy, Greece and Serbia account for more than one third of imports. Albania shares a common border with each of these countries, but it shares much more than just geographic proximity. The majority of Albania's economic emigrants live in Italy and Greece, and Kosovo, now an independent republic (though statistically part of Serbia), is close to Albania as part of the same nation. Furthermore, Albania has preferential trade agreements with Italy, Greece, Serbia, and Kosovo.

Distance, shortened by integration processes, has affected trade among countries that traditionally traded less. The EU's policy of support to trade integration in the Balkans region, and its policy of offering these countries the prospect of EU membership, has facilitated and motivated trade in the region which exceeds that expected only from the geographic proximity between Albania and these countries. Serbia, Federal Republic of Yugoslavia, Croatia, and Bosnia and Herzegovina, four countries which just a few years ago all belonged to the FYROM, are now among the ten top Albania's trading partners, though the trade flows are still not very large. This evidence challenges the pessimistic forecasts of only 10 years ago, which identified Albania as the region's least integrated country, and doubted whether it would ever really be part of the region in an economic sense (Christie, 2001). There were similar findings by Mançellari (2005) and Pere (2009).

Increases in trade flows between Albania and Balkan countries were anticipated by Xhepa and Agolli (2005), who posited that reducing distance costs will significantly increase trade volumes in less time. By 'reduced' distances, they did not mean only improvements of the physical infrastructure. Market access measures, border procedures, movement of people, development, and dissemination of information were also important dimensions of the distance costs that they deemed needed improvements. Important improvements in all dimensions are now in place: Albania's trade infrastructure has substantially improved, people move freely across borders, market access has improved, border procedures are now much lighter, and information flows better.

Size as a factor explaining Albania's geographic distribution of trade

The gravity model predicts that economic size, measured by national income, is an important factor explaining trade among countries. A given country—let us say Albania—is expected to export more to a larger country because this country has higher purchasing power. Complementarily, the given country is likely to buy more from a larger economy because the larger country has more to offer.

Economic size in the framework of gravity model is less important in explaining Albania's agricultural geographic distribution of trade than distance. Out of the ten top world agricultural exporters, based on

International Trade Center (ITC) rankings (ITC, 2012), only three overlap with the ten top countries from which Albania imports. These countries are Italy, Brazil, and Germany, meaning that Albania buys mainly from countries other than export 'superpowers'. The overlap in terms of Albania's exports includes four countries, namely Italy, Germany, the USA, and France, but exports to the USA and France are modest, at slightly lower than US$2 million with France and US$4 million with the USA (UN COMTRADE, 2012), meaning that Albania exports mainly to countries other than import 'superpowers'. Albania's imports from only three agricultural export 'superpowers', and exports to only two import 'superpowers', indicate the low influence of size in explaining the country's agricultural trade. The low predictability of the gravity model size factor in explaining Albania's geographic distribution of trade calls for alternative theories that may be likely to explain trading patterns. Theoretical trade models that need to be considered are factor endowment theory, labor productivity theory, and intra-industry trade theory.

9.4 Trade Patterns

9.4.1 Exports and imports by product and trading partner

The main destinations of Albania's exports are Italy, Germany, Austria, Serbia, and Greece (Table 9.3). Albanian exports to Italy include fresh and processed fish, and different kinds of animal products, including eggs and honey. Fruits, including watermelons and other melons, as well as vegetables, are mainly exported to Serbia and Greece; and MAPs are exported to Germany, the USA, and Turkey.

Albanian imports reflect both the consumers' diet and the emerging shift away from land intensive products. The country imports large quantities of cereals, meat and meat products, fruit, edible oils, sugar and confectionery, and

Table 9.3. Exports of agricultural products to Albania's top three trading partners, 2010, Values in US$ (thousands) and shares. Source: UN COMTRADE database.

HS[a] code	Product	Value of exports	Export shares to three top trading partners		
			First	Second	Third
'16	Meat, fish, seafood food preparations nes[b]	24,896	Italy (92%)	Spain (5%)	Croatia (2.7%)
'12	Oil seed, oleagic fruits, grain, seed, fruit, etc., nes	20,948	Germany (38%)	USA (19.1%)	Turkey (8.3)
'03	Fish, crustaceans, mollusks, aquatic invertebrates nes	12,382	Italy (55%)	Greece (38%)	France (2.4%)
'05	Products of animal origin nes	4,597	Austria (63.9)	Italy (36%)	Serbia (0.1%)
'07	Edible vegetables, and certain roots and tubers	4,445	Serbia (27%)	Bulgaria (24%)	Greece (21%)
'24	Tobacco and manufactured tobacco substitutes	4,220	Greece (98.6%)	Serbia (1.4)	–
'08	Edible fruit, nuts, peel of citrus fruit, melons	3,844	Serbia (40.8%)	Italy (21.8%)	FYROM[c] (8%)
'20	Vegetable, fruit, nut, etc. food preparations	3,419	Greece (84.4%)	Serbia (9.4%)	FYROM (2.4%)
'04	Dairy products, eggs, honey, edible animal product nes	2,984	Italy (75.9%)	Greece (14.6%)	Serbia (9.5%)
'22	Beverages, spirits and vinegar	1,668	Serbia (46%)	France (29.6%)	USA (7%)

[a]HS, Harmonized Commodity Description and Coding System, an international nomenclature designed to classify traded goods on a common basis, developed and maintained by the World Customs Organization; [b]nes, not elsewhere specified; [c]FYROM, Former Yugoslav Republic of Macedonia.

live animals. Tobacco, beverages, and spirits are also important import items (Table 9.4). The imports of cereals are related to both the Albanian diet (wheat bread is the staple food) and to the move away from land intensive products (wheat is a land intensive product because it has a higher land to labor ratio than fruit and vegetables). Imports of meat and fruit are mainly explained by increased consumption, because their domestic production has also increased.

The main suppliers of cereals, which are Albania's most important imports, are the Russian Federation, France, and Serbia. Turkey, Italy and Bulgaria are the country's main suppliers of processed grains, such as pasta, biscuits, and breakfast cereals. Albania imports meat from remote countries, including Brazil and the USA, but also from Serbia. The import of cereals from the Russian Federation, a country rather distant from Albania using the definition of 'distance' presented in this chapter, suggests that factor endowments rather than distance better explain trade. Russia is a country with large land resources (Table 9.5) favoring wheat production—an intensive use of land resources. While the same factor is likely to explain the trade with Serbia, imports from France are most probably due to a combination of factor endowment and labor productivity theories.

9.4.2 Theoretical trade models—which best explains Albania's agricultural trade?

Factor endowment and Albania's agricultural trade

Factor endowment theory—or Heckscher–Ohlin theory—predicts that a country which has a large supply of a particular resource relative to its supply of other resources will tend to produce more goods that use its abundant resources intensively. The result is that countries tend to export goods that are intensive in the factors with which those countries are abundantly supplied (Krugman and Obsfeld, 1994).

Albania is a labor abundant country. Evidence in Table 9.5 confirms that its labor

Table 9.4. Imports of agricultural products from top three trading partners, 2010. Values as US$ (thousands) and shares. Source: UN COMTRADE database.

HS[a] code	Product	Value of imports	Import shares from three top trading partners: First	Second	Third
'10	Cereals	103,198	Russia (37.1%)	France (13.4%)	Serbia (13.3%)
'24	Tobacco and manufactured tobacco substitutes	99,449	Poland (33.1%)	Greece (28.8%)	Germany (21%)
'22	Beverages, spirits, and vinegar	89,542	Italy (31.8%)	Austria (22.8%)	Greece (14.2%)
'02	Meat, and edible meat offal	70,172	Brazil (33%)	USA (23.6%)	Italy (18.3%)
'08	Edible fruit, nuts, peel of citrus fruit, melons	62,389	Greece (42.8%)	Italy (27%)	Ecuador (18.9%)
'19	Cereal, flour, starch, and milk preparations and products	57,193	Turkey (34.6%)	Italy (23.1%)	Bulgaria (14.4%)
'15	Animal and vegetable fats and oils, cleavage products, etc.	54,582	Ukraine (31.3%)	FYROM[b] (16%)	Hungary (11.5%)
'21	Miscellaneous edible preparations	38,004	Poland (18.7%)	Italy (17.5%)	Greece (14.3%)
'17	Sugars and sugar confectionery	37,824	B&H[c] (24.2%)	Brazil (18.6%)	Serbia (17.2%)
'01	Live animals	34,578	Bulgaria (43.3%)	Greece (29.8%)	FYROM (7.3%)

[a]HS, Harmonized System; [b]FYROM, Former Yugoslav Republic of Macedonia; [c]B&H, Bosnia and Herzegovina.

Table 9.5. Labor to land ratio of Albania's main trading partners, 2008. Source: FAO and ILO, 2012, and authors' own calculations.

Country	Arable land and permanent crops (1000 ha)	Employment in agriculture, hunting and forestry (1000 people)	Labor/land ratio
Albania	697	542	0.778
Austria	1,440	227	0.158
Brazil	68,700	16,207	0.236
Bulgaria	3,272	251	0.077
Croatia	946	216	0.228
France	19,408	770	0.040
Germany	12,133	866	0.071
Greece	3,709.1	519	0.140
Italy	10,042	860	0.086
Poland	12,970	2,206	0.170
Russian Federation	123,442	5,994	0.049
Serbia	3,602	707	0.196
Turkey	24,505	5,016	0.205
(Former) Yugoslav Republic of Macedonia	460	119	0.259

to land ratio of 0.778 is the highest among the country's main trading partners. As factor endowment theory predicts, it is expected that Albania would allocate more resources to the production and export of low skilled and labor intensive products.

Trade data (Tables 9.3 and 9.4) clearly reveal that Albania exports labor intensive products, namely fish and processed fish, MAPs and vegetables; and imports land intensive products, such as cereals and meat. The main markets for labor intensive products are countries where agricultural labor is a scarce resource. The main sources of imports for land intensive products are countries where agricultural labor is a scarce resource compared with land. Albania imports cereals from the Russian Federation, France, Serbia, and Brazil, all of which have a much lower labor to land ratio than Albania. Hence, the factor endowment theory of trade or Heckscher–Ohlin theory explains quite well an important part of Albania's agricultural trade.

Labor productivity and Albania's agricultural trade

Labor productivity and comparative advantage, or the Ricardian model, suggests that a country's comparative advantage depends on its labor productivity and wage level. If the relative productivity of a country in producing a good is higher than its relative wage compared with a foreign country, then that country has a comparative advantage over the foreign country in producing that good.

A sound discussion on labor productivity as a factor giving rise to Albania's agricultural trade would require reliable data on product yields/ha, labor used/ha, and wage data for a large number of countries. Such data are difficult to find, especially labor/ha and agricultural wage labor. Therefore, the following discussion does not pretend to provide a solid insight into labor productivity as a factor of agricultural trade; rather, it aims at giving a partial understanding of the issue, which then needs be further probed with more in-depth analysis.

At this time, Albania's two main exports, fresh fish and MAPs, are more exports of natural resources than of processed products. The activity is based on harvesting of the natural resources, very simple processing, and exporting. So in order to understand the capacity of the labor productivity model to explain Albanian trade, we focus on vegetable products and wheat.

It can be shown that Albania has higher relative yields in watermelons, tomatoes, and cucumbers than do importing countries, except for cucumbers from FYROM (Table 9.6). This supports the argument that Albania has higher relative productivity than does its trade partners. Does Albania expend similar amounts of labor/ha of vegetable products? It is difficult to fully determine owing to lack of data. However, because tomatoes and cucumbers are mainly produced in protected areas (greenhouses), the technology adopted requires similar labor inputs/unit area cultivated. Additionally, expert assessment (A. Balliu, Tirana, Horticulture Department Head, Agricultural University of Albania, personal communication 23 January 2012) suggests that Albania is well advanced in greenhouse technology, technical knowledge and labor skills, which are quite high in the cultivation of greenhouse vegetables and watermelons. In terms of relative wage, Albania has relatively lower wages than its vegetable trading partners, despite wage increases in recent years.

In conclusion, it seems that labor productivity explains some export flows, but not the import of wheat. Albania has higher relative yields than its trading partners in watermelons, tomatoes and cucumbers (except FYROM for cucumber in 2010; refer to Table 9.6). This supports the contention that Albania is likely to have higher labor productivity than its trading partners. Following the logic of the factor endowment theory, Albania has chosen to produce labor intensive products, with a further investment in modern technologies, which has most probably resulted in labor productivity increases. In the case of wheat, while Albania has higher yields than the Russian Federation and Serbia, it imports the commodity from these countries, which is quite contrary to predictions of the labour productivity model. The stronger alternative theory to explain Albania's import of wheat is the factor endowment theory, which has already been discussed above.

Intra-industry trade measurement and Albania's agricultural trade

Intra-industry trade arises when a country simultaneously imports and exports similar types of goods or services (Box 9.1). Does intra-industry trade explain Albanian agricultural trade?

Using the Grubel and Lloyd Index of Intra-industry Trade (Grubel and Lloyd, 1975), the sectors exhibiting a certain level of intra-industry trade are: (i) meat, fish, and seafood food preparations; (ii) lac, gums, resins, vegetable saps, and extracts; (iii) fish, crustaceans, mollusks, and aquatic invertebrates; (iv) products of animal origin; and (v) vegetable, fruit, nut, etc., food preparations (Table 9.7).

Trade data reveal that the intra-industry trade measure explains only a tiny fraction of Albanian trade in agricultural and processed agricultural products. The trade flows for the

Table 9.6. Albania's relative yields for selected goods. Source: FAO, and authors' own calculations.

		Year				
Product	Yield ratio	2006	2007	2008	2009	2010
Watermelons	Albania/Serbia	1.4	1.9	2.0	2.1	2.6
	Albania/FYROM[a]	1.1	1.2	1.5	1.5	1.5
	Albania/Slovakia	1.5	1.7	2.6	2.4	4.2
Tomatoes	Albania/Serbia	2.5	3.3	3.7	3.4	3.4
	Albania/FYROM	0.9	1.1	1.4	1.3	1.1
	Albania/Slovakia	1.2	1.4	1.7	1.8	2.4
Cucumbers	Albania/Serbia	2.9	3.5	4.1	3.9	4.0
	Albania/FYROM	0.8	0.9	1.0	1.0	0.8
	Albania/Slovakia	1.6	1.8	2.3	2.3	2.6
Wheat	Albania/Russia	1.6	1.7	1.6	1.7	2.1
	Albania/France	0.5	0.6	0.6	0.5	0.6
	Albania/Serbia	0.9	1.1	0.9	1.1	1.2

[a]FYROM, Former Yugoslav Republic of Macedonia.

Box 9.1. Intra-industry trade measure.

In the empirical literature on intra-industry trade, the measurement technique most commonly used to identify the presence of intra-industry trade is the Index of Intra-Industry Trade (IIT):

$$B_i = 1 - \frac{|X_i - M_i|}{X_i + M_i}$$

where B_i is the Grubel–Lloyd measure of IIT for industry i, X_i is the country's exports in industry i, and M_i is the country's imports in industry i. This measure of intra-industry trade reaches its maximum of 1 when exports equal imports, and its minimum of 0 when either exports or imports equal zero.

Source: Qasmi and Fausti (2001).

Table 9.7. Intra-industry trade in agriculture products. Source: ITC, and authors' own calculations.

HS[a] code	Product at 2-digit HS level	Index[b]	HS code	Product at 4-digit HS level	Value in US$ (1000) for 2010	
					Exports	Imports
'16	Meat, fish, and seafood food preparations	0.94	'1604	Prepared/ preserved fish and caviar	24,701	13,863
'13	Lac, gums, resins, vegetable saps and extracts	0.92	'1302	Vegetable saps and extracts	230	271
'03	Fish, crustaceans, mollusks, aquatic invertebrates	0.90	'0302	Fish, fresh, whole	4,305	4,018
			'0303	Fish, frozen, whole	310	797
			'0305	Fish, cured or smoked, and fish meal fit for human consumption	1,056	3,074
			'0306	Crustaceans	392	1,616
			'0307	Mollusks	6,295	4,659
'05	Products of animal origin	0.67	'0504	Guts, bladders, and stomachs of animals other than fish	4,590	1,981
'20	Vegetable, fruit, nut, etc., food preparations	0.30	'2005	Prepared or preserved vegetables (excluding frozen)	2,996	5,538
	Total value				44,875	35,817

[a]HS, Harmonized System; [b]Intra-industry trade index (see Box 9.1).

sectors affected by intra-industry trade amount to US$80,692 (US$44,875 plus US$35,817; Table 9.7), which represents less than 0.01% of agricultural trade flow.

The factor giving rise to intra-industry trade is product differentiation, mainly at the processing level rather at the farming level. Albania exports special processed fish products (anchovies) to niche markets in the EU, and fresh fish to EU countries, but imports canned and other processed fish and frozen fish from these countries. The same factor gives rise to trade in processed vegetable products. Albania exports processed vegetable products (canned) that are required on the export market, and imports processed vegetables required by Albanian consumers. Though intra-industry trade is currently negligible, the phenomenon suggests that product differentiation, and exports based on product differentiation, are also feasible in Albania; this may also be an export enhancing alternative to be explored for other sectors as well.

9.5 Trade Policy

Until recently, Albania had a non-preferential trade regime. However, in 2002, the country signed its first free trade agreement with FYROM. Currently, two trade regimes are in use: a non-discriminating trade regime within the framework of WTO; and a preferential trade regime represented by a number of free trade agreements. Both trade regimes will co-exist: the preferential trade regime will cover trade with EU and neighboring countries, and the non-discriminating trade regime will cover trade with the rest of the world. Because the EU and neighboring countries are Albania's main trading partners, the preferential trade regime will govern most of Albanian trade.

9.5.1 General trade regime

Tariff protection

While Albania has adopted a liberal trade regime, tariff protection granted to the agriculture sector for all WTO countries is higher than the overall tariff protection. While the latter was 5.2% in 2009, the average tariff applied for all WTO countries on a 'most favoured nation basis' (MFN) (see Box 9.2) for agriculture was 8.8% (Table 9.8). It becomes clear then that agriculture in Albania is moderately more protected than the rest of economy.

Albania has a moderate and decreasing tariff protection for agriculture. The average tariff rate applied in Albania in 2010 was 7.9%, lower than that of 8.8% in 2009. Albanian tariff protection was lower than that for other transition and other countries, namely Bosnia and Herzegovina (30.3%), FYROM (13.1%), Croatia (9.9%), the EU (12.8%), and Japan (17.3%), to name only a few (WTO, 2010).

Relatively higher tariff protection is granted to products that intensively use the country's abundant production factor, labor. The most protected products are fruit and vegetables, and the least protected products are cereals. The average applied tariff rate for fruit and vegetables is 11.8%, while that for cereals is only 2%. As already discussed, Albania is a country with a high labor to land ratio. Fruit and vegetables use labor intensively, which in the case of Albania is the abundant production factor. As the Heckscher–Ohlin model suggests, a country should specialize in products that use

Box 9.2. WTO non discriminating principle.

Non-discrimination in trade is one of the key principles of the World Trade Organization (WTO). It becomes operational by applying two rules: the rule of the 'most favoured nation' and the rule of 'national treatment.'

- **Most-favoured-nation** (MFN) treatment is the rule of not discriminating amongst one's trading partners
- **National treatment** is the rule of not discriminating between domestic and foreign business, or the principle of treating imports no less favorably than the same or similar domestically produced goods once they have passed customs.

Source: WTO, 2012c. Glossary: A guide to 'WTO speak'.

Table 9.8. Summary analysis of Albania's MFN (most favoured nation) tariff, 2009. Source: WTO Secretariat estimates, based on data provided by the Albanian authorities.

Description	MFN (no. of lines)	Average (%)	Range (%)	Coefficient of variation (CV)	Final bound average (%)
Total	9569	5.2	0–15	1.1	6.6
HS[a] 01–24	2251	7.6	0–15	0.8	9.3
HS 25–97	7318	4.5	0–15	1.3	6.1
WTO Agriculture[b]	1987	8.8	0–15	0.6	10.1
Animals and products thereof	322	9.4	0–15	0.3	10.6
Dairy products	151	9.7	2–15	0.2	10.3
Coffee and tea, cocoa, sugar, etc.	293	9.3	0–15	0.5	12.1
Cut flowers, plants	54	6.8	2–15	0.7	9.6
Fruit and vegetables	428	11.8	2–15	0.2	12.3
Grains	55	2.0	2–2	0.0	5.0
Oil seeds, fats and oils, and their products	164	2.0	0–10	1.9	3.3
Beverages and spirits	271	11.1	0–15	0.6	14.9
Tobacco	20	7.3	2–15	0.8	9.4
Other agricultural products nes[c]	229	5.2	0–15	0.8	7.1

[a]HS, Harmonized System; [b]WTO agriculture refers HS chapters 1 to 24 (excluding fish and fish products) and a number of manufactured agricultural products; [c]nes, not elsewhere specified.

its abundant production factor intensively. It follows that Albania should support and specialize in fruits and vegetables, and give up cereals.

The protection of promising industries, and tariff revenue concerns, explain the relatively high tariff rates. The high tariff rates for horticultural, and animal and dairy products, are explained on the grounds of trade protection; they represent promising industries. The tariff rates for coffee, beverages, and tobacco are kept relatively high as well, as they represent important budget income sources.

Domestic support measures

Domestic support to the agricultural sector in Albania takes two main forms: measures to stabilize the agricultural markets, and direct payments to producers. Domestic support must meet the following criteria: it should not have the effect of supporting market prices received by farmers; the support given to a specific product should not be more than 5% of the total value of annual production for such a product; and total support that is not product specific should not be more than 5% of the total value of agricultural production during the respective year (Albanian Parliament, 2007). These criteria are in full conformity with WTO principles regarding domestic support.

The domestic support policy has, therefore, been very market friendly. Albania has reported that it did not provide any product specific support (Box 9.3) for the period 2000–2008 (Table 9.9). Non-product specific Aggregate Measurement of Support (AMS) (Box 9.3) has consisted largely of subsidized irrigation and, since 2007, also gave support to tree plantations, which is an investment support policy, and not a price support policy. The data in Table 9.9 show a sharp increase in non-product specific AMS, starting from 2007. Financial assistance or incentives to farmers falling under non-product specific AMS also includes loan schemes for investments in production technologies, and subsidized interest rates for private bank loans. The main target for 'Green Box' (Box 9.3)—or non-market distorting support spending—has been strengthening inspection and infrastructural services (WTO, 2010).

Box 9.3. Product specific and non-product specific support for agriculture.

- **Product specific support** is government support for specific agricultural products. Although it distorts trade, it is allowed up to a certain level, called 'De minimis'. The 'De minimis' goes up to 5% of the value of production for developed and transition countries, and 10% for developing countries. If the domestic support for specific products goes beyond 'De minimis,' it becomes subject to reduction commitments. Support beyond 'De minimis' is technically calculated as the Aggregate Measurement of Support (AMS).
- **Non-Product specific support** is the domestic support for agriculture that is allowed without limits because it does not distort trade, or, at most, causes minimal distortion. WTO classifies this as Green Box.

Source: WTO, 2012c. Glossary: A guide to 'WTO speak'.

Table 9.9. Albanian domestic support, 2000–2008. Value in thousand lek (ALL). Source: WTO (2010).

Year	Product specific AMS[a]	Non-product specific AMS	Green Box[b]
2000	0	385,000	207,000
2001	0	397,000	195,400
2002	0	243,000	1,025,200
2003	0	277,000	768,000
2004	0	328,000	949,500
2005	0	610,000	970,500
2006	0	500,000	1,286,000
2007	0	976,000	1,777,000
2008	0	854,000	853,800

[a]AMS, Aggregate Measurement of Support; [b]Green Box, non-product specific domestic support.

Government support is mainly support at the farm level. Since 2007, Albania has provided moderate financial support to farmers investing in new plantations of fruit trees, olive trees, vineyards, greenhouse vegetable production, milk production, olive oil production, honey production, organic production, and raising horses and rabbits. The direct financial support to farmers has increased from US$4 million in 2007 to US$12.6 million in 2011 (Ministry of Finance, 2011). While support is needed at several value chain levels (input supply, farm products consolidation, processing, and marketing), support to the farm is a high priority. Small supplies of agricultural produce from individual farms, unreliable and inconsistent supplies in terms of cultivar, standards and quality, and their high price represent major obstacles for the agro-processing industry. It follows that government support to farmers is a justified policy option.

Government support has been designed to give support to products deemed to have a comparative advantage. The products receiving the majority of the financial support are fruit trees, vegetables, and milk production. These are products considered to have a comparative advantage and, therefore, are a priority sector in the Sector Strategy of the Agriculture and Food section of the strategic document of the Albanian government, prepared by the Ministry of Agriculture, Food and Consumer Protection (MAFCP, 2007).

The support policy is mainly a scheme of support to investment rather than to price or production. Farmers are supported with a grant to plant new fruit tree plantations and invest in greenhouse industry. The scheme is not totally free of price support though. For example, it supports the price of cow's milk at US$0.1/l. Additionally, payments per animal head (cows and small ruminants) are included in the scheme, with the objective of increasing herd size. The last two deviations from a market oriented policy show some lack of consistency in the government support policy.

A new support scheme following the EU model is going to be implemented in Albania once the country has obtained EU candidate status, which it has applied for and it is expected to have in the coming two years, depending on further political issues. In this framework, the country has established a Rural Development Programme (MAFCP, 2010), which supports four sectors: fruit, vegetables, meat and milk, at both farming and processing levels. The eligible investments are mainly buildings and machinery. The scheme is expected to support businesses that have the potential to become competitive, while establishing food safety and environment protection standards. Establishing EU-like standards raises, however, a genuine concern regarding the private sector absorptive capacity; Albania has small farmers who will have hard time meeting EU standards, and therefore it will be difficult for them to benefit from EU support.

Other charges affecting imports

Value added tax (VAT) is the main source of tax revenue in Albania, and contributes 38% of government revenues (Ministry of Finance, 2011). Persons and companies involved in any economic activity, including exporting or importing of agricultural and processed agricultural products, but excluding farming, must be VAT registered if they have an annual turnover of US$50,000/calendar year. The VAT rate is 20% for domestically produced and imported goods and services, and is applied on a national treatment basis, which means treating imports not less favorably than domestic products (WTO, 2010).

Excises taxes also represent a substantial source of government revenue (12% of government revenue; Ministry of Finance, 2011). Excise taxes are levied on several agricultural products, including coffee; fruit juices and other non-alcoholic beverages; beer, wine, alcohol, and other spirits; and tobacco and tobacco products. Excise tax is also applied on a national treatment basis.

Import prohibitions and licensing

According to the WTO Trade Policy Review (WTO, 2010), Albania is fully in line with WTO requirements on import prohibitions and licensing. This is important because it shows that import prohibitions and the licensing regime is not used as a disguised instrument of trade protection and generation of non-tariff barriers to trade.

Import prohibitions relate to products considered to be hazardous to or threaten public health, and apply to animal products from certain countries. The Law on Veterinary Service and Inspectorate prohibits the importation of: bio-products and veterinary drugs recognized as dangerous to human and animal health by relevant international organizations; imports from countries or areas within a country infected by a contagious disease; and live microorganisms that are pathogenic to animals. The Law on Plant Protection Service prohibits the importation of plants, plant products, and other objects that are carrying pests.

Albania requires the importers of certain products to be licensed, for reasons including sanitary and phytosanitary (SPS) objectives, national security, protection of the environment, and compliance with obligations under international conventions. Documentation and other information must be presented in order for the license to be granted. If these requirements are met, the license is granted automatically. Exceptions apply to narcotic and psychotropic drugs, and to certain flora and fauna, for which quantitative restrictions apply, as well as to medicinal products, military goods, and fish and fish products, where licensing is not automatic.

9.5.2 Preferential trade regime and its likely outcomes

The conclusion of the Free Trade Agreement (FTA) between Albania and FYROM on 29 March 2002 marked Albania's shift to a preferential trade regime. Now, Albania has four FTAs: with the EU as part of a Stabilization and Association Agreement; the CEFTA (Central European Free Trade Agreement)-2006 FTA with Bosnia and Herzegovina, Croatia, Kosovo,

FYROM, Moldova, Montenegro and Serbia; the FTA with Turkey; and an FTA with European Free Trade Association (EFTA) countries. The driver behind the shift toward the preferential trade regime has been rather political. It has been motivated by the EU policy of trade integration within Balkan countries, and their perspective of becoming part of EU economic space when integration criteria are met.

Albania grants tariff preferences to EU Member States. According to the latest Albania Trade Policy Review (WTO, 2010), agricultural products or groups 1 to 24 of HS face an average tariff of 4.4% which is lower than the average MFN tariff rate of 7.6% (Table 9.10). Similarly, agricultural goods (WTO classification; refer to note at Table 9.8) from the EU are subject to an average rate of 4.9% compared with an average MFN tariff rate of 8.8%. Albania has agreed to eliminate customs tariffs for products defined as non-sensitive products, gradual liberalization for moderately sensitive products until 2010, and constant customs protection for highly sensitive products, and duty free quotas for common wheat and wine. The access granted to Albanian agricultural products is asymmetric in favour of Albania. The Interim Agreement (Council of the European Union, 2006) and the Stabilization and Association Agreement grant duty and quota free access of Albanian agricultural products

Table 9.10. Albania's preferential customs tariffs, 2009 (%). Source: WTO (2010).

Description	MFN[a]	EU	Turkey	CEFTA[b]	Bosnia	Croatia	Kosovo	FYRM[c]	Moldova	Montenegro	Serbia
Total	5.2	1.0	1.8	1.3	1.3	1.4	0.0	1.1	1.3	1.6	1.6
HS[d] 01–24	7.6	4.4	7.4	5.4	5.4	5.8	0.0	4.5	5.4	6.8	6.5
HS 25–97	4.5	0.0	0.2	0.1	0.1	0.1	0.0	0.0	0.1	0.1	0.1
WTO											
Agriculture[e]	8.8	4.9	8.4	6.3	6.3	6.8	0.0	5.2	6.4	7.9	7.5
Animals and products thereof	9.4	7.8	9.5	7.0	7.1	7.8	0.0	7.3	7.1	9.3	9.3
Dairy products	9.7	6.5	9.8	9.7	9.7	9.7	0.0	8.1	9.7	9.7	9.7
Coffee and tea, cocoa, sugar	9.3	3.2	9.0	5.1	5.1	3.6	0.0	3.5	5.1	6.0	5.1
Cut flowers, plants	6.8	1.3	7.1	0.9	0.9	1.9	0.0	2.9	0.9	2.2	0.9
Fruit and vegetables	11.8	5.7	11.3	10.2	10.2	11.1	0.0	10.8	10.4	11.3	11.0
Grains	2.0	0.5	2.0	1.7	1.7	0.0	0.0	1.4	1.7	0.4	0.4
Oil seeds, fats and oils, and their products	2.0	0.4	1.8	1.0	1.0	0.0	0.0	0.9	1.0	1.8	1.1
Beverages and spirits	11.1	9.5	10.7	5.5	5.5	11.1	0.0	0.0	5.5	11.1	11.1
Tobacco	7.3	2.6	5.5	7.3	7.3	4.9	0.0	7.3	7.3	7.3	7.3
Other agricultural products	5.2	1.0	4.3	4.2	4.2	2.1	0.0	2.2	4.2	4.4	3.6

[a]Most favoured nation; [b]CEFTA, Central European Free Trade Agreement; [c]FYROM, Former Yugoslav Republic of Macedonia; [d]HS, Harmonized System; [e]WTO agriculture refers HS chapters 1 to 24 (excluding fish and fish products), and a number of manufactured agricultural products.

to the EU market, with the exception of live bovine animals and meat of bovine animals, and sugar and wine; for sugar and wine, tariff quotas were agreed.

The preferential trade regime does not seem to have had any substantial impact on Albania's trade with the EU. A close look at trade data highlights nearly the same trend before and after the Interim Agreement entered into force. While trade flows increased 1.4 times from 2001 to 2005, they increased 1.5 times from 2006 to 2010 (ITC, 2012). It becomes obvious, therefore, that there is no visible change in the trend between before and after the preferential trade regime become effective.

A rather moderated preferential agricultural trade regime has been granted to CEFTA-2006 countries. Goods classified as agricultural face an average tariff rate of 6.3%, which is moderately lower than the MFN average tariff rate of 8.8% for all countries. Only trade with Kosovo is fully duty free (WTO, 2010). For its sensitive agricultural products, Albania has granted tariff quotas. Similar preferences—both tariffs and tariff quotas—have been granted by CEFTA-2006 to Albanian agricultural products (CEFTA, 2006). The preferential trade regime with—and among—CEFTA-2006 countries has resulted in more trade among these countries (Table 9.11), and its impact has been important for these countries. Trade flows during the period when CEFTA-2006 took effect (2006–2010) increased significantly, reaching US$132 million in 2010, which is nearly ten times the 2001 flow of US$13.8 million. Agricultural trade with CEFTA-2006 countries has increased faster than trade in agricultural products with the rest of the world (Table 9.11). The process of a larger integration perspective (free movement of people, services and capital) among CEFTA-2006 countries and with the EU has most probably reinforced the impact of the preferential trade regime.

A less preferential trade regime has been granted to Turkey. Agricultural imports from Turkey enter Albania at an average tariff rate of 7.4%, which is only slightly lower than the average MFN rate for all countries of 7.6% (Table 9.10). The FTA has not had any positive impact on trade between the two countries, which even contracted right after the preferential trade regime entered into force in 2008 (ITC, 2012).

While Albania trades now more with the countries of the Balkan region (CEFTA-2006 countries), it is also experiencing a growing trade deficit with these countries. Data show that while Albanian exports with the Balkan region have stagnated, ranging from US$3 to

Table 9.11. Trade flows with CEFTA[a]-2006 countries before and after free trade agreement. Values in US$ (millions). Source: ITC, and our own calculations.

Trade flows/exports/imports	Year									
	2001	2002	2003	2004	2005	2006	2007	2008	2009	2010
Exports to CEFTA-2006 countries	3.4	3.7	3.9	10.8	10.3	11.1	15.8	10.7	5.1	10.4
Imports from CEFTA-2006 countries	10.4	13.7	10.9	24.8	24.5	42.2	55.6	74.7	71.2	121.5
Trade flows with CEFTA-2006 countries	13.8	17.4	14.9	35.6	34.9	53.3	71.4	85.5	76.2	131.9
Albania's agricultural trade balance with CEFTA-2006 countries	−7.0	−10.0	−7.0	−14.0	−14.2	−31.0	−39.7	−64.0	−66.1	−111.1
Agricultural trade flows worldwide	288.3	329.2	407.3	498.2	514.0	609.2	776.8	960.1	867.2	925.8

[a]CEFTA, Central European Free Trade Agreement.

US$15 million in the last 11 years, Albanian imports from the same region have increased drastically. They reached more than US$120 million in 2010, a 12-fold increase compared to the US$10 million of imports in 2001 (Table 9.11). The very negative trade deficit shows that Albania has yet to exploit its agricultural export potentials.

9.6 Government Strategies to Improve Trade Balances

Albanian agriculture has contributed heavily to the country's trade deficit. Addressing the negative impact inflicted by agriculture on the national economy should be a genuine concern for any government. Therefore, compiling a working trade balancing strategy is a 'must' for the Albanian government. Evidence seems to show that while there are no countries that are both trade protective and developed, trade liberalization alone is not sufficient for economic development. Trade liberalization might even be detrimental if a solid strategy for improving trade balances were not designed and implemented. Therefore, trade liberalization should be complemented by striking the right balance between export enhancement and import substitution. Additionally, reducing the costs of border crossing or trade facilitation is critical to realizing the potentials of free trade.

9.6.1 Trade liberalization versus trade facilitation

Trade liberalization has been considered as a necessary condition for Albania's development, but it is not a sufficient condition. The high costs faced by businesses in border crossings (costs to move merchandise from one country to the other) motivated the Albanian government to participate in a concerted effort with countries in South East Europe (SEE) aimed at reducing border crossing costs through a trade facilitation policy.

Albania has adopted a liberal trade regime from the very beginning of its economic transition toward a market economy. Several factors contributed to the Albanian government making this choice. The first factor was related to the ways in which the transition was conducted; the 'shock therapy'—wiping up everything considered as a legacy from the former communist regime—promoted by the first democratic government, resulted in the collapse of almost all inherited businesses, including destruction of farm buildings, agricultural machines, and fruit tree plantations. There was therefore not much to be 'protected'. The second factor was related to the pressure from the International Monetary Fund (IMF). The IMF has played a crucial role in Albanian economic policy since the beginning of transition. With its liberal philosophy, and its program of structure adjustment, the IMF challenged the Albanian government to gradually lower custom tariffs and remove the others barriers to trade. The third factor was institutional. Albanian's young customs institutions were not initially able to justly implement a high rate of customs taxation. There is evidence that high rates of taxation led to smuggling and, therefore, to unfair competition for businesses. The accession of Albania to the WTO in 2000 was a major step in the trade liberalization process—it completed the liberalizing 'picture', with trade in goods, services, and intellectual property rights, and added the 'rules of the game' of international trade. Finally, the CEFTA FTA and the FTA with the EU deepened the trade liberalization of Albania. Trade liberalization was therefore justified on political, economic, institutional, and geopolitical grounds.

While Albania's recent history has imposed a liberal trade regime, trade flows were often interrupted by administrative barriers. Business surveys have indicated the difficulties that small and medium sized enterprises (SME) were facing in complying with rapidly changing laws and regulations, particularly as SEE countries progressively aligned their legal framework to European Commission (EC) requirements (Ollivier and Gotcheva, 2003). These changes, combined with insufficient information channels, led to significant increased transaction costs at borders, and lack of predictability in transit times toward the EC market. Increased transaction costs often took the form of delays and frequent 'facilitation' payments, sometimes justified, sometimes not.

In order to fight corruption and reduce transaction costs, Albania was one of the six SEE countries which benefited from the Trade and Transport Facilitation for South Eastern Europe (TTFSE) project, which aimed to reduce non-tariff costs for trade and transport, reduce smuggling and corruption at border crossings, and strengthen and modernize the customs administration and other border control agencies. Main project activities included support for physical improvements at border crossings, technical assistance to strengthen the customs administrations, computerization of procedures at the border crossings, electronic filing of customs declarations, and improved exchange of information between the border control agencies and the business community. The project resulted in rehabilitation of the border crossing point infrastructure. Additionally, the implementation of the Integrated Customs Information System contributed to accelerating the speed in processing customs clearance, and the faster transmission of statistics and risk analyses (World Bank, 2005).

The Albanian Trade and Transport Facilitation Body (ALBAPRO) was established in 1998 to foster cooperation and constructive public–private sector dialogue on international trade and transportation issues. All trade-related government agencies ensured high-level participation in ALBAPRO, thereby emphasizing the importance of cooperation and dialogue among stakeholders. Initially, ALBAPRO was hosted and chaired by the Albanian government. Later, the ALBAPRO's leadership role was transferred to private sector, as in other regional 'PRO' committees.

TTFSE supported ALBAPRO in achieving its objectives, but once TTFSE ended, ALBAPRO was unable to carry out all of its statutory objectives, such as designing specialized projects, formulating trade facilitation programs, conducting professional studies, and disseminating concrete measures for trade facilitation. Currently, ALBAPRO is rather nonexistent, mainly due to lack of commitment on the side of the private sector. The fading of ALBAPRO has left Albanian business without a much needed 'forum', whose main aim was to improve coordination among all actors involved in border crossing.

9.6.2 Export promotion versus enhancement of export competitiveness

Export promotion is a strategy for economic development that stresses expanding exports, often through incentive programs designed to attract more firms into exporting by offering help in product and market identification and development, pre-shipment and post-shipment financing, training, payment guaranty schemes, trade fairs, trade visits, foreign representation (Web Finance, Inc., 2012). It is a relevant strategy when a country already has a rather consolidated export oriented sector, but this does not seem to be the case for Albania. This is the reason why the Albanian government has focused more on agricultural competitiveness enhancement than on export promotion, while including in this strategy some policies related to export promotion.

Based on its export strengths (Box 9.4), Albania has the potential for developing exports in the following agricultural and food products: early and late season fruit and vegetables, nuts, preserved products such as canned and dried fruits and vegetables, fresh fish and processed fish products, and MAPs. Albania has favorable natural and climatic conditions, such as fertile land, and suitable climatic conditions for growing early and late vegetables, which, combined with labor costs that are still low, make labor intensive activities quite beneficial. The proximity to the EU market, and the existing infrastructure for sea transport, make export really feasible in terms of transport costs. Good reputation as business partner in selected sectors (MAPs) also represents an 'asset' to capitalize on. Albania's proximity to the EU market, and the adoption in Albania of the *acquis communautaire* (approximating legislation to that of the EU), makes European export oriented inflow of foreign direct investment (FDI) a real opportunity.

However, a number of important weaknesses (Box 9.4) in production and logistics at the private actors' level, in business oriented services, and at the policy level prevent Albania

Box 9.4. Albania's export potential: strengths and weaknesses.

Strengths	Weaknesses
• Favorable natural and climatic conditions (e.g. for early and late vegetables)	• Small farm size, and small business size in general
• Good reputation as business partner in selected sectors (medicinal and aromatic plants—MAPs)	• Inadequate and often obsolete technology
• Proximity to the EU markets for technology transfer, subcontracting, foreign direct investment (FDI), and exports	• Low though increasing labor and capital productivity
• Existing facilities for storage and/or processing	• High unit production cost
• Existing infrastructure for sea transport	• Insufficient knowledge of modern methods of production
• Liberal regulations for FDI in agro-processing	• Inadequate skills for strategic management of companies and marketing
• Low labor costs	• Low product quality lagging behind international standards
• Under-exploited possibilities for cost reduction and capacity utilization	• Inadequate entrepreneurial culture in market and export orientation
	• Limited financial resources for investment and running costs
	• Limited access to information on modern production technologies and markets
	• Limited access to credit

from turning its export potentials into reality. The small farm size represents a major constraint for the development of Albanian agriculture, including the development of agricultural exports. The average farm size in Albania is only 1 ha (MAFCP, 2011). The small farm size affects technology adoption, production costs, produce consolidation costs, and reliability of export supply. Farmers lack knowledge on new production and post harvesting technologies, marketing, and management, and the public extension service does not have the human, infrastructural or financial capacities to support a large number of very small farms in terms of both advice and information. Access to capital is extremely difficult as well.

There is enough understanding on the part of the Albanian government that, nowadays, there are the systems that compete with one another rather than individual companies. The system in which individual companies operate is critical for their export performance, especially when the export sector is represented by small to medium sized companies, which is the case of Albania. It is important, therefore, that for Albania to improve its exporting capacity, it has to invest in the three pillars of systemic competitiveness, namely, at the level of: (i) private actors'—product design, quality and presentation, realistic price calculation, continuity and reliability of supplies, etc.; (ii) business services—consultancy and financial services, research and development technologies, marketing information, etc.; and (iii) policy and administrative framework conditions—legislation, law enforcement, and bureaucratic procedures and infrastructure. Such investments would facilitate the development of the domestic industry.

For the first pillar, the Albanian government (Ministry of Economy, Trade and Energy, METE) has designed an export enhancement strategy (METE, 2007) that aims at improving competitiveness by supporting actors and business services, and improving the conditions of policy and administrative frameworks. Support to private actors includes: capacity building aiming at improving marketing systems and product quality management; support in setting up agricultural producers' groups according to the EU model; assistance in establishing partnership schemes between local and international companies; support to develop promotional products and business contacts; and the introduction of information and communication technology (ICT), and e-business models.

The second pillar of competitiveness enhancement—after support to private actors—is support to business services. Strengthening the Albanian Investment Promotion Agency (AIPA) is considered critical in terms of provision of high quality consultancy, and tailored training and information services for export oriented businesses. While support to the AIPA is a rather cross-cutting policy, because it is supposed to serve all sectors, setting up collection centers of agricultural produce close to farmers directly affects agricultural exports. Investing in market infrastructure (collection centres) is a recognized priority because they are instrumental in improving the marketing of products, and reduce the cost of supply consolidation from many small farms. Further, programs to strengthen the capacities of institutions (both public and private) that deal with the certification of product quality are important to address the weaknesses that have been identified (Box 9.4) in terms of raw materials, production processes, final products, and storing and marketing conditions. Finally, improving access to credit through a loan guarantee scheme is considered a key factor to improve export financing.

The third pillar of competitiveness enhancement—improving policy and administrative conditions—is critical for a country in transition. Albania's main trading partner is the EU. Additionally, its main trading partners are also neighboring countries, all of which are adopting EU legislation. Therefore, harmonizing domestic legislation in terms of export financing and standards with EU legislation is indispensable for increasing Albania' exports both to the EU and to other EU candidate countries. Harmonizing national legislation with that of the EU is closely related to establishing institutions according to the EU model. Institutions that directly affect agricultural exports are those in charge of the registration of food production units, namely animal farms, crop farms, and food processing entities. Structural improvement is, however, needed to increase Albania's exports. To deal with structural improvement, two major policies have been designed: namely, establishing technology transfer and innovation services in order to enable the modernization of the exporting manufacturers' production technology; and establishing the Albanian Competitiveness Fund, which is designed to support potential exporters.

9.6.3 Enhancement of export competitiveness versus import substitution

Albania's export performance for the whole transition to a market economy has been quite poor. Despite the Albanian government's export competitiveness enhancement policies, the country's exports are quite modest. As previously discussed, the import to export ratio for the whole transition period has been high, reaching 9.4:1 in 2010. Therefore, a policy of trade balance improvement cannot rely on export enhancement only, at least in the mid-term perspective.

The Albanian government has favored an import substitution policy emphasizing replacement of some agricultural imports by encouraging local production for local consumption, rather than producing for export markets. Though not explicitly formalized in strategic documents, the government has clearly adopted an import substitution policy. The priority sectors included in the Strategy of Agriculture and Food (MAFCP, 2007) are horticultural products, meat and milk, grapes, and olives, and related processing industries. Additionally, the four sectors to support in the Rural Development Programme (MAFCP, 2010) are fruit, vegetables, meat, and milk. While only the horticulture sector obviously has an export potential (early and late vegetables, and nuts), the remaining priority sectors are intended to produce for the domestic market.

An agricultural import substitution policy for Albania is, therefore, more feasible than an export enhancement policy. Production costs for products sold in the domestic market are lower than imports as they avoid transport, taxation, and other costs associated with the export market. Further, while supplying export markets would require large quantities of a uniform quality of product, and timely provision, supplying the domestic market is easier because of the smaller quantities required at each type of market (roadside, city markets, and larger outlets) owing to buying and consumption habits. Urban Albanian consumers, for example, buy smaller quantities, and more often. Finally, the domestic market is less demanding in terms of standards in terms of quality, safety and marketing. Government support policy in favor of products intended for the domestic market has been rather effective, and Albania has substantially reduced import dependence in major agricultural and food products.

9.7 Conclusions

Elements of several trade theories, namely factor endowment, labor productivity, and product differentiation, all contribute to explaining Albania's agricultural trade. The country currently exports products using its abundant production factors. The main export products (MAPs, fish, and vegetables) use unskilled labor and natural resources. Additionally, investment in new technologies has resulted in increased labor productivity in the greenhouse vegetable industry, whose exports are increasing. Finally, relatively sizable exports in the fish processing industry suggest that product differentiation may represent an important factor in an export enhancement strategy.

Albania remains a country with low agricultural exports, and comparatively high food imports. An import to export ratio of 9.4:1 in 2010 suggests that the country's agricultural export strategy requires further development. An export promotion strategy should be best built on revealed comparative advantages. Kokko (2002), summarizing the experience of the export-led growth of East Asian countries, concludes that 'targeting based on current comparative advantages is a more reasonable task than the identification of future comparative advantages'. Horticulture will very likely be an export oriented sector. Albania has gradually shifted to labor intensive products (vegetables and fruit) away from land intensive products (wheat). This justified shift has led to investment in modern technologies in vegetable and fruit tree production. The country's exports are still low, but there are increasing quantities of vegetables exported to neighboring and other European countries. Additionally, Albania has a growing tourism sector favored by its long, beautiful coast, and its Mediterranean climate. Supplying fresh and cooked fruit and vegetables to the Albanian tourism market may represent important export and domestic potentials.

MAPs and the fish processing industry will most probably remain key sectors in the export development strategy. Today, they represent two of the most important export sectors. The MAP sector is expected to develop further, mainly as a result of emerging MAP cultivation, and the potential for attracting FDI in the MAP processing industry. Realizing the limits of harvesting wild plants, farmers have already started to cultivate MAPs. This strategy is favored by high business profitability, a suitable land and labor force in the appropriate areas, and already established export channels. Additionally, fish processing has important development potential because of important natural resources and an already successful product differentiation strategy. It is interesting to highlight that product differentiation in the fish processing sector has been combined with FDI. MIGA (the Multilateral Investment Guarantee Agency of the World Bank Group), in assessing the FDI opportunities for Albania (MIGA, 2006), found that there are important opportunities for attracting FDI in simple processing in the MAP sector, and also in the processing of fish products, including anchovies and sardines. Recent studies have found that FDI inflows lead to more exports. Kutan and Vukšic (2007), studying the relationship between FDI and export performance for 12 European transition countries (central and Eastern European and Baltic countries), found that FDI positively affects export performance. Bucevska (2009) confirmed the positive relationship between FDI and exports in the case of EU candidate countries.

The nut tree sector (walnuts, hazelnuts, chestnuts, and almonds) is another important sector identified by the Albanian government as having export potential. This sector seems to have a number of advantages, including suitable climate and land conditions, high profitability, a large unmet international demand, and relatively low marketing and storage requirements.

While the nut tree sector represents a good opportunity for export development, a gradual approach should be considered, as no prior experience exists in this sector.

Applicable lessons learned from the Asian miracle economies, such as market research and information on export opportunities, training in export related issues, technology transfer, and export financing and insurance (Kokko, 2002), are quite in line with the Albanian export development strategy. These measures need be implemented, however, for the strategy to generate results.

Import substitution is considered by the Albanian government as a policy complementary to export enhancement; both are supposed to improve the agricultural trade balance. For products with less export potential (i.e. with fierce competitors, and requirements for high standards), but with important development potential (factor endowments, and tradition), an import substitution policy is being pursued by the Albanian government. This policy has already produced tangible results. The country has improved its self-sufficiency in fruit, milk, meat, olives, and olive oil. Finally, Albania will continue to import large quantities of food, mainly cereals. As discussed, this is related to both a production and a consumption model; Albania is expected to produce less land intensive products like fruits and vegetables (cereals are relatively more land intensive than fruit and vegetables), and still consume large quantities of cereals.

References

ACIT (2009) *Albania 2008 Trade Report*. Albanian Center for Competitiveness and International Trade, Tirana, Albania.

ACIT (2010) *Albania 2009 Trade Report*. Albanian Center for Competitiveness and International Trade, Tirana, Albania.

Albanian Parliament (2007) *Law No 9817, 22 October 2007, On Agriculture and Rural Development*. Tirana, Albania.

Balisteri, E.J. and Hillberry, R.H. (2006) Trade frictions and welfare in the gravity model: how much of the iceberg melts? *Canadian Journal of Economics* 39, 247–265.

Bank of Albania (2011) Time Series Database of the Bank of Albania. Available at: www.bankofalbania.org (accessed 14 January 2012).

Bucevska, V. (2009) The impact of FDI on export performance: empirical evidence from EU candidate countries. Presented to STCPMs (Special Topic Contributed Paper Meetings) Sessions, STCPM 115 at: *57th Session of the International Statistical Institute, 16–22 August 2009, Durban, South Africa*. Available at: http://www.statssa.gov.za/isi2009/ScientificProgramme/IPMS/0754.pdf (accessed 8 March 2012).

CEFTA (2006) Agreement on Amendment of and Accession to the Central European Free Trade Agreement. Central European Free Trade Agreement Secretariat, Brussels. Available at: http://www.cefta2006.com/cefta-agreement (accessed 9 September 2012).

Christie, E. (2001) *Potential Trade in Southeast Europe: A Gravity Model Approach*. Working Papers 011, The wiiw (Vienna Institute for International Economic Studies) Balkan Observatory, Vienna. Available at: http://balkan-observatory.net/wp/2001%2012%20wiiw%20bo%20wp%20011.pdf (accessed 9 September 2012).

Council of the European Union (2006) *Interim Agreement on Trade and Trade-Related Matters between the European Community, of the One Part, and the Republic of Albania, of the Other Part*. Available at: http://ec.europa.eu/enlargement/pdf/albania/st08154.06_en.pdf (accessed 9 September 2012).

Grubel, H.G. and Lloyd, P.J. (1975) *Intra-Industry Trade: The Theory and Measurement of International Trade in Differentiated Products*. Macmillan, London.

IMF Albania (2011) *Albania: Staff Report for the 2011 Article IV Consultation*. International Monetary Fund, Washington, DC. Available at: http://www.imf.org/external/pubs/ft/scr/2011/cr11313.pdf (accessed 9 September 2012).

ITC (2012) Trade Map. International Trade Centre, Geneva, Switzerland. Available at: www.trademap.org/Bilateral_TS.aspx (accessed 9 September 2012). Available at: http://www.docstoc.com/docs/34757214/Interim-Report-IR-01-050October-Export-Led-Growth-in-East-Asia (accessed 9 September 2012).

Kokko, A. (2002) *Export-led Growth in East Asia: Lessons for Europe's Transition Economies*. Interim Report IR-01-050/October. International Institute for Applied Systems, Laxenburg, Austria.

Krugman, P. and Obsfeld, M. (1994) *International Economics. Theory and Policy*, 3rd edn. Harper Collins College Publishers, New York.

Kutan, A.M. and Vukšic, G. (2007) Foreign direct investment and export performance: empirical evidence. *Comparative Economic Studies* 49, 430–445.

MAFCP (2007) *Sector Strategy of Agriculture and Food 2007–2013*. Ministry of Agriculture, Food and Consumer Protection, Tirana, Albania.

MAFCP (2010) *Rural Development Programme 2011–2013. Under the Instrument for Pre-Accession Assistance (IPA)*. Ministry of Agriculture, Food and Consumer Protection, Tirana, Albania.

MAFCP (2011) *Vjetari Statistikor 2010*. Ministry of Agriculture, Food and Consumer Protection, Tirana, Albania.

Mançellari, A. (2005) *Trade Liberalization and FTAs: The Case of Albania*. Working Paper, European Institute, Sofia, Bulgaria. Paper originally presented at the Conference "Trade Liberalization in the Western Balkan Countries", held in Tirana, 21 October 2005. Available at: http://www.westernbalkans.info/upload/docs/1b_Albania_trade.pdf (accessed 9 September 2012).

METE (2007) *Business and Investment Development Strategy (2007–2013)*. Ministry of Economy, Trade and Energy, Tirana, Albania. Available at: http://www.dsdc.gov.al/dsdc/pub/business_and_investment_development_strategy_155_1.pdf (accessed 10 September 2012).

MIGA (2006) *Investment Horizons: Western Balkans. Benchmarking FDA Opportunities*. Available at: http://www.miga.org/documents/Benchmarking_study_Balkin.pdf (accessed 9 September 2012).

Ministry of Finance (2011) *Buxheti ne Vite*. Tirana, Albania. Available at: http://www.minfin.gov.al/minfin/Buxhetinevite1268 1.php. (accessed 9 January 2012).

Ollivier, G. and Gotcheva, K. (2003) TTFSE.ORG: adding value for business through transparency. In: Cosgrove-Sacks, C., Apostolov, M. and Tomkova, J. *Sharing the Gain of Globalization in the New Security Environment: The Challenges to Trade Facilitation. Second International Forum on Trade Facilitation, 14–15 May 2003*. UN Economic Commission for Europe, Geneva, Switzerland, pp. 253–258.

Pere, E. (2009) The significance of economic trade relation[s] for regional integration in the West Balkans. *Geographica Pannonica* 13, 159–165. Available at: http://www.dgt.uns.ac.rs/pannonica/papers/volume_13_4_4.pdf (accessed 9 September 2012).

Qasmi, B.A. and Fausti, S.W. (2001) NAFTA intra-industry trade in agricultural food products. *Agribusiness* 17, 255–271.

UN COMTRADE (2012) United Nations Commodity Trade Statistics Database. Available at: http://comtrade.un.org/db/ (accessed 9 September 2012).

WebFinance, Inc. (2012) BusinessDictionary.com. Available at: http://www.businessdictionary.com/ (accessed 9 September 2012).

World Bank (2005) *Implementation Completion Report (IDA-34270 PPFI-Q2290) on a Credit in the Amount of US$8.6 Million to Albania for the trade and transport facilitation in Southeast Europe project, October 12 2005*. Report No: 31710, World Bank, Washington, DC. Available at: http://www-wds.worldbank.org/external/default/WDSContentServer/WDSP/IB/2005/10/17/000160016_20051017161308/Rendered/PDF/31710.pdf (accessed 9 September 2012).

WTO (2010) Trade Policy Review. Report by the WTO (World Trade Organization) Secretariat, Albania.

WTO (2012a) Trade Profile: Albania. World Trade Organization, Geneva, Switzerland. Available at: http://stat.wto.org/CountryProfile/WSDBCountryPFView.aspx?Country=AL&Language=F (accessed 9 September 2012).

WTO (2012b) Trade Profiles: Albania, Bulgaria, FFYR Macedonia, Hong Kong, China, Singapore, Slovenia. Available at: http://stat.wto.org/CountryProfile/WSDBCountryPFView.aspx?Language=E&Country=AL,BG,MK,HK,SI,SG. (accessed 10 September 2012).

WTO (2012c) Glossary. A guide to "WTO speak". World Trade Organization, Geneva, Switzerland. Available at http://www.wto.org/english/thewto_e/glossary_e/glossary_e.htm. (accessed 9 February 2012).

Xhepa, S. and Agolli, M. (2005) *Albania's Foreign Trade through a Gravity Approach*. Discussion Paper, Albanian Centre of International Trade, Tirana, Albania.

Acronyms and Abbreviations

ACIT	Formerly the Albanian Centre of International Trade, now the Albanian Centre for Competitiveness and International Trade
ALBAPRO	Albanian Trade and Transport Facilitation Body
AMS	Aggregate measurement of support
CEFTA	Central European Free Trade Agreement
EC	European Commission
EFTA	European Free Trade Association
EU	European Union
FDI	Foreign direct investment
FYROM	Former Yugoslav Republic of Macedonia
GDP	Gross domestic product
HS	Harmonized Commodity Description and Coding System
ICT	Information and communication technology
ILO	International Labour Organization
IMF	International Monetary Fund
ITC	International Trade Centre
MAFCP	Ministry of Agriculture Food and Consumer Protection, Albania
MAPs	Medicinal and aromatic plants
METE	Ministry of Economy, Trade and Energy, Albania
MFN	Most favoured nation
MIGA	Multilateral Investment Guarantee Agency of the World Bank
SEE	South East Europe
SME	Small and medium sized enterprises
SPS	Sanitary and phytosanitary
TTFSE	Trade and Transport Facilitation for South Eastern Europe
UN COMTRADE	United Nations Commodity Trade Statistics Database
USA	United States of America
VAT	Value Added tax
WTO	World Trade Organization

10 Albania's Agriculture Sector: Trends and Projections

Luciano Leonetti
Development Solution Associates, Tirana, Albania

10.1 Main Features and Trends in the Evolution of the Agro-food Sector

When analyzing the evolution of the Albanian agro-food sector over the last 20 years, one often has the impression of looking at consecutive waves of revolutionary changes and recurrent reforms, which keep changing the structure of the sector and its governance, but eventually do not have much impact on the final result. This is a much fragmented sector characterized by low productivity, few investments, and focus on maximizing short-term marginal profitability. In fact, in spite of favorable natural resources potential, the proximity of countries which are major markets for exporters of Mediterranean food products and long-lasting support from international development organizations and projects, the sector does not succeed in achieving even a neutral balance in international food trades, which could be considered a modest objective for a prevalently rural country. Looking at the sub-sectors, the picture becomes more complex and even more puzzling, as the most successful business models are those food processing industries that are totally disconnected from domestic primary production, and those enterprises focused since the very beginning on export to a demanding and extremely competitive international market.

To understand the deeply entrenched constraints that hamper sector development, and the factors that prevent the establishment of complex supply chains within the country, this chapter will try to show how the substantial changes experienced by Albania in the last 20 years have decisively conditioned the trends of the agro-food system in terms of production, consumption, and trade balance.

In terms of trends, the period from 1991 to the present can be roughly divided in two segments. The first, from 1991 to 2005, saw dramatic socioeconomic changes and recurrent crises, which created the conditions for the present trends. In the second period, from 2005 to present, total supply trends became more flat and stable, with demand starting to follow patterns previously recorded in other European markets. The transition from the first to the second period was gradual, and the first decade (1991 to 1999) witnessed major fluctuations and sharp inter-annual changes in the production, consumption and trade of different food products, while the years from 2000 to 2005 have seen the gradual establishment of a more stable growth pattern, setting the conditions for the more stable growth in the following period. In the second period, from 2005 on, production in all agro-food sub-sectors grew, although at a different pace among sectors. Demand increased and changed, too, as the market became more segmented and lifestyles evolved. The production of some domestic products kept the pace of these changes, while that of others lagged behind. Imported products maintained, and, in some cases, reinforced their dominating position in the end-up segment of many food products. In non-food products, such as wood, domestic producers have so far generally managed to be competitive only in some segments, leaving to imported products the most valuable segments of the market.

A few major events occurred in the first years of transition; also, some dynamics that started in these years had a lasting and decisive impact on the evolution of the food sector and food market in Albania. The first of these was the collapse of the supply chains established during the planned economy, immediately followed by a fast, but chaotic, process of land privatization. Next, the internal market was opened quite early to international trade. These issues were compounded by the rapid and substantial changes in the social structure of the country, characterized by major migration flows—the total population decreased from 3.18 million to 2.83 million inhabitants between 1991 and 2011 (censuses 1991, 2001, and preliminary results of 2011 census from INSTAT—the Albanian Statistical Institute), by urbanization (between 1991 and 2011, the share of the urban population increased from 35.7% to 53.7% of the total population), and concentration of the residents in the Tirana–Durres urban and peri-urban areas. These factors contributed to the creation of a weak and fragmented food production sector, induced deep changes in consumption lifestyles, and made Albanian consumers knowledgeable and accustomed to a much wider range of imported food products and brands. Among the above mentioned events, privatization had the most immediate and disruptive impact on the agro-food sector. In fact, the end of the socialist system and a hasty privatization of lands, processing units, and other assets prompted the collapse of all the previously established supply chains, leading to a system based on self-sufficiency farming and elementary trades of food products. Organized food processing was discontinued, so that consumers had to use either imported products or basic processed products, self-produced or purchased from farmers. A handful of processing plants and logistics facilities somehow managed to continue their activity on a much smaller scale, in most cases dealing with imported half-finished products.

The first value chain segments that succeeded in recovering in the years between 1991 and 1997 can be divided in three categories: (i) the simplest primary production, including vegetable production, small ruminant breeding and, up to a limited degree, dairy cattle breeding at subsistence level; (ii) those agro-industrial activities which could rely on a good domestic demand, and could easily access cheap imported raw food products, such as the milling industry (based on imported flour and grains), and the meat processing industry (mostly based on frozen beef and pork meat); and (iii) the supply chains that could source raw materials directly from natural resources, such as marine fish catching, and the harvesting of wild medicinal and aromatic plants (MAPs). These last supply chains were generally export oriented.

Recovery in primary production was the result of the uncoordinated activity of a very large number of farms, generally very small and poorly equipped in terms of financial, technical and human resources, and usually having the following characteristics: (i) a short and simple supply chain; (ii) able to provide basic food products suitable for consumption with little or no processing; and (iii) being competitive even under conditions of minimum capital and a small area available for cultivation and/or pasture. Semi-commercial poultry breeding was the only supply chain dealing with primary production and requiring more capital and organization, and this rapidly rebounded after 1991.

Agro-food processing activities using imported raw materials or half-finished products and export-oriented supply chains based on natural resources require more capital and skills than basic farming. These barriers to entrance were not very high in the first three to four years of transition, as there were no more dominant players (because the large state food processing industries collapsed after privatization) in the market, and the regulatory framework was reduced to a minimum and weakly enforced. However, insiders had a definite advantage in getting into complex businesses, so that the first entrepreneurs dealing with the above mentioned activities were usually specialists in the relevant sectors, who had already gained experience and established trade relations as high-ranked civil servants or directors of state companies.

A second sharp economic downturn occurred in 1997, with the collapse of 'pyramids', i.e. a large-scale financial Ponzi scheme that affected most households, resulting in turmoil, and collapse of the political establishment and the economy. This event had relatively limited effects on primary production, but heavily affected the food industry, destroying most of the investments made in the early 1990s. Once again, the companies that survived this crisis relatively well were those that were export oriented (such as those involved in exporting fish and MAPs), and those able to produce cheap food while having good profit margins, as was the case for some meat processing enterprises, broiler farms, and the milling industry. Most of the few food processing industries which survived the 1997 turmoil are now among the market leaders in their group of products.

The real development of the food market started with and immediately after the Kosovo war (i.e. in the years 1999–2000), when political stability increased and purchasing power started to grow. The large pool of emigrants played a major role in the evolution of market and lifestyles, through the large flows of remittances, and the introduction of new consumption patterns and of a much larger variety of food products and preparation recipes. Urbanization also led to more homogeneous consumption patterns; in the past the daily diet had been largely conditioned by the range of locally available products.

The trends emerging in the structure of demand included an initial and generalized increase of the per capita consumption of food, followed, after 2007, by the gradual leveling up of total quantities, and increasing awareness of the health risks related to excessive consumption of fats of animal origin, meat, and sugar. At the same time, a process of market segmentation was started, as consumers became selective, thanks to increasing revenues (giving more possibility to choose) and to the possibility of access to a wider range of products. These changes in consumer behaviors and focus followed the same patterns of postwar food consumption trends in southern Europe, but at a much faster pace.

In parallel with the increase in total demand, the composition of the diet changed, too. Consumption of fish and seafood increased more quickly than that of meat. Preferences for types of meat also changed, with increased demand for beef and poultry meat and cold cuts, and stagnating demand for lamb and mutton meat. At present, demand for chicken meat, beef, and processed meat is still growing. Demand for fruit and vegetables and dairy products is also quite high, but the range of products in demand is still mostly limited to the most common fresh produce (apples, oranges, tomatoes, etc.), and to products which are part of the traditional diet, such as okra. The successful introduction of a few new products, such as rucola (arugula, or rocket) and kiwi, does not change this overall picture.

Consumption of bread always was and remains high, though demand for other staple foods is changing. Consumption of pasta and potatoes is increasing, demand for rice is stable, and consumption of other kinds of cooked cereals (whole wheat grains, corn porridge, etc.) has almost disappeared. The use of wheat flour-based staple foods is also changing. The traditional burek (a kind of pie) is still extremely popular, but is now facing competition from other very popular preparations, such as pizza.

In parallel with the evolution of the structure of consumer preferences, market segmentation started as well. This development was the result of increasing purchasing power, especially in urban areas, and of the possibility of access to a much larger choice of products of different quality and price. In the past, high-end products largely coincided with imported ones. Segmentation is now more articulated, with widening price differences for different qualities, a wider variety of prices and brands for the same product, and also different prices based on product orgin. The market segmentation process makes import substitution more difficult, as Albanian producers provide, in most cases, only a limited range of basic products, so that most products with higher value added are still imported products. However, this process could also eventually prove to be a drive for the development of Albanian niche products, and for the growth of small-sized, quality oriented domestic producers, as, for comparable qualities, local products are usually preferred to imported products. Moreover, in general, all the consumer surveys done in the last few years show that Albanian consumers have a strong preference for local products (e.g. Chan-Halbrendt *et al.*, 2010b, in a study of wine preferences; DSA, 2010, in a study of lamb meat preferences).

A powerful drive for change is given by the expansion of supermarket chains, which have substantially widened the range of choices for consumers and have also pushed traditional retail toward specialization. The first large supermarket (i.e. with an area greater than 1000 m^2) was opened in 2006, when the first commercial center was built. At present, there are two main supermarket chains, both foreign owned. Other smaller local supermarket chains and smaller chains belonging to international groups are also present. Supermarkets are providing a substantial contribution to widen consumer choice, introducing a much larger variety of products and brands, contributing to the growth of demand for frozen food and other food processed products, and also pushing the food industry towards a more clear-cut dualism between enterprises that supply the supermarkets and the others, which do not have the size, the financial means or the quality standards to become stable partners of supermarket chains. Even if the share that supermarkets have of total retail sales is still small, their increasing importance is contributing to a consolidation of the food processing sector faster than in the recent past. The development of fast foods and restaurants has also contributed to changing the structure of demand, and to increasing demand for frozen products and for products such as seafood, that most consumers are not accustomed to preparing properly at home, as they were not part of traditional diet in most of the country.

The process of European Union (EU) integration has also been a driver for the evolution of the Albanian food sector. In the past few years the influence of this driver has been mostly felt in the changes introduced in the Albanian legal framework on food safety, labeling and control of food-related health risks. In the near future, the influence and focus of the EU integration process will increasingly shift toward the improvement of competitiveness of the largest farms and food processing units, as the EU Instrument of Pre-Accession Assistance for Rural Development (IPARD) will be made available. IPARD is a financial facility for the development and modernization of commercial-sized primary producers and food processing enterprises, and its approach consists of supporting a clear-cut separation between commercial agro-food enterprises and subsistence farms and artisan food processing, which are supported using other tools, such as regional development funds and rural livelihood support programs.

The introduction of EU-harmonized laws on food safety and the availability of financial instruments to increase compliance of commercial enterprises with such standards represent a major challenge for small producers and processing industries. Food producers will be increasingly pushed to speed up the process of scaling up their size and improve their standards, the alternative being to be forced out the market.

After years of gradual and relatively slow adaptation to the changing markets, Albanian operators in the food sector are now increasingly feeling the pressure for consolidation into larger and more formalized businesses. In terms of consumption trends, the driver for expansion is rapidly shifting from increasing quantities of per capita food consumption toward the value added per unit of food product (introducing higher quality food products and partially processed food) and demand for a wider range of products. This shift from quantity to quality and diversification started later than in other transition countries, but is now gaining momentum.

The capacity of Albanian food producers to keep up with the pace of changes occurring in consumer demand and in the retail sector, and their ability to face increasing competition from foreign producers, will be the main challenges of the future. So far, the available information shows both signs of weakness in these aspects, such as the growing gap between food imports and exports and the stable share of imported products in the high-end segment of the market (showing inability to substitute the import of high value-added products), and signs of vitality and strength, which are coming from increasing investments in the agro-food sector and the growth of commercial-sized operators. Notwithstanding several common trends and features, the situation and perspectives are not the same for all the sectors. So the rest of the chapter (Sections 10.3–10.6) presents a more detailed analysis of the main food and non-food sectors of agribusiness in Albania.

10.2 Impact of the Collapse of Planned Economy Supply Chains and Land Privatization on the Evolution of Agribusiness Sectors

The end of the socialist regime that ruled Albania from 1945 to 1991 led to a complete and sudden collapse of state and cooperative enterprises, with substantial changes in the structure of primary production and a regression of food processing to the stage of cottage-level basic processing. The land was almost immediately (July 1991) distributed to rural households resulting in the establishment of a huge number of very small (1.1 ha on average) subsistence farms. Moreover, each beneficiary received more than one plot, located in different and often distant areas, thus making impossible any cost-effective production and use of mechanization. Finally, dissatisfaction with land distribution criteria and conflicts over property rights prevented any private investment in agriculture. This was perhaps of less consequence, however, as the much impoverished population had no resources to invest in more efficient farms.

As a result of the above, production was focused on the self-sufficiency of the rural family, and on labor intensive production requiring minimal inputs and land. The area and output of all industrial and field crops were drastically reduced. In particular, the production of wheat dropped from 621,000 Mt (average output 1988–1990) to 252,000 Mt in 1992, and the production of maize decreased from 256,00 Mt (average output 1988–1990) to 156,000 Mt. Production of cotton (10,200 Mt in 1990) and sugar beet (267,000 Mt in 1990) was phased out. Commercial-scale breeding disappeared and rural families tried to keep a number of animals sufficient to provide milk and eggs for family needs.

Altogether, production of potatoes and vegetables increased: this was entirely due to increased productivity, as the cultivated area remained stable. Fruit and vegetable productivity increased in spite of a drastic reduction in the use and quality of mechanization and agricultural inputs (seeds, fertilizers, pesticides), showing how big the productivity and efficiency problems were that affected state farms and cooperatives. Different factors negatively affected fruit production. Large numbers of fruit trees and vineyards were cut, and production decreased accordingly. Short-term increases in productivity were soon offset by lack of agronomic services and the spread of plant diseases.

The dramatic reduction in the output of cereal crops and the termination of industrial crop production experienced between 1990 and 1992 proved a permanent change, while recovery in fruit production of different fruit crops after 1992 followed different trends. This rebound depended on various factors, including the resilience of different fruit processing activities (for example, jam production was never totally discontinued after 1991, while the production of fruit juices was terminated and not resumed), the impact of plant diseases, which in some areas was devastating, the trends in demand, and the severity of property rights conflicts.

In general, the recovery of production after 1992 followed different trends for different groups of products. The reasons affecting these different trends are still in force and are still valid for estimating future developments (Table 10.1). The main factors affecting these different trends in agricultural food and non-food products are discussed in the following sections.

10.3 Supply Chains Based on Crop Production

10.3.1 Cereals and industrial crops

Cereals

Primary production of cereals and industrial crops is affected by a general lack of competitiveness, owing to the fragmentation of the production base and to the natural conditions being more suitable for other crops. As a consequence, production suffered a sharp drop after 1990 and never recovered, or recovered almost 20 years later, as in the case of maize (Table 10.2).

In the case of cereals, the early opening of the domestic market to international trade exposed local producers to competition with much more efficient producers and traders. The scarce competitiveness of the domestic production of cereals is due to several structural constraints, ranging from the quantity and quality of soils, to the lack of economies of scale caused by a much fragmented production, to the better cost/profit opportunities offered by alternative uses (other agribusiness activities, industrial and commercial establishment, residential use, etc.) of the best land. The factors making domestic production of cereals scarcely competitive in an open market have not changed over the years. As a result, the production of wheat is still 53% less than in the period 1988–1990, and the production levels of maize achieved in the period 1988–1991 were reached only in 2009.

In particular, the production of wheat followed an irregular pattern, characterized

Table 10.1. Main groups of agricultural products and relevant trends. Source: Elaborated FAOSTAT[a] and national statistics.

Product	Main production trend features
Wheat	Declining trend
Maize	Growing; pre-1991 levels reached in 2009
Potatoes	Constant growing trend
Beans	Stable; production higher than pre-1991 since 1992
Vegetables and melons	Constant growing trend
Olives	Constant growing trend
Citrus	Growing; pre-1991 levels reached in 2010
Grapes	Growing; pre-1991 levels reached in 2002
Fruit	Growing; pre-1991 levels reached in 2003
Tobacco	Declining trend
Sunflowers	Very low production after 1997
Cow's milk	Growing until 2005, then constant
Sheep's/goat's milk	Growing until 1995, decline in 1997, then constant
Eggs	Growing; pre-1991 levels reached in 1997
Skins/hides	Growing faster in value after 2005
Fish	Growing; pre-1991 levels reached in 2000
Aquaculture	Growing; pre-1991 levels reached in 2002
Medicinal and aromatic plants	Growing trend until 2008, then declining
Wood	Declining

[a]FAOSTAT, Food and Agriculture Organization Statistical Division Databases.

Table 10.2. Trends in production of main cereal and industrial crops in Albania (thousand Mt). Sources: FAOSTAT[a], INSTAT[b].

Crop	Average 1988–1990	1992	1997	2000	2005	2010
Wheat	621	252	420	341	260	295
Maize	257	156	193	206	220	362
Tobacco	15	13	4	6	2	2
Sunflowers	19	9	1	3	2	3

[a]FAOSTAT, Food and Agriculture Organization Statistical Division Databases; [b]UNSTAT, UN Statistics Division (UNSD) Statistical Databases.

by a sudden drop (from 621,000 Mt average production in 1988–1990 to 252,000 Mt in 1992), followed by a steady recovery (from 252,000 Mt in 1992 to 420,000 Mt in 1997), a second, longer declining trend (from 420,000 Mt in 1997 back to 260,000 Mt in 2005), and a more recent, weak recovery (from 260,000 Mt in 2005 to 295,000 Mt in 2010). The causes of these changing trends are strictly linked to the general evolution of Albanian agribusiness. Early recovery between 1992 and 1997 was first generated by the work of a large number of cash-strapped subsistence farmers willing to satisfy their immediate needs for staple food, and then speeded up in the framework of the short lived investment boom of 1995–1997, which was linked to the rise and fall of the 'pyramids' schemes, i.e. the Ponzi schemes that involved well over two thirds of the total Albanian population, and ended in a financial hole worth more than half of the current Albanian GDP (gross domestic product). The crash of the pyramids in 1997, and the subsequent economic crisis, severely affected all economic activities, including agribusiness. The recovery of agribusiness in the following years did not include the scarcely competitive wheat production sector, with the result that wheat production continued to decline until 2005, when output was still 60% lower than in the period 1988–1990. The limited growth experienced after 2005 does not change the overall picture, as 2010 output was 295,000 Mt, i.e. less than half the average production of the period 1988–1990. Domestic production of maize keeps increasing, mostly in line with the needs of animal breeding; however, maize imports keep growing too.

At the same time as these domestic production problems were occurring, the opening of the market to international trade allowed the establishment of a solid and competitive milling industry, based on imported flour and grains, which evolved into an oligopolistic industry structure, dominated by three large (by Albanian standards) companies (DSE, 2011). These companies gradually gained flour market shares at the expense of imported flour, but most of their production is made with imported wheat, with the consequence that the quantity of imported flour kept decreasing from 1997, while the quantity of imported wheat increased until 2005.

Smaller regional mills did not manage to expand, finding a market niche as traditional service facilities (i.e. milling wheat into flour for local farmers), and as minor competitors in the local markets. More traditional consumption patterns in rural areas (rural dwellers consume more bread than urban consumers) have also contributed to keeping profitable the business of small local mills, whose core business is still the traditional production of wheat flour, while the three main players in the milling industry are seeking new sources of profit, having started to produce a wider range of products, such as pasta.

Imports supplied the missing quantities of cereals and industrial crops, with the value of wheat imports growing from US$14.7 million in 1997 to US$46.6 million in 2005 and US$68.5 million in 2010, and maize imports growing from US$1.5 million in 1997 to US$18.1 million in 2010. (Table 10.3). Exports have been always very low or negligible. The import data also prove the gradual consolidation of the domestic milling industry, as imports of flour decreased over the time

Table 10.3. Trends in imports of cereals and products of the milling industry (million US$). Source: UNSTAT[a].

Cereal import/milling product	1997	2000	2005	2010
All cereals	21.3	26.9	60.4	103.2
Wheat	14.7	17.6	46.6	68.5
Maize	1.5	4.0	6.1	18.1
Products of milling industry	41.3	23.9	11.6	17.4
Wheat flour	39.6	21.1	8.5	12.7

[a]UNSTAT, UN Statistics Division (UNSD) Statistical Databases.

(from US$39.6 million in 1997 to US$12.7 million in 2010), while the import of whole grains increased. The parallel growth of both production and import of maize is linked to the gradual intensification of animal production, which required more animal feed.

The analysis of imports as quantity (Table 10.4) tells a slightly different story, and helps to explain the reversal of trend in wheat production experienced after 2005. In 2005, domestic production covered less than 40% of net supply of wheat and wheat flour. In 2010, this share was more near to 50%. This is partially due to the growth of international prices of wheat, which made it more profitable to produce in Albania, and to the development of consumer preferences, which witnessed a stabilization of urban demand for bread, and an increasing demand for a wider range of products of the milling industry (pasta, biscuits, cakes, etc.).

Future developments will see the total consumption of wheat and wheat flour remaining stable or growing very slowly, with an increasing share of imports of durum wheat, as the local pasta industry will start to substitute parts of the sizable imports from Italy. Imports and production of maize will continue to grow, pushed by demand for animal feed. Domestic production of maize will still rise, but domestically produced maize cannot be competitive with maize from the central European plains, so it is expected that imports of maize will continue to cover 20–25% percent of a growing total demand.

Industrial crops

The production of industrial crops, such as sugar beet, oilseed crops and tobacco, has been drastically reduced (oilseed crops) or practically phased out (sugar beet and tobacco) since 1990, mostly as a result of the collapse of the processing industries. The decline of tobacco production has been slower, as the processing industry survived the first wave of privatization. Imported raw materials (oilseeds) and finished products (raw and refined vegetable oil) are now covering all the needs for oilseed edible oil (excepting olive oil). A couple of relatively large processing plants are processing the oilseed oil (using imported raw material) required by the domestic market, for which demand seems rather stable. Imports are now stabilized at around US$13 million/year. Tobacco is almost completely imported, and sugar is completely imported, both as finished products. Imports of sugar reached US$36 million in 2010 and keep growing at a rate of US$2 million USD/year, while imports of tobacco scored US$48.7 million/year, also showing a steady growth rate of 3%/year. The growing import trend for both products is likely to continue.

10.3.2 Vegetables, potatoes and beans

The production of vegetables and potatoes kept growing after 1991. Bean production, after increasing immediately following land privatization, has remained stable over the last 20 years. Production of vegetables for the fresh market suits well the typical structure of the Albanian small farm, so there is still potential for growth, even if competitiveness in the international market is, and will remain, limited in the present business environment conditions. At the same time, the production of vegetables specifically oriented to industrial processing (i.e. the use of apposite varieties and production practices) did not catch on, owing to farm fragmentation, and lack of interest and capacity in building up producers'

Table 10.4. Trends in imports of main cereals and products of the milling industry (thousand Mt). Source: UNSTAT[a].

Imports of cereals/milling products	1997	2000	2005	2010
Wheat	74.8	190.8	315.0	270.6
Maize	8.5	36.8	51.6	66.3
Wheat flour	140.5	113.0	41.5	40.2

[a]UNSTAT, UN Statistics Division (UNSD) Statistical Databases.

associations and cooperatives, and/or establishing effective contractual agreements between fruit and vegetable processing industries and the large number of small farmers.

At present, Albanian vegetables are almost totally sold on the domestic market. International flows (i.e. imports and exports) are quite limited, representing 5% or less of the sector supply balance (for each product and in all years), with limited differences between different products (5% for potatoes, around 2% for most of the other fresh produce). Table 10.5 shows development in the production of vegetables, potatoes and beans.

The production of potatoes has increased constantly since 1991, even if the pace of growth has been irregular, with a relatively small increase between 1992 and 1997 (from 79,000 to 89,000 Mt) and between 2000 and 2005 (from 161,000 to 169,000 Mt), and faster growth between 1997 and 2000 (from 89,000 to 161,000 Mt) and from 2005 to 2010 (from 169,000 Mt to 208,000 Mt). There is no single reason for the changing speed of potato production growth, rather it is a combination of several factors, such as changes in international prices of potatoes (making domestic production more or less competitive), growth in domestic demand for fresh potatoes (a rather stable trend), changing competitiveness of other crops in key production areas (beans from Korça became less competitive in international markets, thus making potato production more attractive), and increases in production and reduction of losses in different production areas, such as Fier and Dibra.

Bean production has remained almost constant, but this apparent stagnation is the result of a reduction of exports (beans were the most important vegetable exported from Albania until 2005), and an increase in domestic demand.

The production of other vegetables and melons is constantly increasing, with an acceleration after 2005. Over 20 years (1990 to 2010), total output more than doubled. Production increased from an average in 1988–1990 of 364,000 Mt to 565,000 Mt in 1992, 620,000 Mt in 2000, and 860,000 Mt in 2010. In spite of the impressive growth of domestic production, imports of vegetables are growing too, even though they represent a small share of the total supply of vegetables. The vegetable international trade deficit in 2010 was over US$20 million USD, and keeps growing (Table 10.6).

An important feature of Albanian vegetable production is the slow development of greenhouse production. Most of the existing

Table 10.5. Trends in the production of vegetables, potatoes and beans (thousand Mt). Source: INSTAT[a], MAFCP[b] Yearbooks.

Crop	Average 1988–1990	1992	1997	2000	2005	2010
Potatoes	79	79	89	161	169	208
Beans	19	25	18	25	24	24
Vegetables and melons	364	565	590	620	685	860

[a]INSTAT, Albanian Institute of Statistics; [b]MAFCP, Albanian Ministry of Agriculture, Food and Consumer Protection.

Table 10.6. Trends in the international fruit and vegetables trade balance (million US$). Source: UNSTAT[a].

Fresh/processed fruit/vegetables	1997	2000	2005	2010
Vegetables	–3.1	–6.0	–14.6	–22.0
Fruit	–13.0	–26.5	–50.6	–58.5
Processed fruit and vegetables, excluding beverages		–8.3	–14.7	–16.1

[a]UNSTAT, UN Statistics Division (UNSD) Statistical Databases.

greenhouses are concentrated in a relatively small area of central Albania, in the Lushnje and Berat regions. These greenhouses are mostly low-tech plastic greenhouses, operated without a heating system (even if some of them do have such a system). The use of heating systems in existing greenhouses is not economically viable, owing to the rising cost of the gas oil needed for the burners that are commonly available on the market, and to the absence of demand for and supply of alternative heating systems (based on multi-fuel burners, biomass burners, hybrid systems with solar panels and fuel, etc.). Also, the very favorable climatic conditions in some areas of the country make the use of heating really necessary only for 3 months of the year, so most farmers do not find it profitable to invest in heating systems. At present, without an agro-food system strongly organized for exporting in wintertime, and a very efficient distribution system, it is more cost-effective and less risky to have low-tech greenhouses producing for nine months/year than a high-tech greenhouse production system giving high yields all year round, with higher investments and running costs. As a result, even if Albania could have a competitive advantage in greenhouse vegetable production, there is still a supply gap in the winter months for some vegetables (such as tomatoes), which are very important in Albanian customary diet. This gap, presently filled by imports, will be filled by domestic production only if the distribution and export system becomes much more efficient, making it profitable to invest in greenhouses with higher technology and more complex (and expensive in terms of investment) heating systems.

The slow trend in the growth of low-tech greenhouse surface is therefore likely to continue, but the patterns (high concentration in the areas with better climatic conditions and at the same time near to the main wholesale markets, low technology, absence or non-use of heating systems) are also likely to remain unchanged in the short-term. In a longer term perspective, the gradual improvement of the trade and logistics system, and the availability of subsidies for investments in more cost-effective heating systems and higher technology greenhouses are likely to foster investments and production, so it is possible to expect that in a 7–10 year perspective, Albania will fill the supply gap in winter vegetables, and also start to export at regional level. Actually, for some periods of the year, Albanian tomatoes and cucumbers are already competitive in regional markets; for example, Albanian tomatoes produced in non-heated greenhouses can be delivered to regional markets (e.g. Serbia) two to three weeks before those from Macedonia (which, when available, have a better price/quality ratio), while favorably competing in terms of price and transport costs with those from Greece and Turkey.

10.3.3 Citrus, stone and pome fruits

The production of citrus, stone and pome fruits follow similar trend and development patterns, even if there are many aspects specific to each group of fruit. The production of all of them suffered much in the 1990s, as a consequence of conflicts among farmers unsatisfied by the outcomes of privatization, and because of quarantine diseases that fatally affected many trees (citrus plantations were almost wiped out). With the start of the transition, the production of stone and pome fruits decreased from 76,000 Mt/year (average 1988–1990) to 34,000 Mt (1992). It started a slow recovery after 1997 (55,000 Mt were produced in 1997), but only in 2003 did the output exceed pre-1991 levels. After 2005, the pace of growth in production increased, as output rose by 77,000 Mt (+85%) in the five years from 2005 to 2010, more than in the 13 years between 1992 and 2005, when production increased by 56,000 Mt (Table 10.7).

For citrus, new investments have been made only in the past few years (from 2005 onward) and have been very focused in terms of products (almost exclusively on clementines), and location (a small area in the district of Saranda, near Greece). The output of citrus fruit reached pre-1991 levels only in 2010 (Table 10.7), but in terms of type of fruit, the production of oranges and lemons is still lower than in 1991.

The main factors hampering recovery and development of fruit production have been the lack of post-harvest facilities and

Table 10.7. Trends in the production of citrus, stone and pome fruits (thousand Mt). Source: INSTAT[a], MAFCP Statistical Yearbooks[b].

Fruit	Average 1988–1990	1992	1997	2000	2005	2010
Pome and stone fruits	76	34	55	65	90	167
Citrus fruit	13	13	3	3	5	13

[a]INSTAT, Albanian Institute of Statistics; [b]MAFCP, Albanian Ministry of Agriculture, Food and Consumer Protection.

investment, which made it impractical to establish commercial-scale production of fruit far from the main markets (which practically coincided with the area of Tirana–Durres). Another factor was the parallel and mutually reinforcing weakness of the fruit and vegetable production and processing sub-sectors.

Scarce accessibility, the cost of storage facilities, and high marketing risks related to the activity of storing, discouraged investments in post-harvest facilities in the few large production clusters that survived to the transition. This has been the case for apples in Korça, where before 2009 the share of losses of harvested apples exceeded 40% (DSA, 2009a), mostly due to the fact that traders considered it too risky to invest in storage facilities and in buying apples for storage. At the same time, farmers never succeeded in establishing associative organizations (i.e. producers' associations or cooperatives) to establish and manage fruit storage facilities.

The conditions for overcoming these constraints, especially for pome and stone fruits, started to ease from 2003 onward, with the opening of the first wholesale markets in Lushnja, which contributed to improved logistics and access to market in the main fruit production areas. The wholesale markets network has been gradually extended, and now covers all the main production areas of the country. From 2007 onward, a government support scheme has been established to subsidize new fruit plantations and drip irrigation systems. From 2009, another support scheme was provided, which subsidized loans taken to establish post-harvest facilities. These state support schemes, together with the improvement of marketing infrastructures and a more favorable business environment, have eventually proved successful in stimulating investments in new pome and stone fruit orchards and cold storage facilities.

In the case of orange and lemon production, it is quite difficult for small Albanian orange producers to compete with the larger and more organized Greek producers and exporters. It is easier however, for Albanian farmers to compete for the soft citrus domestic market (which in Albania, as in Greece, is almost totally dominated by mandarins and clementines), for which Greek production is relatively small (so not leaving much for export) and not very competitive, as it largely consist of original mandarins (*Citrus reticulata*). In marketing terms these fruit are considered an obsolete variety as other seedless varieties of mandarin and hybrids such as clementines are much more popular. The import of oranges, almost totally from Greece, is now the main source of trade deficit among subtropical fruit.

In spite of the increase in domestic production, the international trades deficit of fresh fruit remains large and structural and keeps growing (US$58.5 million in 2010, compared with US$26.5 million in 2000 and US$13 million in 1997 – Table 10.6), even if at a slower pace, as compared with the first years of the 21st century. The main trade imported items are oranges, bananas, and apples.

10.3.4 Fruit and vegetable processing industry

Even if many fresh fruits and vegetables are among the most competitive Albanian agro-food products, the fruit and vegetable processing industry is very weak, being made up of less than 40 formally registered enterprises, of which all except two are small or very small (Leonetti *et al.*, 2010). The late and difficult development of this industry, as compared with other agro-food sub-sectors, shows once again the main feature that has so far characterized the development of food supply chains in Albania: those agro-industries that manage to grow and consolidate better are those that can rely on cheap and easily accessible imports of raw materials, and that have good processing margins without an excessively sophisticated technology (World Bank, 2007). This is not the case for the fruit and vegetable processing industry, as fruit and vegetable supply chains in neighboring countries are quite competitive and well organized (and to be competitive, it is necessary to rely on large-scale investments and modern and efficient processing technologies), and there are not many half-finished products that can be profitably imported and processed. At the same time, as in other sectors, it has proved quite difficult to get a regular flow of domestically produced raw vegetables and fruit from a highly fragmented primary production sector, which is oriented to the fresh market. This means offering to processing industries the products that cannot be profitably sold in the fresh market, rather than raw materials specifically grown for industry needs. Only one or two companies have managed to negotiate with farmers production agreements for tomatoes and cucumbers specifically grown for industry needs. In the other cases, processing plants must rely on local collectors/wholesalers and on some relatively larger farmers with whom they have more consolidated trade relations.

As a result of the above, fruit and vegetable processing companies are small, all offering more or less the same range of basic products, and competitive with imported products only for basic products, such as pickles and some jams. Even when the quality of the domestic product is better than that of cheap imports, poor marketing skills keep domestic products in the low end of the market range. Low economies of scale, scarce suitability of raw material for processing, and ineffective logistics, have proved to be insurmountable obstacles to the establishment of processing facilities, which need a very large production base to be competitive; examples are juice production (existing juice factories are reconstituting imported concentrated juices), and, with a couple of exceptions, tomato processing (tomato sauce produced in Albania is mostly based on imported tomato paste).

In the fruit and vegetable processing sub-sector, the establishment of supermarket chains has proved a real game-changing factor, making available to consumers a range of imported processed food which is much wider (larger variety of products) and more far ranging (different qualities, brands and prices for similar products), thus creating a much harsher competitive environment for Albanian fruit and vegetable processing industries. Smaller companies have now few options: (i) getting larger and offering a wider range of products to provide the quantities and the range of products that the supermarkets require; (ii) focusing on niches in which they can be more competitive; or (iii) limiting their outreach to local markets and traditional retail, a market that will gradually shrink.

The gap between the few companies that will manage to grow and consolidate, and the majority of enterprises that will eventually be marginalized is going to increase, as both commercial trends and developing policies are pushing in this direction. Even the EU-funded IPARD facility is crafted to support the growth of the largest companies, able to make investments in the range of hundreds of thousands of US dollars. This exceeds the capacity of most of the Albanian fruit and vegetable processing industry. As a result, it is likely that no more than four or five companies will manage to become large enough to remain, or to become, national-level competitors, while the rest will be gradually forced into a marginal market of basic processed products, sold in local markets and traditional retail.

However, the overall picture and the perspectives for Albanian fruit and vegetable processing industries is not negative: the largest companies are investing to enlarge their range of products, and the core of the Albanian market of processed fruit and vegetables remains that one of pickles and jams, where domestic products are competitive. As a result, the trade deficit of processed fruit and vegetables has been relatively stable since 2005, in the range of US$15–16 million/year, after experiencing a fast growth between the late 1990s and 2005 (it was US$8 million in 2000). In the future, it is possible to expect a substantial consolidation of the sector, with a few companies growing larger, and the others remaining as they are, shrinking or even disappearing. Total output will grow very slowly, mostly as a result of an increased range of products made in Albania rather than as the outcome of growing volumes of traditional products.

10.3.5 Grapes and wine

Grape growing

Grape production recovered to pre-1991 levels in 2002. New plantations mainly consisted of vineyards planted with varieties specialized for winemaking, especially after 2005. The trigger for a faster expansion of grape production was represented by the fast-growing demand for wine and table grapes in the period 2000 to 2006, which pushed the farm gate value of table grapes to levels of around US$1/kg, and opened new opportunities for informal winemaking at the cottage level. The increasing specialization of grape growers is confirmed by the faster rate of increase in yield and output of vineyards compared with pergolas, whose production is primarily for self-consumption. Between 2000 and 2010, the output of vineyards more than tripled, whereas pergola output increased by only 80%. As a consequence, the share of vineyards and pergolas in total production has substantially changed, with the vineyard share growing from 41.5% to 55.3% of the total and the pergola share decreasing from 58.5% to 44.7% (Table 10.8). Also, the area devoted to vineyards has grown faster than the number of vines in pergolas (in statistics, vineyards are recorded by ha, while pergolas are recorded by the number of vines). The surface cultivated with vineyards increased by 68% between 2000 and 2010, while over the same period the number of vines in pergolas increased by 36%.

Since 2008, the availability of state aid for new vineyard plantations has contributed to a limited degree to the faster expansion of semi-commercial and commercial grape-growing plantations over pergolas (the aid was given for new vineyards larger than 0.2 ha). However, the expansion was well defined before the provision of such aid. In fact, output from vineyards in 2004 was already higher than that from pergolas.

So far, domestic production has mostly succeeded in substituting imports of table grapes, and grapes for processing and cheap wines. The trade deficit in fresh grapes now slightly exceeds US$3.3 million, after reaching

Table 10.8. Trends in areas cultivated with grapes, yields and output, 2000–2010. Source: processed data from MAFCP[a] Statistical Yearbooks.

Grape area/yield/output	2000	2005	2010
Vines/areas in production (2000 = 100)			
Pergolas (000 vines)	100.0	115.7	118.6
Vineyards (ha)	100.0	137.3	166.8
Yields (2000 = 100)			
Pergolas	100.0	103.4	136.1
Vineyards	100.0	127.8	168.2
Output index (2000 = 100)			
Pergolas	100.0	121.6	180.1
Vineyards	100.0	183.9	314.7
Output share			
Pergolas (%)	58.5	48.3	44.7
Vineyards (%)	41.5	51.7	55.3
Yield in quantity			
Pergolas, yield (kg/vine)	11.9	12.3	16.2
Vineyards, yield (00 kg/ha)	70.5	90.1	118.6

[a]MAFCP, Albanian Ministry of Agriculture, Food and Consumer Protection.

a level of US$4.7 million in 2005. According to UNSTAT (the UN Statistics Division (UNSD) Statistical Databases), imports of fresh grapes in 2010 were 5018 Mt, worth US$3.2 million, figures that are comparable to those for 2000 in both quantity (5300 Mt/year) and value (US$1.5 million), but much lower than in those for 2005–2006, when imports of table grapes peaked in both quantity (14,900 Mt in 2005, 10,900 Mt in 2006), and value (US$4.7 million in 2005, US$5.8 million in 2006). Meanwhile, domestic demand continued to grow (almost doubling between 2000 and 2010), so that the share of total supply covered by imports has decreased. Considering the available figures for output from pergolas (mostly used for fresh consumption), and findings from different researches and studies (Agrotec, 2008; Leonetti *et al.*, 2010), it is possible to estimate that, at present, no more than 3–5% of the total consumption of table grapes is covered by imports.

In future, the expansion of supply is expected to slow down, as per capita consumption levels are approaching the levels of those in other countries in the region, and yields are already in line with those of other countries in the region. For table grapes, the domestic market as estimated by the author is 45,000 to 48,000 Mt, and is almost completely satisfied by domestic production, so the growth prospects for Albanian table grape growers are limited to growth in the domestic market. The possibility of exporting sizable quantities of table grapes is remote, at least until substantial progress is achieved in post-harvest services, logistics, and marketing skills of Albanian fruit and vegetables wholesalers and exporters, as these are essential pre-conditions for competing in foreign markets. Prospects for growth in the production of grapes for wine are not particularly good either, as the low-value segment of the wine market (which is the one into which most Albanian wines fall) is now saturated. This situation was reflected by farm gate values of grapes for wine, which experienced increasing downward price pressures in 2010 and 2011. Further growth will depend on the natural growth of the market and on the gradual introduction of higher quality Albanian wines to compete with imported wines in the medium and high-value segments of the market.

Wine production

The largest wine factories were privatized and continued production after 1991, though on a much smaller scale than before. The only positive exception was the winery 'Skanderbeu', in Durres, which always maintained its market leadership at the national level, at least in terms of quantity and turnover. Its bottled red wine of the 'black shesh' grape variety is still considered the benchmark product in the market, and represents the threshold between cheap and quality Albanian wines.

As it is exposed to tough competition from wines with a very good price/quality ratio from Italy, Montenegro and Macedonia, the Albanian wine industry found it difficult to start a real process of qualitative improvement and size growth. This was due to a number of factors, including the limited size of the domestic market, the preference of farmers for trying to process and sell autonomously the grapes that they could not sell in the fresh market, and the amount of investment and know-how required to start a vertically integrated business to improve quality production. However, the overall business environment from 2000 on has remained positive, being characterized by growing domestic demand and a general preference of consumers for domestic products (Chan-Halbrendt *et al.*, 2010a). In this framework, some entrepreneurs started to invest in wine production, initially with modest results in terms of quality and market response.

Changes and developments started to unfold more quickly from 2003 onward. Several factors triggered this more dynamic phase of wine development. The first of these factors was the high prices of fresh grapes and wine, which fostered investments in grape growing and winemaking. This was coupled with increasing evidence of the shift of consumer preferences from raki to wine, which made it increasingly risky for farmers to distill unsold grapes to produce raki (a simple and traditional process, requiring minimal investment and needing limited and widely available skills), and more profitable (even if more difficult) to produce wine. Finally, the increased income of urban consumers created a new middle class with enough disposable income to boost demand in neighborhood and rural restaurants, thus also increasing demand for cheap wine (the type of wine normally served not bottled, but by quantity, in that kind of restaurant). This demand for non-bottled wine from neighborhood and rural restaurants (usually not regularly invoiced) strongly contributed to make wine production an attractive alternative for farmers.

In the years 2003–2005, the opportunity created by the changes described above became evident enough to raise the interest of grape growers, who were already processing excess grapes into raki at the farm level. These farmers started to increase the areas cultivated with vines, and to make some minimal investments to convert grape processing from raki distillation to winemaking. The positive market trend also attracted the interest of other subjects, often external to agribusiness, who had financial resources to invest in larger scale investments than did farmers. Also the existing wineries, after years of minimal investments, started to invest again. The increase of demand of grapes for wine from previously existing and newly established wineries very soon met a major obstacle in the difficulty of ensuring adequate supplies of wine grapes at affordable prices. Wine processing companies soon realized that most farmers expected to sell grapes for wine at the same price as fresh grapes, preferring, as an alternative, to start processing grapes into wine on their own, and sell the wine directly. In addition, the quality of grapes for wine was quite poor, as the priority of farmers was still the fresh grapes market. A lax enforcement of fiscal and food quality laws left the door open to farmers who moonlighted as wine producers. For some years, they found a good demand in local markets, while the demand from traditional retail and high-end restaurants was satisfied by the long-established domestic wineries and by imported wine.

Thus, the first attempts at integrating vertical production and processing were made by growers rather than by processors. In terms of quality, the results were very modest, as these initiatives were little more than attempts made by semi-subsistence farmers to cope with a growing production of low quality fresh grapes (because of lack of post-harvest services; inappropriate agronomic practices, all oriented to maximize yields; and limited new investments in the table grape varieties that were more valued in the market), and a declining demand for raki. Further, the scale of investments made by these farmers was much smaller (thousands or a few tenths of thousands of US dollars) than that required to establish an integrated production of quality wine (several hundreds of thousands of US dollars).

The reaction of professional winemakers to the difficulty in finding quality domestic grapes for processing at a reasonable price led to importing concentrated must from Italy, and buying grapes from Macedonia, which proved to be a reliable source of good quality grapes for red wine (World Bank, 2007). The substantial improvement of road connections between Albania and Macedonia also played an important role in making it economically and technically feasible to import large quantities of grapes from Macedonia, to be processed in Albania. During this period (2004–2007) some larger sized investments (i.e. over US$0.5 million) were also made in establishing new quality oriented modern wineries, often based on integrated grape production and processing business models. These investments, in most cases, were made by entrepreneurs not previously involved in wine production or grape growing. So from an industry that had been limited to the products of a couple of wineries, a more competitive industry with higher quality wines, started to emerge.

The next step in the evolution of the sector started at the end of 2007, when the government started to take action to reduce the quantity of wine produced in a totally informal way (which at that time was quite sizable), and to stem the increasing flow of grapes from Macedonia. The authorities imposed a regime of licenses for importing grapes, while starting to carry out fiscal controls on the largest informal winemakers. At the same time, the most active wineries managed to consolidate their businesses. During this period, competition among winemakers for the market of bottled wines was very dynamic, with a large number of small companies and trademarks quickly gaining and losing market shares.

The ever-increasing production of low quality wine in rural areas eventually led in 2010 to a sharp reduction in the prices paid for grapes and low quality wine compared with that in previous years. This change triggered the next step of the evolution of the sector, which is still ongoing, and is characterized by a swift slowdown of investments in new vineyards and wineries, and the beginning of a much needed period of consolidation and rationalization. The general slowdown of the country's economic development contributed to the reduction of new investments. At present, the most turbulent and dynamic phase of the process of rebuilding the Albanian wine industry after the long stagnation that followed the collapse of the large wineries inherited by the planned economy can be considered mostly over. A more professional and structured industry is emerging, characterized by a larger number of serious competitors. However, the companies that were market leaders in terms of turnover (the Skanderbeu winery) and reputation for quality (the Çobo winery) 10 years ago, when the process of sector renovation and growth became more dynamic, have still managed to keep their leadership.

At present, Albanian wine is competitive in the segment of cheap wines, especially those which are sold in bulk containers for retail sale by quantity (i.e. not sold by the bottle) and, to some degree, in the medium to high-end market, as affluent Albanian consumers often prefer products from reputed local wineries to imported wines, even when the price/quality ratio is less favorable. This is related to a general preference of Albanian consumer for domestic products (Chan-Halbrendt *et al.*, 2010a). In contrast, Albanian wines are not yet managing to pose a real challenge to imported wines in the market segment of lower priced fine wines (bottled quality wines with a price range equivalent to US$4–6), where wines from Italy and Montenegro are still very competitive.

Contrary to expectations, the establishment of supermarket chains did not lead to an invasion of cheap table wine from Italy, specifically produced for price, quantity and packaging for supermarket outlets. There could be different explanations for this poor market performance. A first reason could be linked to the wine consumption pattern of the typical supermarket client, who is an urban, educated and relatively well-off consumer, more accustomed to buy higher quality wines for occasional consumption in supermarkets, while using other alcoholic and non-alcoholic beverages for daily consumption. The practice of buying wine in bulk directly from producers, or obtaining wine from relatives living in production areas is common, too, thus reducing the appeal for cheap bottled wine purchased in supermarkets. More specifically, a cheap wine in Tetra Brick (Tetra Pak) is sold at the equivalent of US$1.4–2.5/l, a price which is still excessive for the daily food budget of the average Albanian consumer.

The changes in international trade of grapes and wine reflect the different stages of the sector development in Albania, as outlined above. Imports of grapes and the trade deficit in fresh grapes and must reached a maximum in around 2005 (US$4.7 million and US$0.1 million, respectively; see Table 10.9), when the import of grapes for winemaking from Macedonia was very high and the import substitution of table grapes was still incomplete. After this date, grape imports started to decline as a result of government measures and the increasing competitiveness of domestic producers. Wine imports remained stable in quantity, but continued to grow in value, as cheap imported wines were out marketed by domestic

Table 10.9. Trends in the international trade balance in grapes and the wine supply chain,1997–2010 (million US$). Source: UNSTAT[a].

Grapes/juice and must/wine	1997	2000	2005	2010
Fresh grapes	1.4	−1.6	−4.7	−3.3
Juice and must	0.0	0.0	−0.1	−0.1
Wine	0.0	−0.2	−0.4	−6.7

[a]UNSTAT, UN Statistics Division (UNSD) Statistical Databases.

production, but more valuable imported bottled wines succeeded in dominating the expanding segment of medium-priced wines, i.e. wines sold for a retail price of US$3–7/bottle) (Agrotec, 2008). The wine trade deficit is not particularly high, but its growth in the past few years has been quite fast, having risen from US$0.4 million in 2005 to US$6.7 million in 2010.

When considering the changes and developments that have occurred in the last 10 years, it is possible to conclude that the Albanian grape and wine sector underwent major changes and has made great progress since 2000. But the transition toward a modern and competitive industry is still far from complete, even if the trends and development scenarios are now more clear and stable than they were 10 years ago. Because of changes in consumption patterns, the development of the food distribution system, the competitive environment, and government policies for the sector are now more consolidated and clearly defined.

Future trends will necessarily see a rather long process of consolidation in fresh grape production and distribution, and in the wine industry, characterized by a relative stagnation of investments in plantations and fewer, but larger, investments in post-harvesting facilities for fresh grapes and in quality-oriented wineries. There will be more competition in the market for low priced fine wines. Demand keeps growing, as middle-aged consumers shift from raki to wine, and a large share of younger people have now become customary wine consumers. However, the growth in demand is strictly linked to the growth of urban income and further changes in lifestyles. As rural demand is more than satisfied, so it is expected that the rate of growth will slow down.

All factors considered, wine remains one of the most promising sectors for both domestic producers and foreign exporters. The next 10 years will not see the same rate of expansion as has been experienced in the last 10 years, when domestic production of grapes more than doubled and imports grew 30 times larger, but it will still provide tangible rewards to quality wine producers and traders. Notwithstanding the expected slowdown in demand growth and investments due to the need for consolidation of the sector, forecasting an overall further growth of the sector is a safe bet, even though it is not included in those that will get support from the European Commission (EC) co-financed IPARD rural development facility.

10.3.6 Olives and olive oil

During the planned economy, yield and productivity of the olive production sector were low, even compared with traditional (i.e. non-commercial) farming practices in other Mediterranean countries. After 1991, olive production has been consistently higher than in previous years, with a few exceptions due to particularly adverse weather conditions. In absolute quantities, though, production is small when compared with the potential of the country and with the output of other Mediterranean countries with a comparable production base. In 2010, total output was 70,000 Mt in 2010, while for most of the previous decade it was 30,000–60,000 Mt. Domestic demand for olive oil was always good, and olive trees continued to produce even with minimal or no agronomic services, so the continuity of production was not disrupted. Also, olives were less affected by the epidemics of quarantine diseases that contributed to the decline in production of other fruit crops.

Production increased by 50% between 1992 and 1997, remained stable until 2005, and then started to grow again, as a result of new plantations commencing production. Productivity was, and remains, very low, because, as already mentioned, few agronomic services are provided to the olive groves (World Bank, 2007). In the case of olives, the trigger event for faster growth consisted of two years of scarce harvest (2006 and 2007), which coincided with the peaking of an olive oil price cycle in international markets. This situation prompted a sharp rise in prices of raw olives in Albania, which lasted even when prices in international markets abruptly decreased in 2008. Favorable prices pushed farmers to exploit their olive groves better, and also to plant new trees. From 2007, there has been a state development support scheme in place as well, co-financing new olive tree plantations. The industry was ready to process this larger flows of olives, as earlier in the first part of the decade, most of the oil mills had been renovated. This was the result of a large development aid project, although until 2006 these mills had been hugely underused. However, processors were not ready to pay a high price for the raw olives, often preferring to work as a service facility, processing the olives into oil for a fee, and leaving to the farmer the risk and the task of selling the oil.

The price of raw olives remains high in Albania, so that there is apparently room for further expansion of production without affecting the financial sustainability of the current plantations. International prices of olive oil remain low, though, and beyond a certain level, it becomes profitable to import olive oil and bottle it in Albania. This previously happened in 2009, when the price of olives was particularly high (DSA, 2009b). Such a threat is particularly strong considering that the main players in the olive oil industry are bottlers, rather than processors: they mostly buy olive oil in bulk from oil mills, but they could switch to importing olive oil in bulk without making major changes to their organization or business model.

At present, the olive oil industry is roughly divided in three segments: (i) small and medium sized oil mills, making almost half of their turnover from processing olives into olive oil on behalf of farmers who then sell the oil on their own, and the rest by buying olives and selling olive oil; (ii) small quality oil mills, specializing in a high-end segment of the market, selling to a small number of loyal buyers, and the only producers able to export; and (iii) industrial processors/bottlers. The enterprises in this last group are the only ones with the size, organization and financial means to work with supermarket chains. In most cases, these largest operators also have a limited processing capacity (DSA, 2009b).

Out of the above mentioned categories, stands a single company, established in 2007 with support of an IBRD (International Bank for Rural Development) loan, which could really represent a game changer, considering its size (by far the largest in Albania) and ambitions (to become a major regional level competitor in olive oil production and export). The first years of activity of this newcomer proved quite difficult, as the plant, due to its size needed process a large quantity of raw olives to be profitable, and it proved impossible to guarantee a sufficient supply of raw olives at affordable prices. As a result, it was decided to reorient the business toward an integrated production system, in which the company produces the olives it will process and sells the olive oil in a full range of retail and wholesale packaging solutions. Nevertheless, the reorientation of the business continues to require huge additional capital investments which are difficult to find, so to date (2011), the company is still struggling to shape a fully sustainable business model. This experience is important because, on the one hand, it shows the importance of establishing a vertically integrated enterprise, which seems the most appropriate business model to overcome the recurrent raw olives supply problems. On the other hand, shows the difficulty in establishing a company large and efficient enough to generate enough cash flow to recover in a reasonable time the huge investments that are required.

The Albanian government considers olive oil to be one of the most potentially competitive Albanian products, and has made the development of the olive sector the main priority of its agribusiness development aid scheme since 2007, with some results in terms of increased cultivated area and output, but, so far, without many results in terms of increased efficiency among the sectoral enterprises, import substitution, or increase of exports; in fact, both autonomous investment decisions of farmers and processors, and state incentives (provided in the form of a contribution to investments in new olive groves) in the last 10 years have mostly addressed to the expansion of the production base (i.e. a greater cultivated area, more oil mills), and not improving the productivity of existing orchards or improving quality along the supply chain. Furthermore the allocation of a small budget for incentives to produce extra-virgin olive oil does little to change the overall orientation of the government development policy to quantity.

So far, domestic production has succeeded in satisfying all the growth of domestic demand, but olive oil imports are stable in quantity and increasing in value, showing that the high-end segment of the olive oil market remains dominated by foreign products. In particular, the olive trade deficit increased from US$3.9 million in 2005 to US$10.2 million in 2010 (Table 10.10). A small, but increasing trade deficit is also recorded in table olives (US$0.5 million in 2010).

Future trends are not particularly positive: domestic demand is growing and will continue to grow in parallel with income growth, as customers who can afford the price are keen to replace consumption of other vegetable oils with olive oil. However, the structural imbalances affecting competitiveness of the Albanian olive oil supply chain have been barely addressed by the recent investments and public development schemes, as merely an increase in the size of the production base (i.e. more olive trees) did not parallel an increase in efficiency and quality at the production and processing levels. As a result of poor efficiency along the supply chain, and high prices of raw olives, the wholesale price of Greek and Spanish virgin olive oil is 50% lower than on Albanian. Also, support schemes for quality olive oil production are not backed by a bold policy of quality controls to prevent commercial frauds. Without serious controls on food quality, the available incentives to the production of extra-virgin olive oil should be considered as subsidies to the olive oil industry, rather than incentives for quality improvement.

Production has nearly doubled in the last 10 years and will further increase by more than 40% in the next five years, considering that new plantations with almost 2 million trees will enter into full production, bringing the production base from 4.3 million trees in 2005 to 6.3 million trees in 2010 (Table 10.11). In the last 10 years, farm gate prices of raw olives in Albania also grew, in spite of a steep decrease in international prices of virgin olive oil after 2008.

When the new plantations established after 2007 enter into full production, the supply will be much larger than domestic demand. This scenario will be incompatible with present prices, and a raw olives price crisis will inevitably occur, as happened in a similar scenario with wine grapes. The experience of the past shows that the reaction of most farmers to a price crisis will be to minimize cultivation inputs, or even abandon crop, rather than invest further to increase efficiency. This realistic scenario poses a threat to the sustainability of the investments made in the past few years to expand both the production base and processing capacities.

The strongest segments of the olive oil supply chain remain the bottlers (who are mostly traders) and those enterprises that can somehow develop an integrated supply chain, producing at least part of the olives that they process, or those that can rely on particularly favorable supply conditions at the local level. Downwards pressure on international prices of

Table 10.10. Trends in the olive and olive oil trade balance (million US$). Source: UNSTAT[a].

Olives/olive oil	1997	2000	2005	2010
Olives	0.0	−0.3	−0.3	−0.5
Olive oil	−0.9	−1.2	−3.9	−10.2

[a]UNSTAT, UN Statistics Division (UNSD) Statistical Databases.

Table 10.11. Trends in the olive and olive oil production base. Source: MAFCP[a] Yearbooks.

Olive/olive oil production	2000	2005	2010
Olives trees (000 plants)	3,611	4,264	6,255
Olive trees in production (000 plants)	3,256	3,488	4,298
Olives production (000 Mt)	36,200	30,200	70,000
Estimated olive oil production (000 Mt)[b]	5,700	4,800	11,200

[a]MAFCP, Albanian Ministry of Agriculture, Food and Consumer Protection; [b]Author's estimate.

virgin olive oil is going to continue, as well as the expansion of supermarket chains, so Albanian olive oil producers will find it increasingly difficult to compete in this market environment, while bottlers will have increasing bargaining power. As a result of the above, it is possible to conclude that the phase of consolidation and development of the Albanian olive and olive oil supply chain has still not really started, as the expansion of supply and production base is only one of the factors that could trigger an increase in the competitiveness of Albanian olives and olive oil.

Olives and olive oil are one of the strongest and potentially more competitive agribusiness sectors in Albania, in which the country has a real potential competitive advantage. However, without readdressing investments in improving efficiency at all stages of the existing supply chains, the supply chain operators will not manage to exploit this advantage. According to present trends, the country will produce more than 100,000 Mt of olives by 2015, but yields and productivity along the value chain will remain low. The same production base, more efficiently managed, could easily produce twice as much, and the processing and marketing costs could be lower too.

10.4 Livestock Production

10.4.1 Broilers and egg production

Poultry breeding in Albania has continuously grown since 1991. Commercial production initially consisted of broiler production, but very soon commercial-level egg production started as well. The avian flu outbreak had a major impact in 2005 and 2006, so the output of commercial farms in 2007 was considerably lower than that in 2004. However, comparing figures over a longer interval, the impact of the avian flu crisis does not seem so severe, and the overall growth trend was not interrupted. Avian flu caused major short-term financial damage to the broiler farms, but had no impact on the infrastructures and the market chains, so those farms that survived the financial crisis induced by the disease were able to swiftly restart their businesses. In spite of the temporary disruption of the production cycle, overall broiler production increased from 4000 Mt in 2000 to 16,000 Mt. in 2008 and 17,000 Mt in 2009, which represents about 11% of the total domestic meat production (Mautner Markhof *et al.*, 2010).

Table 10.12 shows the changes in the number of poultry commercial farms in the years from 2005 to 2009, which includes the years of the avian flu outbreak and the following recovery. The impact of the avian flu crisis has been higher on the smaller commercial poultry farms (those with 1000 to 5000 birds), whose number declined between 2005 and 2007 from 11 to 7 units, showing their sensitivity to financial and commercial shocks. In contrast, the number of poultry farms with more than 5000 birds increased between 2005 and 2007 from 12 to 22 enterprises, showing the financial steadiness and resilience of these larger companies. At present, there are at least five broiler factories with a production capacity of over 600,000 broilers/year.

Based on the experience of and indications provided by the avian flu crisis, it is possible to foresee that broiler factories with at least 5000 head are better positioned to survive the process of consolidation that will start when the domestic market becomes saturated. Still, domestic demand is low

Table 10.12. Number of poultry farms with more than 1000 head of birds, 2005–2009. Source: MAFCP[a] Yearbooks.

Poultry meat farm	2005	2006	2007	2008	2009
1,000–5,000 head	11	8	7	17	16
5,000–10,000 head	3	3	6	5	5
Over 10,000 head	9	9	16	14	17
Total	23	20	29	36	38

[a]MAFCP, Albanian Ministry of Agriculture, Food and Consumer Protection.

compared with demand in neighboring countries, imports are high (US$32 million in 2010), and domestic production has proved the capacity to out compete imported products, so this activity offers good growth prospects for the next years. There are no particular competitive advantages in poultry breeding in Albania, so it does not seem realistic to expect that once saturated, the domestic market Albanian broiler factories will be able to export significant quantities of poultry meat.

When analyzing the trends in the poultry stock, it is also possible to obtain further confirmation that the avian flu crisis affected the growth of the business, but that the sector enterprises showed strong resilience. Table 10.13 shows changes in poultry stock and egg production at the national level. The stock almost doubled between 1992 and 2010 (from 2.6 million birds to 5.2 million birds). In fact, when analyzing changes in production over a 20-year perspective, the impact of the avian flu crisis is limited, resulting in a decrease of 5% in the 2005 stock of birds compared with the stock in 2000 (4.7 million birds in 2005 versus 4.9 million birds in 2000).

The impact of the avian flu crisis on the production of eggs was even lower, and is hardly visible in the time series, as production kept growing from 270 million eggs in 1992 to 530 million eggs in 2000 (+96% in eight years), and to 846 million eggs in 2010 (+60% in 10 years).

The growth in egg production is likely to continue over the coming years, even if the pace of the growth will be probably lower. From 2007 onward, Albania has also exported a certain quantity of eggs, whose value in 2010 was US$3.4 million (see Table 10.14). This market channel is becoming more consolidated, so it is possible to expect some increase in exports.

10.4.2 Cattle production

Bovine production in Albania mostly consists of dairy cattle. Meat cattle breeding farms are a rare exception, and bull calves are slaughtered very early, when their weight is 150 kg or less. The breeding of other bovines—Mediterranean buffaloes—has been almost totally phased out (110 head left).

Milk production and cattle stock show a consistent trend since 1992, characterized by increasing specialization and productivity in dairy cattle breeding. In fact, from 1992 to 2010, the number of dairy cattle has decreased from 840,000 head to 493,000 head (Table 10.15), while milk output has increased from 486,000 Mt in 1992 to 930,000 Mt in 2005, and remained stable after that date (though yields per head continued to grow).

After a fast increase in per capita consumption between 1992 and 2000, the demand for milk grew slowly over the last 10 years, and has actually been stable since 2005. Over the same period the cattle population slightly decreased in number, as same period the breeders focused on increasing productivity per cow. Since 1992,

Table 10.13. Trends in poultry breeding and egg production, 1988–2010. Source: MAFCP[a] Yearbooks.

Poultry stock/eggs	Average 1988–1990	1992	1997	2000	2005	2010
Poultry stock (000)	2392	2624	3358	4870	4671	5245
Eggs (millions)	318	270	337	530	738	846

[a]MAFCP, Albanian Ministry of Agriculture, Food and Consumer Protection.

Table 10.14. Exporte of eggs, 2003–2010 (millions US$). Source: MAFCP[a] Yearbooks.

2003	2004	2005	2006	2007	2008	2009	2010
0.8	0.6	0.7	0.1	1.6	3.3	1.8	3.4

[a]MAFCP, Albanian Ministry of Agriculture, Food and Consumer Protection.

milk production has increased by 90%, while the cattle stock has decreased by 40%. At present, the production of milk/cattle head is in line with that of some Western Balkan countries (Bosnia and Herzegovina, Macedonia, Montenegro) but still very far from the average in Western European countries, and 25% lower than that other comparable, but more advanced, Balkan agricultural systems, such as those of Greece, Croatia, and Bulgaria. The figures on meat production are largely the result of estimates, as most cattle are slaughtered without proper records (indeed, documentation on live animal movements is seldom produced and no regular slaughtering registers are kept). Slaughtered cattle are almost entirely used for fresh meat consumption, as the Albanian meat processing industry, one of the strongest food processing sectors in the country, mostly uses imported frozen meat.

The apparent stagnation in the production of both meat and milk in recent years is the result of a swift reduction in the profitability of cattle breeding that occurred from 2008 onward. This is due to the growth of input costs, first of all fuel (a key cost component for self-production of animal feed), and then animal feed. Table 10.16 shows how the growing trend recorded in the years before 2005 changed between 2006 and 2008. It is still early to judge the long-lasting impact of the rise of input costs on the production trends for cow's milk and cattle meat. Even if figures from the MAFCP (Albanian Ministry of Agriculture, Food and Consumer Protection) and FAOSTAT (Food and Agriculture Organization Statistical Division Databases) do not fully match, in both cases the change of trend occurred between 2006 and 2008, and is the same.

Considering the above, it is not possible to extrapolate a clear future trend for the total output of cattle meat and cow's milk. The available information does, however, allow us to draw a couple of conclusions. Firstly, the trend toward increasing of yields in milk production is likely to continue, although at a slower pace; there is still much room for improvement. So far, yield improvements have been in line with those in the neighboring countries of the former Yugoslavia. The second conclusion which can be drawn is that changes in production

Table 10.15. Cattle breeding and production of milk and meat, 1988–2010. Source: MAFCP[a] Yearbooks.

Cattle/milk/meat production	Average 1988–1990	1992	1997	2000	2005	2010
Cattle (000 head)	616	840	771	728	655	493
Milk production (000 Mt)	379	486	791	807	930	930
Meat production (000 Mt)[b]	Not available	23.4	32.5	35.9	40.8	40.8

[a]MAFCP, Albanian Ministry of Agriculture, Food and Consumer Protection; [b]Food and Agriculture Organization Statistical Division Databases (FAOSTAT).

Table 10.16. Trends in production of cattle meat and cow's milk in 2005–2010 (thousand Mt). Source: MAFCP[a], FAOSTAT[b].

Milk/meat production	2005	2006	2007	2008	2009	2010
Milk production						
MAFCP	930			895	907	930
FAOSTAT	430	420	396	360	353	355
Meat production						
MAFCP	68			66	66	68
FAOSTAT	41	49	50	40	39	41

[a]MAFCP, Albanian Ministry of Agriculture, Food and Consumer Protection; [b]Food and Agriculture Organization Statistical Division Databases (FAOSTAT).

levels will depend on the ability of local breeders to obtain animal feed at affordable prices. The low productivity of maize and other fodder crops, and the high cost of concentrated feed and agricultural inputs, have greatly affected the cattle industry value chain. The capacity to improve pasture management, make use of agro-industrial byproducts as components for animal feed, and increase efficiency in the production of fodder crops will all be factors that will decide whether cattle meat and the production of cow's milk will stagnate or recover a growth path.

10.4.3 Sheep and goat production

Sheep and goat production in Albania is characterized by some limited adaptations to the changing social and economic environment of a very tradition-oriented primary production sector. A first characteristic of this production sector is that production and consumption in the southern regions have been traditionally higher than in the remaining part of the country. In particular, consumption patterns in southern Albania are more similar to those recorded in the neighboring Greece (where the consumption of lamb meat is particularly high), while patterns in other areas of the country are more similar to those of other Balkan countries. Partly as a consequence of this, lamb from Gjirokaster and sheep dairy products from the south (and in particular white cheese from Saranda, and kashkaval cheese from Gjirokaster) are traditionally reputed by consumers to be the best in Albania.

Dairy products from sheep's milk are particularly valued, while all goat products are mostly for self-consumption and local markets. This factor played in important role in the sustainability of some risky investments made in the first decade of 2000, as consumers proved ready to pay a premium price higher than expected for quality sheep's milk dairy products, so the investors eventually proved to be right in their optimistic assumptions. At the same time, the availability of larger processing facilities in the south has contributed to reorienting production, which is now more sheep oriented and milk oriented than in the recent past.

The consumption of sheep and goat meat is traditionally stronger in rural areas, where consumption of bovine and poultry meat is much lower than in urban areas. Consumption of fish in the country's more inland areas is minimal, too. Urban consumer preference is shifting toward beef and poultry fresh meat, and processed meat. Among sheep and goat meat, urban consumption of lamb meat is increasingly linked to celebratory and convivial consumption (Bayram, End of the Year, Easter, marriages, restaurants), as lamb meat is perceived as tasty, but too fatty and unhealthy. Goat kid meat is considered less fat, but consumption is always tied to a seasonal (spring) niche. As a result of the above, the consumption of lamb meat is stagnating or decreasing (no clear trends are identified) and the consumption of goat kid meat is subject to a seasonal peak between April and May during any given year. The demand for goat kid meat, which in urban areas almost disappeared in the mid 2000s, is now recovering, thanks entirely to this seasonal peak.

Internal migration toward urban areas after 1991 contributed to mitigating the decline in demand for lamb meat in urban areas (as new urban dwellers continued for some time to keep their consumption habits), but has also negatively affected production and consumption in rural areas, in parallel with the depopulation of the inner and mountain regions. Also, an increasing number of immigrants from northern regions to Tirana, and decreasing supplies of lamb meat from the south, have contributed in redirecting the supply flows in the main urban market of the country. Now many of the largest butchers specializing in lamb meat in Tirana are from the northern regions, and keeping solid trade links with the areas from which they come. Also, consumers with family links to the north do not consider lamb meat from the south as the best.

This market environment and these trends create conditions for more intensive milk-oriented sheep breeding, especially in southern regions, and for a gradual reduction in goat breeding. In fact, this is what is happening, as the milk yield of sheep slowly increased (from 55 l/head in 1992 to 70 l/

head in 2000 and 77 l/head in 2010; Table 10.17) and the number of goats has constantly declined over the last 20 years (from 1.6 million head in 1992 to 0.77 million head in 2010; Table 10.17).

Notwithstanding the favorable market conditions for milk oriented sheep breeding, the number of sheep has remained more or less stable between 1997 and 2010 (1.86 million head in 1997 versus 1.80 million head in 2010), and yield/head has not increased as fast as could be expected considering the high price paid for sheep's milk, which should have stimulated investments in commercial, more intensive, and specialized sheep breeding. In fact, the recovery and expansion of sheep breeding after 1991 has been and remains constrained by the scarcity and expense of animal feed, a problem that has affected sheep production more severely than other animal productions. This condition is mostly linked to the prevalent sheep breeding pattern, based on transhumance and access to public pastures. During the planned economy, state forests and pastures were scarcely productive and not very effectively managed, although pastures were relatively well maintained (in particular, water points were regularly maintained) and overgrazing prevented through a strict control of animal movements. Investments in pasture maintenance and rigid controls were discontinued with the collapse of the regime, with the result that there was a sudden increase in the number of small ruminants (both sheep and goats) held by individual breeders. Widespread depletion of pasture followed, caused by overgrazing due to the contemporary reduction of accessible pastures (lack of maintenance of water points made otherwise good pastures unusable) and overuse of the still usable pasture resources. This situation first caused a reduction in the number of small ruminants, after an excessive growth between 1992 and 1995, and then a stabilization in the numbers of livestock that was compatible with the quantity and quality of available pastures. Exploiting the favorable market conditions for sheep's milk would require the use of more productive breeds and/or an increase in the number of animals. Owing to the limited quantity of available pastures, this would be possible only by making increasing use of animal feed. However, using purchased animal feed to substitute for pasture feeding requires a substantial change in the small ruminants supply chain, with fewer, larger and more specialized commercial-sized

Table 10.17. Trends in sheep and goat stock, milk production and yields, 1988–2010. Source: MAFCP[a] Yearbooks.

Stock/milk production	Average 1988–1990	1992	1995	1997	2000	2005	2010
Stock (000 head)							
Sheep	1796	2480		1858	1939	1760	1806
Goats	1234	1650		1148	1104	941	775
Total stock	3030	4130	–	3006	3043	2701	2581
Milking stock (000 head)							
Milking sheep		1232	1736	1372	1448	1312	1337
Milking goats		857	1150	840	800	701	576
Total milking stock		2089	2886	2212	2248	2013	1913
Milk production (000 Mt)							
Sheep's milk		55.3	81.6	68.3	70.0	75.0	77.0
Goat's milk		70.4	95.5	74.1	71.0	71.0	63.0
Total milk	94.0	125.7	177.1	142.4	141.0	146.0	140.0
Annual yield (l/head)							
Sheep		44.9	47.0	49.8	48.3	57.2	57.6
Goats		82.1	83.0	88.2	88.8	101.3	109.4

[a]MAFCP, Albanian Ministry of Agriculture, Food and Consumer Protection.

breeders, a change which did not occur over the last 20 years and is not likely to occur in the near future. This is because the vast majority of traditional breeders do not have the means or the willingness to become involved in intensive breeding, while profit-oriented agribusiness entrepreneurs find it more profitable to focus on dairy cattle breeding, as the market for bovine milk and bovine dairy products is much larger than that for sheep and goat dairy products.

As a consequence, the Albanian supply chain of small ruminants is still mostly shaped in function by traditional transhumant breeding, characterized by minimal cash costs in production, relatively low productivity and small total revenues (but high margins for production units). In this market environment, service and input providers (such as animal feed suppliers and traders of purebred animals for intensive breeding) consider poultry and cattle breeding as their core business and are not much interested in offering new and better services and products or particularly competitive prices.

Considering the business environment and market trends described above, it is easy to understand the forces that have influenced changes in the size and structure of sheep and goat stock, and in the structure of production, and to identify future development trends. A first key trend is relevant to the gradual rationalization of small ruminant breeding activity, with a decreasing head number and increasing productivity. The total stock of small ruminants experienced a large increase immediately after the collapse of the planned economy, partially as an effect of the surge in personal consumption and of a much weaker application of the rules in planning and controlling access to public pastures. As a consequence, the total small ruminants stock grew from 3 million head in the period 1988–1990 to 4.1 million heads in 1992. After this peak, a slow process of rationalization started, with increasing milk yields per animal, although the decline in the number of goats has been faster than the increase in milk yields, with the result that the total output of goat's milk has declined.

The second key trend is related the market preference for dairy products made from sheep's milk, and the Albanian consumer's parallel lack of interset in goat's milk and dairy products. Scarce demand in the urban market for milk and dairy products based on goat's milk, and no interest by the processing industries in promoting such products contributed to the weakness of the goat's milk market, such that the interest of breeders increasingly shifted towards sheep breeding. The demand for sheep's milk from processing factories remains high, even if the farm gate prices of sheep's milk have much increased, especially since 2005. Strong demand for sheep's milk and weak demand for goat's milk resulted in an increasing ratio of sheep to goats (from 1.5:1 in 1992 to 2.3:1 in 2010), as well as an increasing difference in prices between sheep's and goat's milk. The price and demand shifts also produced a gradual specialization of sheep breeding towards milk-oriented breeding. This last trend was first recorded in the southern regions (Vlora, Gijrokaster), whose dairy products were traditionally more reputed among consumers. As sheep's milk production in these regions did not grow over the years, raw sheep's milk prices also increased in neighboring regions, and processing plants in other regions (such as Korça) started to produce dairy products based on sheep's milk on a larger scale.

A third important issue is the level of investments, which remains low, as investments in commercial breeding were oriented in the past to the more profitable activity of dairy cattle breeding. Indeed, in the last few years investments are low for any type of breeding, owing to the increased cost of inputs and decreased profitability. In the case of small ruminants, the traditional transhumant breeding pattern is not much affected by the increasing cost of feed, but the limited availability of quality pastures is a major constraint. In particular, investments in improving sheep and goat breeds have always been low and concentrated in the south. The MAFCP extension service center of Korça (CATT Korça) is the only extension center in the country with some capabilities for supporting sheep and goat breeding improvement. As a result, the presence of the traditional 'Ruda' sheep breed in the north is more common, while in the south the most common sheep breed is the 'Tzigaya', which was introduced during the planned economy and is lighter in weight (and therefore has a lower fat content and is preferred by urban consumers), and more productive in terms of milk yield. More recently, some efforts have been made to introduce new varieties, all milk oriented. However, the genetic potential of most flocks is rather depleted, as a systematic policy for sheep/goat breed improvement has never been pursued. Furthermore, support from development projects has been limited to introducing few new breeds, which worked well in the countries that financed the projects (the 'Awassi' sheep breed and 'Des Alpes' goat breed) at CATT Korça.

A fourth important factor influencing the present situation and future trends is that investments in dairy processing plants specialized in cheese based on sheep's milk have been also concentrated in the south (because of larger number of sheep and better reputation among consumers of sheep dairy products). Sheep breeding in northern and central Albania is more meat oriented as demand from dairy plants and prices of sheep's milk are much lower there than in southern Albania.

Finally, options for the development of sheep breeding in Albania have always been focused on meat or milk production. Wool production oriented breeding has never been considered as an option. The existing breeds are not suitable for quality wool production, the domestic textile industry is based on semi-processed imported fabric, and the international demand for wool is weak, except for high quality wool. As a consequence, the importance of wool production is negligible.

10.4.4 Meat consumption and the meat industry

Meat consumption in Albania is rising. Per capita consumption has doubled in the last 10 years and is now in line with that of most Western Balkan countries, but is still over 30% lower than in Greece (FAOSTAT 2007, the last data available). Meat is a major component of the traditional Albanian diet and, for the older generations, daily consumption of meat was synonymous with familial wealth. In the last years of the planned economy and the first years of the transition, first scarcity, and then budget constraints, reduced consumption, which began to grow substantially only after the year 2000. Meat is mostly sold in butchers' shops. The practice of door-to-door meat trading has disappeared, while supermarket chains find it difficult to capture a sizable market share, as consumers give the most value to meat 'freshness' (which means meat of an animal slaughtered the same day), and strongly prefer to see the butcher cutting the meat according to consumer requests.

The trends of different meat types reflect both changes in family budgets, and changes in consumption patterns and lifestyles (Table 10.18). Beef consumption has been always influenced by price. In principle, most Albanian consumers prefer beef and veal to other types of meat, but the price difference from poultry meat and the large availability of sheep and goat meat in rural areas was and is an issue. There is, though, a direct relation between increase in income and increase in the consumption of bovine meat, and this trend is likely to continue. Even if data

Table 10.18. Evolution of yearly meat consumption (kg per capita), 2000–2007. Source: FAOSTAT[a]; also LSMS[b] 2002 and 2005, and HBS[c] 2007 data from Mautner Markhof *et al.* (2010).

Annual meat consumption	2000 (FAO)	2002 (LSMS)	2005 (LSMS)	2005 (FAO)	2007 (HBS)	2007 (FAO)
Total meat (without offal)	29.0	19.3	33.4	41.7	43.0	47.6
Beef and veal	12.2	13.7	17.9	15.4	17.8	18.6
Pork	4.5	0.2	0.5	9.4	3.1	11.9
Poultry	5.8	3.0	8.5	9.8	7.9	9.9
Sheep and goat meat	6.5	1.9	1.7	7.0	2.0	7.2

[a]FAOSTAT, Food and Agriculture Organization Statistical Division Databases; [b]LSMS, Living Standards Measurement Survey; [c]Household Budget Survey.

from different sources are not fully comparable (FAOSTAT data include meat imports for processing, while data from the Living Standards Measurement Survey (LSMS) and HBS (Household Budget Survey) provide data recorded by consumers about their purchases of fresh meat, and record processed meat separately by weight after processing), the growing trend is clear: a 52% increase between 2000 and 2007 according to FAOSTAT, compared with a 30% increase based on LSMS 2002 and HBS 2007 data.

Sheep/goat meat consumption is stagnating. Lamb meat is consumed more in the south and in rural areas, but urban consumers prefer other types of meat, because of price, taste and presumed health issues (lamb meat is considered more fatty and less healthy). As already noted, different sources (FAOSTAT 2000, 2005 and 2007; LSMS 2002 and 2005; HBS 2007) show quite different data, but the trend is similar: FAOSTAT shows a modest annual growth in consumption from 6.5 to 7.2 kg per capita between 2000 and 2007, while data from LSMS 2002 and HBS 2007 (which have similar methodologies) show an almost totally flat annual trend (1.9 kg per capita in 2002 against 2.0 kg per capita in 2007). The most recent qualitative research (Leonetti and Kristo, 2010) shows that consumption of sheep/goat meat is increasingly becoming seasonal and related to celebratory or convivial consumption. There is no a reason to believe that this trend will change in the future.

Pork consumption is definitely increasing, even though an increasing number of consumers are becoming regular clients of halal meat shops, which do not sell pork. However, data on domestic production (+60% in 10 years), imports (+300% in 10 years), and the little data on consumption, clearly show an increase of demand. This trend is likely to continue. Supply is largely based on imports, as domestic production is growing slowly, while import growth is growing every year with an additional trade deficit of US$1–2 million.

The consumption of poultry meat mostly consists of chicken meat, as the demand for turkey meat is highly seasonal. Production has increased more than fourfold in the last 10 years and import value threefold. Demand keeps growing, and the trend is likely to continue. As for the other meats already discussed, the available data are from different sources, but the trend is the same: FAOSTAT data shows a substantial increase in annual poultry meat consumption between 2000 and 2005 (from 5.8 to 9.8 kg per capita, i.e. +69%), and a stagnation in consumption between 2005 and 2007 (also the years of the avian flu); data from LSMS and HBS also show a substantial growth in annual consumption from 2002 to 2005 (from 3.0 kg to 8.5 kg per capita), while a comparison between the LSMS 2005 data and HBS 2007 data show a decrease of annual consumption, from 8.5 to 7.9 kg per capita.

The latest data available on consumption are from 2007, and they reflect the impact of the avian flu crisis (consumers were scared about the risk of eating poultry meat and eggs, even though the health risks were associated only with live birds), so it is not yet possible to say how much consumption has grown after 2007, and whether the growth in consumption of poultry meat will accelerate again in the future. Data on the growth of poultry stock (Table 10.13) lead to the conclusion that the prospects are good.

The growth in meat demand has outpaced the growth of meat production, so meat imports have increased and are likely to continue to grow, thus increasing the meat trade balance deficit (Table 10.19). While for beef this situation and trend is a direct consequence of the orientation of bovine breeding toward milk production (see Section 10.4.2), the fact that the trade deficit in poultry meat tripled in value between 1997 and 2010. This spite of an increase of domestic production by 57% in the same period (Table 10.13), shows how much poultry meat has increased its importance in the daily diet of Albanian consumers over a few years, mostly thanks to its low price compared with other types of meat.

Another interesting aspect is related to the increasing imports of pork, which tripled in value from 2000 to 2010 (US$19.9 million of imports in 2010 versus US$6.5 million in 2000). The resulting trade deficit coincides with imports as no pork exports are recorded from Albania. In this case,

Table 10.19. Trends in the meat trade balance (million US$). Source: UNSTAT[a].

Meat type	1997	2000	2005	2010
Bovine meat	–3.8	–1.2	–6.8	–5.8
Pork meat	–1.2	–6.5	–12.6	–19.9
Sheep/goat meat	–0.1	0.0	–0.9	–1.0
Poultry meat	–10.7	–10.1	–18.4	–32.0
Fat	–1.1	–0.5	–2.6	–4.4
Other meat	–0.6	–0.4	–0.7	–5.6
Total fresh and frozen meat	–17.4	–18.6	–42.1	–68.8
Processed meat	–1.5	–1.6	–4.2	–14.1

[a]UNSTAT, UN Statistics Division (UNSD) Statistical Databases.

increasing demand does not correspond at all to increasing domestic supply, as there is a tradition of small-scale pig breeding only in a small part of Albania (the region of Lac). The large-scale pig breeding facilities—the only ones able to produce pork at internationally competitive prices—were destroyed in the first years of transition and never re-established, as investors are resistant to investing in the production of a good that can be easily imported at very competitive prices. The most recent trends since 2010 show increasing flows of live pigs from Greece and other neighboring countries, which are kept in Albania for finishing (i.e. to take on some additional weight) for a duration generally of less than a month, and then slaughtered. However, it is currently (2012) still too early have consolidated data on these flows.

Meat processing in Albania is characterized by an almost radical separation between fresh meat and processed meat supply chains. Fresh meat comes from domestic and imported live animals, which are mostly slaughtered in totally inappropriate facilities (so-called 'slaughtering points', i.e. very rudimentary and unhygienic facilities or even shop backyards), not complying with any standard. An exception to this situation is represented by broiler factories, as the largest ones do include modern poultry slaughtering facilities. The few public and private investments made in modern slaughterhouses have failed to improve the situation, as norms on public hygiene and food safety in slaughtering were incomplete until 2010 (when additional norms on food safety were approved), and are still not effectively enforced, so that meat traders are not really forced to apply internationally accepted (and more expensive) standards in the slaughtering process. Food safety hazards and the environmental impact of slaughtering are some of the most worrying shortcomings of Albanian agro-food sector.

In contrast, the processed meat industry, which mostly produces cold cuts, is almost totally based on imported frozen meat and is one of the most structured and solid food industries in Albania. This is due to the fact that it is based on a business model that proved solid and profitable, and able to outcompete most of the imported products in large demand in the domestic market. The establishment of supermarket chains has created new challenges for Albanian meat processors, as a larger choice of imported products of all prices and qualities is now available, but so far the largest domestic meat industries have managed to maintain their market leadership, even in supermarket outlets. In traditional retail, imported processed meat products take a marginal share of the market and are considered to be of lower quality.

At present, the industry is structured in two tiers: (i) two national-level players, each of which are a leader in a segment of the market and (ii) specialized regional competitors, very strong in their regions, and in many cases having a certain outreach in Tirana and Durres, through a network of owned shops and distribution networks. It is estimated that 80% of the market is controlled by no more than ten enterprises (Mautner Markhof *et al.*, 2010).

The Albanian meat processing industry (both national market leaders and regional competitors) shows three main characteristics. Firstly, the range of products is limited to cold cuts (cooked hams, wurstel, short-ripening dry cured (fermented) sausages, etc.) and some fresh products for fast food (meat balls, doner kebabs, etc.); no canned meat or other types of processed meat are produced. Second, there is a total reliance on imported frozen meat as raw material, as the domestic meat production and slaughtering system are totally oriented to fresh consumption. Finally, the industry does not include slaughterhouses, as slaughtering is still mostly performed in informal facilities, and existing slaughtering facilities could absolutely not comply with the needs of the meat industry in terms of food safety, efficiency, size, and organization.

The industry is tailored to satisfy the domestic market of cold cuts, has a relatively wide range of products covering most of the needs of Albanian consumers. So far, the industry has no export ambitions, which is realistic, considering that regional and international markets of cold cuts are quite competitive and demanding in terms of food safety standards and quality. In terms of compliance with international standards food safety and quality, the largest meat processing companies are more advanced than any Albanian food industry oriented to the domestic market. Only the fish industry has a comparable level of quality standards, but in that case this was a must from the beginning, as the fish industry is mostly focused on international markets. However, the Albanian meat industry is still far from international standards when dealing with fresh meat processing (the weak point being the slaughterhouses), and almost totally noncompliant with environmental standards, also due to an incomplete and scarcely applied national legal framework.

Prospects for the processed meat sector and the relevant domestic industry are good. An important and increasing share of the meat consumed is processed meat, but even the most recent sector analyses found it difficult to give a precise figure (Mautner Markhof *et al.*, 2010). The leading industries generate enough cash to invest without too much stress in the improvements that will be required to comply with EU safety and environmental standards. At the same time, prospects for the fresh meat market are uncertain, as an analysis of the elasticity of Albanian meat demand to price has never been made, and prices of fresh meat will inevitably rise (in real terms) in the future, when the supply chain will become more structured and compliant with international standards. The enforcement of stricter rules in fresh meat processing will increase the final price of fresh meat at least by 10%, as estimated with reference to the lamb meat supply chain (Leonetti and Kristo, 2005, 2010). In the short-term, there is not much real pressure for investing more in fresh meat safety and better slaughtering waste disposal. Consumers do not perceive the present situation as potentially risky and there is not a real political commitment to decisively curb the now prevalent practice of informal slaughtering. Butchers and meat traders know that sooner or later they will be asked to invest in complying with more demanding norms, but they openly declare that they will do this only when so obliged (Leonetti and Kristo, 2010).

In a framework of economic downturn, an increase in the cost of meat could cause a reverse in the trend of increasing consumption of red meat, in favor of cheaper poultry meat. In particular, a considerable price increase is expected to affect the demand for beef and lamb meat more (as these prices are already higher), with the most likely result a further growth in the share of poultry meat consumption in total meat consumption. The increasing role of the supermarket in food trade is expected to have a marginal impact on traditional retail, as consumers show strong fidelity to their trusted neighborhood butchers.

10.4.5 Milk and dairy consumption and the milk processing industry

The milk and dairy food chain is one of the most problematic in Albania. Consumption is growing slowly in some segments, such as cheese, and stagnating in others, such as fresh milk, and the domestic milk and dairy industry is undergoing a very slow process of consolidation and modernization. Domestic supply and consumption have been more or less stable since 2000, with total supply increasing by 12% from 2000 to 2010, though the growth was entirely achieved in the first half of the decade (Table 10.20). Imports of milk and dairy products grew, but the total size of the trade is small compared with that of other major food sectors. It is estimated than less than 50% of milk produced in Albania is processed, with the rest being used for animal feeding, consumed by the producers themselves (Schröder *et al.*, 2010), or directly sold by farmers to final consumers.

As for most food products, Albania has a growing, structural trade deficit in milk and dairy products (Table 10.21). The milk trade deficit rose from US$1.9 million in 2000 to US$8.7 million in 2010, and the dairy products trade deficit rose from US$2.8 million in 2000 to US$10.6 million in 2010. The trade deficit largely coincides with imports, as exports are negligible.

Overall, milk consumption is already in line with that of other countries of the region (FAOSTAT), while consumption of cheese and butter is comparable with that of other Western Balkan countries, excepting Greece. The dairy industry is starting to provide a wider range of cheese and milk to stimulate consumption and get more value added, but so far, only some types of cheese similar to gouda and mozzarella have succeeded in becoming popular among consumers. Even if domestic demand does not grow much, it provides, at current levels, a stable and sizable market, a favorable condition for domestic producers. There are two main problems affecting the development of the domestic dairy industry. Firstly, the margins in milk processing are low. This is due to the relatively high cost of raw milk, which is in line with that of EU member country levels (of a lower quality), while the price of the final products is lower. This problem could be somehow overcome by increasing the use of imported milk as powder, but in this case the quality of the final product would be lower than that obtained from quality fresh milk (low cost milk in powder is generally safe, but of low quality). Second, there is a high level of informality in milk processing. The informal dairy units produce with no respect for hygiene, labor, and food safety rules, but are able to sell low quality (and often unsafe) dairy products at a much lower price, thus negatively affecting the capacity to compete with the dairy companies that are operating according to the rules and are forced to keep low prices, while having much higher costs than informal dairy units (World Bank, 2007).

So far, the establishment of supermarket chains has not brought many changes to the distribution patterns of milk and dairy products. Supermarkets are more interested in selling imported products, and only one or two Albanian companies have succeeded in working with supermarket chains. Most of the milk and dairy products are sold in traditional retail

Table 10.20. Evolution of milk production (thousand Mt). Source: MAFCP[a] Statistical Yearbook.

Milk type	Average 1988–1990	1992	1997	2000	2005	2010
Cow's milk	379	486	791	807	930	930
Sheep's milk		55	68	70	75	77
Goat's milk		70	74	71	71	63
Total milk		611	933	948	1076	1070

[a]MAFCP, Albanian Ministry of Agriculture, Food and Consumer Protection.

Table 10.21. Trends in the milk and dairy trade balance (million US$). Source: UNSTAT[a].

Milk/dairy	1997	2000	2005	2010
Milk	–0.9	–1.9	–3.6	–8.7
Dairy, including butter	–2.5	–2.8	–8.6	–10.6

[a]UNSTAT, UN Statistics Division (UNSD) Statistical Databases.

shops and in green markets in a very traditional way. Cheese is usually sold unpackaged, and it is estimated that some 10% of fresh milk is still sold directly by farmers in recycled plastic bottles (Schröder *et al.*, 2010). A few larger and modern enterprises are slowly emerging as market leaders, but the overall sector remains quite fragmented. In 2009, 27 processing units out of 418 registered dairy companies had a capacity larger than 10 Mt/day. Many hundreds of informal cottage level processing units are processing less than 500 l/day.

The process of pre-accession to the EU is expected to trigger a much needed process of rationalization and modernization of the sector. The enforcement of EU-harmonized rules on food safety, animal health and environmental management would lead to the dismantling of most of the Albanian dairy industry, which is far from complying with such standards and has not the means to invest in viable dairy plants fully complying with the rules, even considering the possible conditioned grants that can be obtained from the IPARD facility. This process could lead in the short-term to even further growth of informal production, as marginal companies would not find it possible or profitable to remain in the formal sector.

On the demand side, the slow development of consumer preference for products processed in proper dairy industries will continue, as well as the introduction into the market of a wider range of products, but no major changes are expected in the next 10 years. Demand for milk will stagnate or decrease, while demand for cheese should increase, as it is still much lower than in Greece, although comparable with that of other neighboring countries. Supermarkets are expected to play a major role in widening the range of dairy products available to consumers, even if the core market will remain as a few products, namely fresh whole milk, yogurt, a white cheese similar to Greek feta, and kashkaval.

The future sector developments will be conditioned by slow growth of demand in quantity but on one hand to the increasingly urgent need to consolidate and improve efficiency of the dairy industry on the other. The value added should grow faster, as an increasing share of consumers are buying bottled milk produced by dairy industries, rather than raw milk in recycled bottles, directly sold door to door by farmers. At the same time, qualitative evolution of demand will slowly continue, with an increasingly wider range of products in demand, thus reducing the space for unqualified domestic producers. The degree of political will in enforcing the existing legal provisions and the stricter norms that will be gradually introduced in the process of approximation to the EU legal framework will strongly affect the pace of rationalization of the milk and dairy sector. The risk is that most companies will not have the capacity to profit from the financial support that will be made available through IPARD to invest in complying with the EU standards. Furthermore, only a few companies will manage to grow, pushing many small enterprises back to informality, and eventually leaving more room for imported products. To conclude, the larger and strongest enterprises can rely on positive prospects, while the small and middle-sized enterprises will face increasing challenges.

10.4.6 Leather and hides

The collapse of the planned economy also caused the total disruption of the skin–leather supply chain which starts from slaughterhouses (most of which were dismantled after 1991) and ends with the industrial use of leather (footwear, furniture, garments, etc.). With time, the simplest segments of the upstream supply chain were reconstituted, based on a network of local skin collectors and on a few relatively large exporters, all located between Tirana and Durres, as Italy was the main market from 1991 to 2009. A few tanneries working in Lezha were soon closed, because they were causing too much pollution. The only small and complete value chain remained in Korça where the local slaughterhouse managed to survive after privatization, making collection and drying of skins an important part of its business. A local tannery and a shoe factory completed the supply chain.

With time, a relatively dynamic footwear and garment industry based on subcontracted or outsourced garment manufacturing contracts was re-established (most clients being Italian garment and footwear producers), but all these industries are working with imported leather, while dried raw skins are exported. This persistent gap in the supply chain is caused by the complexity and size of investments, and the knowledge required to process raw dried skins into quality leather. Actually, the process itself is not difficult, but the skill of the workforce, and entrepreneurs and the amount of investments are important constraint factors.

Up to 2009, Italy was by far the main trading partner both for Albanian exports (skins) and imports (leather). After 2009, the gap in the supply chain also resulted in a separation of the directions of trade flows. Exports are now largely to Turkey, where the leather industry is growing strongly, while imports of leather are still mostly part of more complex contractual agreements (such as subcontracted or outsourced manufacturing contracts) with Italian partners. This structure of trade flows, characterized by a major value added difference between exports (low value added skins) and imports (high value added leather, often already pre-cut into size for processing into garments and footwear), translated into a high and increasing trade deficit, which after 2005 ranged constantly between US$48 and US$60 million a year. Table 10.22 shows that both imports and exports of skins, hides and leather have increased since 1997, but the trade deficit is growing, in parallel with the expansion of the Albanian footwear and garments industry. The size of the trade deficit in 2010 exceeded US$50 million.

Future trends are likely to remain unchanged, with a growth in exports of hides, and imports of processed skins and leather, and a growing trade deficit, even if in the last few years the growth of both imports and exports has slowed down. For the first time since 1997, the 3 year average figures of both imports and exports in 2008–2010 were similar to the values recorded in 2010, showing a slowdown of growth, if not a stagnation. It is also too early to say whether the redirection of exports towards Turkey will prove to be a permanent change of trade partnerships, or whether this is a temporary condition linked to the financial difficulties of many Italian small and medium enterprises, which used to be the main customers of Albanian skin exporters.

10.5 Fisheries

10.5.1 Fish and seafood industry

During the planned economy, Albania's fishing industry consisted of four state-owned fishing enterprises (one for each of the main fishing ports of the country, i.e. Durres, Saranda, Vlora, and Shengjin), 16 aquaculture enterprises, and two shellfish enterprises. There were also three fish canning factories. In the past, fish did not represent a major staple of the customary diet, except in coastal urban areas. Also, the consumption of seafood was mostly limited to the main production area (Saranda, the southernmost part of the country) and, to a more limited extent, to Durres. Freshwater fish was available, sourced from lakes and from the numerous artificial ponds established in the country, but scarcely in demand, with the exception of Shkoder (carps from Shkoder lake) and Pogradeç (koran, a kind of trout from Ohrid lake), where there was a certain tradition in fish consumption. During the planned economy wild sea fish (i.e. not from aquaculture) resources were not overexploited, also because domestic demand was low. With urbanization and the introduction of

Table 10.22. Trends in international trade of hides, skins and leather (million US$). Source: FAOSTAT[a].

Trade	1997	2000	2005	2010	Average 2008–2010
Exports	7.2	7.7	11.9	22.0	18.1
Imports	13.5	17.5	52.5	75.6	73.4
Trade surplus/deficit	–6.3	–9.8	–40.6	–53.6	–55.4
Index (1997 = 100)	100	157	645	851	879

[a]FAOSTAT, Food and Agriculture Organization Statistical Division Databases.

consumption habits from Italy and Greece, the situation is much changed in the urban areas, where a wide range of fish and seafood is now available and consumed, even if the average consumer is accustomed only to few types of fish and seafood recipes, mostly coming from Italian cuisine.

As for international trade, wild marine fish has been one of the most important Albanian food export items in the years immediately after the end of planned economy, as it was an abundant and untapped resource, requiring little processing, and finding an insatiable market in nearby Italy. Starting from the beginning of 2000, the fishing fleet has been 'modernized' by buying cheap secondhand trawler fishing vessels from EU countries, which were being retired because of the introduction of more stringent fishing rules. These purchases allowed the industry to rebuild the commercial fishing fleet at a low price, but at the cost of introducing unsustainable fishing practices. The once untapped resource of wild fish was gradually depleted, and the amount and quality (size) of catches started to decrease from the first half of the 2000s.

A few larger companies emerged as market leaders in international trading of fish and seafood (UNDP, 2005). As export is by far the most profitable market, these companies were forced to make important investments in processing and logistics facilities for fresh fish and frozen seafood. Such investments included complete cold chains and, in some cases, even distribution warehouses and specialized outlets abroad (DSE, 2011). The cold rooms established in the main fishing ports were good enough to receive a CE number (the CE is a mandatory conformity mark for products placed on the market in the European Economic Area, or EEA), which ensures conformity with EU legislation. The food processing units complying with EU standards can get an identification number, which must be indicated on the labels of all food items sold in EU member countries, to ensure traceability. To directly export a food product into EU Member States, the product must bear a CE mark (i.e. it must come from a processing unit bearing a CE number) thus allowing export directly to EU Member States. Also, processing facilities for salted anchovies, frozen seafood, and pasteurized mussels were established or modernized.

In parallel with this evolution, domestic demand started to grow fast. The development made by the Albanian fish and seafood sector is well explained by the comparison of import and export flows in 2000 (i.e. at the beginning of the commercial fishing fleet restructuring) and in 2010 (Table 10.23). In 10 years, exports of fresh fish grew from US$0.3 million to US$4.3 million, exports of processed fish (mostly salted anchovies) increased from US$4 million in 2000 to US$24.9 million in 2010, and exports of frozen seafood rose from US$1 million to US$6.3 million. Other types of products, such as cured fish, fish fillets and frozen fish, also registered a strong growth in exports between 2000 and 2010.

A more detailed analysis of exports in the decade 2000–2010 shows that most of the growth occurred between 2005 and 2010

Table 10.23. Trends in Albania's international trade of fish and seafood (US$). Source: processed UNSTAT data[a].

	2000		2010	
Item traded/trade balance	Import	Export	Import	Export
Fresh fish	572,661	308,078	4,018,319	4,305,170
Frozen fish	44,624	2,120	796,605	309,564
Fish fillets	27,075	1,779	68,340	23,082
Cured/smoked fish	2,875,752	50,459	3,074,413	1,055,656
Processed fish	2,165,867	4,004,247	28,098,270	24,895,617
Mollusks	509,035	996,741	4,659,469	6,295,421
Total	6,195,014	5,363,424	40,715,416	36,884,510
Trade balance		–831,590		–3,830,906

[a]UNSTAT, UN Statistics Division (UNSD) Statistical Databases.

(Table 10.24), with a dynamic similar to that recorded in many other food sectors. However, in the case of the fish and seafood sector, the growth in production and exports occurred in the 5 years between 2005 and 2010 and was particularly sizable. Exports of fresh fish grew almost fourfold, exports of cured fish grew by 360% and exports of processed fish grew by 52%, rising in 5 years from US$15 million to US$24 million. Substantial growth in export flows was recorded in all other subcategories of the fish and seafood sector. The growth in production and exports is likely to continue in the future; should the export ban to the EU on bivalve mollusks be lifted, exports could easily exceed US$60 million by 2015.

Part of the exceptional growth in the international trade of fish and seafood is linked to re-exports. The largest fish and seafood trading companies are buying locally fresh fish, crustaceans and mussels, but are also importing fish and seafood which is then mixed with local products and frozen again to produce products with higher value added (e.g. seafood salads) to be sold in the domestic market or exported.

In parallel with a substantial growth of exports, imports increased dramatically, too (Table 10.23): fresh fish imports rose from US$0.6 million in 2000 to US$4 million in 2010, processed fish imports rose from US%2.2 million to US$28.1 million, and imports of mollusks rose from US$0.5 million to US$4.7 million. Fresh fish is the only product among fish and seafood in which the trade deficit registered in 2000 was upturned into a trade surplus in 2010, while processed fish, which showed a trade surplus in 2000, recorded a trade deficit in 2010, in spite of a sixfold growth of exports.

The fact that the trade deficit of fish, seafood, and sea products in 2010 is almost four times higher than in 2000, in spite of a sevenfold growth of exports, shows how much domestic demand for fish, seafood, and fish products has increased since the beginning of the transition, and particularly in the last 10 years. This level of demand denotes a major change in the structure of food consumption in Albania, as consumption of fish and seafood before 1991 was mostly limited to the areas near to the fishing ports.

10.5.2 Aquaculture

While the food chain of fresh and frozen fish and seafood succeeded in a relatively early recovery and development, aquaculture suffered a much more difficult process of reorganization and development. Before 1991, aquaculture was rather developed, and included both inland and marine aquaculture. A long-standing cooperation with China has imparted both the needed expertise and infrastructures for the small-scale breeding of cyprinid species (e.g. carp) in artificial ponds built for irrigation in many inner areas. Larger establishments were also built to grow trout (most of them in the southern regions and

Table 10.24. Trends in exports of fish and seafood products from Albania to third countries, by category of product; index, 2005 or first year of recorded exports =100. Source: processed UNSTAT[a] data.

Product/value/share	2005	2006	2007	2008	2009	2010
Fresh fish	100	105	161	187	303	390
Frozen fish	100	97	8	18	193	71
Fish fillets	0	0	0	100	357	93
Cured/smoked fish	0	100	145	236	375	362
Processed fish	100	131	130	136	146	152
Mollusks	100	443	1462	3772	2503	6309
Total	100	132	139	160	177	205
Value (million US$)	18.0	23.8	25.0	28.8	31.8	36.9
Share of fresh fish and mollusks in total export (%)	6.7	6.7	12.9	20.2	18.3	28.7

[a]UNSTAT, UN Statistics Division (UNSD) Statistical Databases.

Ohrid lake), koran (Ohrid lake), mussels (Butrinto lagoon, in Vlora region). This last had a potential output of 5000 Mt/year of mussels, not a major production center by international standards, but very large for Albania, especially considering the consumption rate at the time. In the last period of the planned economy, a major aquaculture production unit was established near Vlora. This plant was designed to be gradually expanded, to become the largest aquaculture establishment in the Mediterranean Sea. Its basins could have been individually fed with seawater or fresh water, and were thus able to raise either saltwater or freshwater fish, according to demand. This structure never entered into full production. After the start of transition, most of the infrastructures were destroyed and only in the past 2 years is a small part of the plant being reconstructed. Only mussel growing in Butrinto lagoon was never discontinued, but output was substantially reduced. The only important investment made before 2005 was in a shrimp farm in Kavaja (Tirana region).

An analysis of the evolution of the aquaculture output confirms that most of the development after 1991 occurred in the period 2005 to 2010 (Table 10.25). Output decreased from 1991 (859 Mt, worth US$0.7 million) until 1997 (minimum output 98 Mt). Since then, output has continuously grown in quantity and value, reaching 1473 Mt of mussels and fish, worth US$3.6 million in 2005, and 2504 Mt of mussels and fish in 2010, worth US$6.7 million in 2010.

Data produced by the Ministry of Environment, Forestry and Water Administration (MEFWA) for the 4 years from 2007 to 2010 are different, showing a decline in production of aquaculture in quantity (no data on values are provided). This is a result of declining output of freshwater aquaculture production, not compensated for by the growth in marine aquaculture production, and stagnation of mollusk production (but the data account only for production in Butrinti lagoon). Furthermore, MEFWA data do not consider the production of the shrimp farm.

The quality of aquaculture products is improving, as better quality fish feed is being used, and also as more efficient quality control systems and procedures are applied. The introduction of these improvements has been instigated by the need to respect the quality standards required by international markets, which are quite important for the profitability of some of the most important aquaculture units in Albania, such as the shrimp farm and the sea bream and sea bass fish farming units.

According to an evaluation performed in the World Organisation for Animal Health (OIE) in 2008 (OIE, 2008), there were then 33 aquaculture production units (Table 10.26), with 2250 fish farmers dealing with inland aquaculture, and 300 with marine aquaculture. However, the revenue generated by marine aquaculture was several times higher than that from freshwater aquaculture. The largest and more important facilities in terms of size and turnover were the shrimp farm of Kavaja (near Tirana), and the mussel growing complex in the lagoon of Butrinto, in Vlora region. This latter was established during the planned

Table 10.25. Trends in aquaculture output in quantity and value, 1991–2010. Source: FAOSTAT[a], MEFWA[b].

Quantity/value	1991	1995	1997	2000	2005	2006	2007	2008	2009	2010
FAOSTAT data										
Quantity (Mt)	859	340	97	307	1473	1970	2008	1858	2182	2504
Value (000 US$)	669.5	251.4	222.7	477.7	3649.2	3536.3	4714.8	5913.9	5750.5	6742.1
MEFWA data										
Quantity (Mt)							2860	2540	2730	2119

[a]FAOSTAT, Food and Agriculture Organization Statistical Division Databases; [b]MEFWA, Albanian Ministry of Environment, Forestry and Water Administration.

Table 10.26. Type, number and location of commercial aquaculture establishments in Albania. Source: OIE, 2008.

Product	Production units	Region
Sea bream/sea bass	9	Lezha (1), Vlora (8)
Koran	1	Korça
Trout	12	Dibra (2), Gjirokaster (4), Korça (2), Shkoder (1), Vlora (3)
Carp	8	Durres (1), Elbasan (1), Fier (2), Korça (2), Shkoder (1), Tirana (1)
Shrimps	1	Tirana (1)
Mussels	2	Lezha (1), Vlora (1)

economy as a state company, and is now still owned by the state, which provides yearly licenses to private fishermen for each mussel growing section (so-called 'lines'), into which the complex is divided.

The most dynamic segment of Albanian aquaculture is marine aquaculture, which is made up of nine fish farms, producing sea bream and sea bass, and three seafood production units, producing shrimps and bivalve mollusks. A large share of the crustaceans, and almost all of the mussels exported (after processing) or sold in the domestic market, come from the shrimp farm and two mussel breeding plants that are active in the country. An increasing share of sea bream and sea bass sold in fish shops and restaurants also comes from aquaculture. Marine aquaculture is providing an increasing share of the total output of the Albanian fish and seafood sub-sector, at the expense of wild fish harvest and freshwater aquaculture. Marine aquaculture's increased share in the total value of Albania's fish and seafood exports can be seen in the growth of fresh fish and processed mollusks from 6.7% of the total in 2005 to 28.7% in 2010 (Table 10.23). This increase comes entirely from marine aquaculture, as wild fish catches are decreasing due to over-exploitation of natural resources. The collection of wild mollusks is also negligible (i.e. almost all the crustacean and mollusk trade come from seafood farming activities) and exports of freshwater aquaculture products are likewise negligible.

Investments in aquaculture for seafood production, and, in particular, for mussel growing experienced a relatively late recovery after 1991, also due to the fact that international markets are closed for Albanian live bivalve mollusks, shrimps and prawns. Exports to EU Member States are banned because of food safety concerns, while importers from other countries in the region, who could legally import live bivalve mollusks from Albania, prefer to remain on the safe side until the ban is released, importing only frozen and processed seafood. The import ban has been issued by European competent authorities because the food safety controls performed by the Albanian authorities on the mollusk production units are not considered sufficiently reliable and effective. The ban will be not released until the inspectors from EC, who visit Albania every year, consider the food safety control system satisfactory and reliable.

The impossibility of exporting live bivalve mollusks from Albania has had the consequence of slowing down investments in seafood aquaculture, while pushing fish and seafood processing companies to find alternative sources of raw seafood to be further processed and exported, as ingredients of products with higher value added. This complex supply chain, made by imported frozen seafood, fish, and fish products, which after import are further processed, mixed with domestic products, and then exported as a product with higher value added, is a specific feature of the fish industry, which is not found in any other food industry in Albania.

In the last decade, aquaculture has generated increasing employment and income, although three key issues are still hampering further development. Firstly, food safety control procedures are not yet considered satisfactory by the EC, thus preventing the lifting of the export ban on seafood. The removal of the ban would boost exports, and create conditions for further investments, as Albanian seafood could be quite competitive in the Italian market. Second, Albania currently depends on Italian and Greek imports of larvae and fingerlings, and, as a result, the price of fish remains relatively high. Finally, all fish feed is imported, thus increasing production costs and the complexity of the business.

Even considering the above issues, the prospects for Albanian aquaculture enterprises are quite positive, especially for marine aquaculture, as domestic and international demand for seafood (shrimps and mussels) and saltwater fish such as sea bream and sea bass remain very good. Prospects for Albanian freshwater aquaculture are less positive, as domestic demand for freshwater fish is growing slower than the demand for saltwater fish, and there is not much international demand for products from Albanian freshwater aquaculture (carp, trout and koran).

However, the growth of exports from marine aquaculture and the fish processing industry is not likely to be sufficient to balance the international trade deficit in fish and seafood products that Albania has constantly recorded over the last 20 years. Imports are rising more than exports, especially in the subsectors of processed, cured, and smoked fish, and this trend is likely to continue in the near future.

10.6 Forestry and Non-timber Forestry Products

10.6.1 Wood

Albania has favorable conditions for an extensive lumber industry, but its natural resources are rather depleted and domestic production is mostly oriented to providing firewood and timber for construction. Illegal logging has been a very serious problem, but its impact is declining, as the valuable wood that is easy to reach has been already cut and mountain areas are increasingly depopulated.

The postwar government invested much in reforestation and in the wood industry, but results were uneven, as sawmills were outdated, far from wood resources, and wasted too much raw material. The paper industry was similarly uncompetitive. The yearly output of the nine state forestry industry complexes between 1977 and 1988 reached 2.3 million m^3 of roundwood a year (Zickel *et al.*, 1994), though FAOSTAT provides a much lower figure of 722,000 Mt during the same period. The 28 sawmills cut about 200,000 m^3 of wood a year (Zickel *et al.*, 1994). In the early 1990s, the main wood production areas were in the central and northern mountain regions. The country's southern part was mostly deforested, as a result of historical events, such as the destruction of woodlands to create pastures, and the wood trade that fed the shipyards of Venice and Ragusa (Dubrovnik) until the 18th century.

Before 1991, wood was the main source of fuel for rural and urban communities, and it still plays a major role in heating houses in rural areas. It is estimated that in 1991, 90% of household fuel needs were covered by wood, an amount that reached 100% in mountainous areas (Zickel *et al.*, 1994). Ten years later, in 2000, the use of wood for heating was lower, but almost a quarter (23.4%) of households in urban areas were still using wood for heating (INSTAT, 2000). The exceptional role (for a European country) of wood as source of energy in Albania is also shown by its contribution to the total demand of energy. In 2000, wood provided one sixth of the country's energy needs (excluding transportation), a quantity comparable to that of hydropower (Table 10.27). In 2004, the share of wood was reduced to 10%, still very high compared with European standards.

In the decade between 1991 and 2001, pressure on natural resources was extremely high, due to the high need for fuelwood, to deforestation linked to small ruminant breeding (the first years of transition until 1997 saw a steep increase in the number of goats and sheep), and to sawmill activity. Most of the sawmills, quickly privatized after 1991, managed to survive and continued their activity. Some of them also started to export logs and semi-finished furniture parts to Italy. At the same time, the local small furniture industry mostly relied on imported wood. Demand for timber from a booming construction sector added pressure to demand.

A conservative estimate indicates that forests, wooded areas, and plantations decreased by 14,000 ha in the years between 1990 and 2000: from 942,000 ha to 928,000 ha. The decision to transfer 40% of the state forests to communes contributed to reducing controls on the sustainable use of the transferred wooded areas. Because forestry assets were, in most cases, not inventoried, the task of making sustainable use of the assets transferred was extremely difficult. Furthermore, local administrations did not have the human resources or the political will to take care of the transferred assets.

After 2000, reduction of logging, depopulation of the mountain areas, and reforestation efforts, led to an increase in the total wooded area, which by 2005 reached 967,000 ha, and since then has continued to grow slowly. Most of the country's forested area was classified as 'modified forest', as pristine primary forests remained unchanged owing to their inaccessibility. Cultivated forestry followed a constant declining trend, decreasing from 106,000 ha in 1990 to 88,000 ha in 2005 (most recent data available).

The stabilization and recovery of forested areas did not bring a parallel development of the forestry industry. Data supplied by Albanian and international sources are mostly based on estimates, but the trend of official outputs shows a clear stagnation, although these changes of output seem to be more a result of different classification than real changes of flow. Even though the available data should be taken cautiously, the image provided is that of a declining quality of production, now mostly made up of charcoal and roundwood (Table 10.28), while the production of veneer logs and sawnwood has decreased over the years.

At present, Albania is continuing to export mostly low value added wood products, except for some furniture parts, and is importing increasing quantities of wood products for construction (sandwich panels, parquet, etc.), and furniture making. As a result of different data classification, data on international trade are also quite different, depending on the source (Table 10.29). However, all available information shows an increasing trade deficit.

In this sector, more efficient governance and incentives to private investments for the overall modernization of the sector could really make the difference. Modernization efforts could range from the retrofit of heating systems in rural areas to use wood byproducts, to the rehabilitation of damaged commercial forests, to support for more extensive forest cultivation, among others. Should the operational environment remain the same as during the last

Table 10.27. Total energy consumption and energy generated by wood and hydropower. Source: Cela and Hizmo (2009).

Energy consumption/source	2000	2004	2007
Total energy consumption (TJ)	76,379	94,550	89,217
Hydroelectric energy consumption (TJ)	19,335	22,634	10,718
Share of hydroelectricity in total supply (%)	25.3	11.9	12.0
Wood energy consumption (TJ)	12,740	11,242	9,002
Share of wood in total supply (%)	16.7	23.8	10.1

Table 10.28. Trends in output of the main wood and forestry industry products (m^3). Source: FAOSTAT[a].

Product	1992	1995	1997	2000	2005	2010
(Other) industrial roundwood (NC[b])	n/a[c]	n/a	n/a	n/a	8,000	39,695
Plywood and particleboard	6,200	6,200	6,200	n/a	0	10,000
Sawlogs + veneer logs (NC)	433,000	40,000	40,000	53,000	39,900	9,638
Sawnwood (NC)	296,000	2,900	2,900	43,000	50,000	4,000
Wood charcoal	n/a	n/a	n/a	9,000	55,100	60,000

[a]FAOSTAT, Food and Agriculture Organization Statistical Division Databases; [b]NC, non-coniferous; [c]n/a, not available.

Table 10.29. Trends in international trade of wood and wood products (million US$). Source: FAOSTAT[a], UNSTAT[b].

Trade	1991	1995	1997	2000	2005	2010
FAOSTAT						
Import	2.1	6.8	19.5	16.8	25.8	25.8
Export	0.8	6.2	6.6	5.9	7.4	7.4
Trade surplus/deficit	−1.3	−0.6	−12.9	−10.9	−18.3	−18.3
Index (2000 = 100)	12	6	118	100	168	168
UNSTAT						
Import			n/a[c]	40.2	53.1	69.7
Export			n/a	10.4	11.1	15.2
Trade surplus/deficit			n/a	−29.7	−42.0	−54.5
Index (2000 = 100)				100	141	183

[a]FAOSTAT, Food and Agriculture Organization Statistical Division Databases; [b]UNSTAT, UN Statistics Division (UNSD) Statistical Databases; [c]n/a, not available.

10 years, the future trends appear to be clearly defined and heavily conditioned by the inefficiency of the sector governance (both at central and local level), the scarcity of private investment, and the orientation of the whole sector toward the provision of low value added products, such as fuelwood and charcoal. Outputs of sawmills and charcoal are expected to remain in the range of 130,000 to 150,000 m^3/ year, while the quantity of fuelwood is set to slowly decline. Exports could grow by 30–40% over the next 10 years, but imports and trade deficits are expected to grow faster, and, in 2020, could easily be more than twice as much as in 2000.

10.6.2 Medicinal and aromatic plants

Medicinal and aromatic plants (MAPs) traditionally constitute one of the strongest subsectors of the Albanian agro-food sector. The range of MAPs (flowers, herbs, leaves, fruits, roots and aromatic plants) that are harvested and traded total 86 species. The main exported MAPs are sage, oregano and thyme. With no cosmetic and pharmaceutical industry, and depending on imports for MAP derivatives used in the food industry, the domestic market is limited to the use of MAPs in domestic cooking. The industrial use of MAP derivatives in the food, cosmetic and pharmaceutical industries, is quite vast and increasing, but all of these uses require several stages of processing, some of which quite complex and expensive. So far, there are only a few relatively small facilities in Albania, extracting essential oil from just a few species of MAPs. Sage (*Salvia officinalis*) is by far the most important one. The output of these processing units is basic essential oils, which are only the first processing stage of much more complex supply and value chains.

During the planned economy, a network of collection points covered the whole country. Part of this system survived and many of the state officers dealing with the collection and distribution of MAPs at regional level went into private business, working as agents or independent collectors for the newly established private exporters. In the first decade of the transition, export was a quasi-monopoly of a joint venture between a multinational and an Albanian partner. Even in the worst years of the early transition period (1991–1997), Albania was exporting large quantities of MAPs, and was one of the ten largest world MAP exporters in terms of value (Schippman *et al.*, 2002).

In recent years, an increasing number of competitors have emerged. Most of these new competitors are former collectors at the regional level, who managed to increase their businesses and to accumulate enough capital to invest in equipment and facilities. These operators also received support from development projects. Meanwhile, the ownership of the joint venture which used to retain the market leadership was totally transferred to the Albanian partner, and this has contributed to reducing its capacity to control the whole market as important trade links, ensured by the foreign partner, have been lost. More recently, another joint venture has been established with Turkish capital, and has immediately become one of the key actors in the business.

Altogether, MAPs constitute an export oriented, healthy, and important business, controlled by Albanian companies. Some of the most dynamic Albanian regional collectors have managed to grow and evolve into exporting companies. Many of these companies have a turnover of some million US$/year, making them large companies by Albanian standards, but very small players for the international MAP markets, which are dominated by relatively few multinational giants. However, the increased number of competitors has led to a certain fragmentation of the exporting companies, so the impact of the possible entrance of international players (which at world level are a few, large companies) could be huge, as is demonstrated by the case of the above mentioned Turkish–Albanian joint venture. The Turkish partner is itself a large company, linked to a US-based multinational dealing with MAPs and essential oils. The joint venture was established in 2006, and almost immediately became the second largest exporter in the country. As a consequence, Albania is a major international player in some products, such as wild sage and thyme, but the structure of its leading players—the exporters—is still fragile, in spite of major improvements achieved in the past 15 years.

As with the fishing industry, the combination of favorable market perspectives, weak governance of natural resources management, and external socioeconomic factors, such as the gradual depopulation of inner areas, has caused a general over-exploitation of wild MAP resources. This was particularly severe in the less depopulated silvopastoral communities, where resources are wildly exploited to maintain and possibly increase the total production volume of previous years, and compensate for the depopulated areas where MAPs are scarcely harvested. Difficulty in keeping up with demand has caused growth in the prices of Albanian wild MAPs, therby making Albanian MAPs less competitive, and, on the other hand, have added pressure to the fading natural resources. The rush to grab as many wild MAPs as possible, and as early as possible (even before harvesting time, to anticipate competitors), is seriously endangering a wide range of wild MAPs. As a result, by 2000, 5% of flowering plants were considered 'critically endangered', and 18% 'endangered' (REC, 2000; Kathe *et al.*, 2003). The three most important Albanian MAPs for international trade—sage, oregano and thyme—are not endangered at the national level, but high quality sage and thyme are increasingly scarce, as the resources near to the silvopastoral settlements are seriously over-exploited. Other herbs of high commercial interest, such as *Gentiana lutea* and *Sideritis* (locally known as 'mountain tea') are endangered in the whole Western Balkan region. In addition to over-harvesting, the mode of harvesting is also often a source of damage, as MAPs are harvested too soon and without attention to the plant survival.

Local governments have the responsibility to set the annual collection quotas for MAPs, and to enforce these quotas. Thus far, the controls on the quotas are not effectively enforced. Furthermore, private stakeholders (exporters, collectors, harvesters) did not manage to agree between themselves on a self-control regime. Both private and public stakeholders have not invested in harvester training and awareness programs, mostly leaving the task to international development projects. As a result, the quantity of harvested wild MAPs has declined since 2002 (when it reached a peak of 9024 Mt), and since 2009 the value of exports has declined as well, exports having reached a record value of US$23 million in 2008 (Table 10.30). The international economic crisis has contributed to redirecting international demand toward cheaper suppliers, thus delivering a big blow to the Albanian exporters.

The reaction of exporters to the increasing cost and scarcity of wild MAPs has been to promote MAP cultivation and to start to invest in MAP conditioning equipment (sorting, grinding, and packaging machines), and first processing (MAP distillation to produce essential oil). So far, these efforts have been quite limited, considering the overall size of the Albanian MAP industry. They were, though, a

Table 10.30. Export of medicinal and aromatic plants (MAPs) in 2000–2010. Source: MAFCP[a], UNSTAT[b].

	Value		Quantity	
Year	000 US$	Index (2000 = 100)	Mt	Index (2000 = 100)
2000	9,463	100	7,519	100
2001	9,652	97	8,588	86
2002	10,387	104	9,024	90
2003	11,527	115	8,672	87
2004	13,290	133	7,073	71
2006	15,761	158	7,964	80
2007	18,748	187	8,110	81
2008	22,995	230	8,367	84
2009	13,264	133	4,855	49
2010	14,001	140	5,565	56

[a]MAFCP, Albanian Ministry of Agriculture, Food and Consumer Protection; [b]UNSTAT, UN Statistics Division (UNSD) Statistical Databases

great effort for Albanian operators, requiring millions of US dollars in investments, the acquisition of new skills, and an overall increase in business complexity.

At present, the exporters willing to remain among the leading players at national level must develop and control a business which is much more complex than in the recent past, when their activity was focused only on trading dried wild MAPs. Now, the core business of the leading entrepreneurs in the MAP sector is still international trade of dried wild MAPs, but to remain competitive, they must be also involved in other activities, such as MAP cultivation and processing to produce essential oils. In fact, they are now involved in three different business segments, which have different business models (i.e. dried wild MAP trading, which mostly consists of mastering logistics and conservation techniques; MAP cultivation, which is similar to the cultivation of any vegetable crop; and MAP processing to extract essential oils, which is an industrial distillation activity), while before they were focused only on one.

This evolution in the scope and complexity of the business of MAP exporters is forcing them to get more directly involved in previously neglected issues, such as the choice of varieties to be cultivated (each species of MAP has many varieties, suitable for different uses) and/or collected only as wild products (and therefore protected from over-exploitation), and the establishment of contractual agreements with farmers for MAP cultivation, with all the related functions (procurement of seeds and seedlings, training, supervision and coordination of production). Quality control is also becoming a more complex and burdensome activity, as it is now necessary to control the quality of three groups of products (cultivated MAPs, wild MAPs, and essential oils). Finally, as wild MAPs become scarcer and more expensive, more time and some investment must be addressed to ensuring that wild MAP harvesting and the subsequent drying process are properly performed. In the past, only a few international development projects provided some training to MAP harvesters, and the issue of MAP drying was completely left to regional collectors (i.e. to the exporters' suppliers). If the quality of wild MAPs was low because of improper harvesting or drying, the exporter simply did not purchase the product, or paid a lower price. Under the present market conditions, the loss of revenues and market shares caused by a wasteful management of scarce wild MAP resources is becoming too high to ignore these aspects, so exporters are forced to get involved in these steps of the supply chain too.

The trends for the future are strongly conditioned by the situation that has been described, leaving limited space for external intervention to modify their future course. One of these trends is that the export of wild MAPs will remain the core business for Albanian exporters, providing each year US$15–20 million in hard currency incomes, for a quantity of around 6000–7500 Mt/year. However, without a bold and comprehensive action of public and private stakeholders to reduce the depletion of natural resources, it is possible that these amounts will decrease to well below 5000 Mt/year. The three leading MAPs will keep their prominence, but other MAPs will be also much in demand, such as *Gentiana*, *Sideritis*, wild apples and *Satureia montana*. The current trends in the cultivation of MAPs, especially sage and oregano, will continue, with cultivated MAPs playing an increasing role in the business. This will create the need for developing a new agribusiness specialized segment, including seeds and seedlings, specialized technical assistance for agronomic practices, improved drying facilities, and specialized greenhouses.

As for the development and introduction of new products, it is likely that some investments will be made in the next few years to introduce some simple processing, to manufacture products ready for final consumption and based on dried MAPs, such as herbal teas, ground MAPs in retail packaging, and milled mixed MAPs. Establishing such production does not require major investments (anyhow, not in the range of several hundred thousand US$), or changes in the existing supply chain, and is fully within the capacity of the enterprises dealing with MAPs in Albania. Growth and development of MAP processing based on essential oils (i.e. the industrial processes that start with the production of raw essential oils) will continue, but slowly, as the size of investments required for establishing competitive facilities is quite large. More complex industrial activities, for which MAP essential oils are only a minor ingredient (such as detergents) will be probably established in the next few years, but is unlikely that a complete supply chain, starting with wild and cultivated MAPs, and up to the final complex products (prepared food, medicines, cosmetics), will be developed in the next 10 years.

The evolution of MAP sector has many elements in common with that of the fish and aquaculture sector. As with the fishing sector, the MAP sector benefited from an early recovery after 1991, because it could rely on a vast pool of high value natural resources. Both MAPs and fish immediately found a very receptive international market. In addition, Albania was already a major international player for MAPs. An early market leader emerged, with sufficient capital, know-how and entrepreneurship to act as a pivot for the whole domestic production, and to operate with success in international markets. With time, more competitors emerged. Most investments (all investments in the case of MAPs) have been addressed to obtaining more value added from the export markets, without investing in the sustainable use of natural resources. When these natural resources started to be depleted, the over-exploitation has not been addressed by investing in more sustainable practices; rather, investments were made to replace wild products with farmed products, even though these last have a lower value.

10.7 Conclusions

The collapse of the supply chains established during the planned economy and the following hastily organized and chaotically implemented privatization has had a particularly disrupting impact on the whole Albanian agrofood system. However, the large number of small self-sufficiency farms and micro-processing facilities that emerged from this disruption resulted, in many fields, as more productive than the late large-scale integrated companies owned by the state, in spite of major inefficiencies.

Segments of several supply chains started to recover immediately after 1991. Among these segments were primary products that could be produced with minimal investment and sold directly to consumers without any processing, such as vegetables. Other segments that experienced an early recovery were those processing units which could rely on cheap and easily accessible imported raw agrofood products, and on the few logistics facilities that kept working; examples of these processing activities are the milling industry and the meat processing industry. Finally, sectors that could rely on abundant and relatively little-exploited natural resources for which there was receptive foreign demand of products not needing much in the way of post-harvest processing or practices, such as wild caught fish, logging, and harvest wild MAPs, did well in the post-transition period. Thanks to their ability to rely on hard currency incomes and a stable market, these latter supply chains were also those in which more investment was possible from the beginning. In contrast, domestic production decreased or even disappeared when a minimal production scale or integration into supply chain systems was required, as in the case of cereals and industrial crops. Moreover, in general, the more a supply chain was complex or required cooperation between different actors (producers, traders, processors, service providers), the slower the recovery and start of a new development process.

From the demand side, consumption by the impoverished population at first contracted, but soon two main trends emerged, destined to have long-term effects on the whole of society. Firstly, a huge migration outflow (the resident population recorded in the 2011 census was smaller than in the 2001 census) in time generated a flow of remittances which proved to be the real engine of the Albanian economy. Second, a flood of urbanization built up along the Tirana–Durres axis, the two cities being 30 km apart. This created an economic and social core in the country, grouping some 40% of the population and an estimated 60% of national purchasing power. According to the 2011 census, more than 50% of the population now resides in urban areas, for the first time in the country's history. The urbanization, the flow of remittances, and the continuous contact with foreign lifestyles favored by proximity with EU member countries, and by the presence of large communities of new Albanian emigrants in many Western countries, contributed in changing the structure of demand of Albanian consumers in a relatively short period. The liberal international trade regime paved the way for an increasing flow of imports, while the flow of remittances sustained demand even if food prices were relatively high.

The second economic dip experienced in 1997 did not change these main trends, mostly having the effect of slowing down the recovery of the most dynamic segments of the supply chains, forcing the closing of non-viable enterprises (including some new investments made during the financial boom of 1996), and entrenching the lack of confidence in cooperation between Albanian economic actors. The remittances, the flow of international aid, and the money inflow related to the management of the Kosovo crisis (in 1999) contributed to gradually putting the country into a more stable growth pattern, which became visible from 2000 onward.

The relatively slow and gradual process of establishment of a network of newly built wholesale markets (the first being the market of Lushja, established in 2003), and rehabilitation of the main roads, provided the possibility for producers far from Tirana to sell in that market. This development, and the improved economic situation, created conditions for increasing investment in primary production. By 2005, most of the 'easy' investment and production growth had been achieved. From that moment on, a gradual saturation of the domestic market for some basic products (mostly products from animal production) was recorded, while other sectors, which had previously experienced a relatively slow growth (e.g. all fruit and vegetable products) continued on a path of sustained growth. In the past few years, signs of domestic market saturation are also emerging in some crop production supply chains, such as those for grapes and wine.

The more demanding skills and the capital required for more complex supply chain development triggered the next phase of growth, characterized by a more stable market, and by an overall process of consolidation and rationalization of an increasing number of supply chains. This process is still ongoing. The establishment of the supermarket chains, started in 2006, hastened this process, as customers are now provided with more product choice, and competition from imported products is fiercer. Also, supermarkets want to deal with large suppliers, thus making the process of consolidation faster for most supply chains, though so far they have not proved a real game-changing factor, except in some sub-sectors, such as the fruit and vegetable processing industry.

The process of consolidation in the food industry, and in commercial and semi-commercial farms, will be fostered by the structure of state support schemes and by the provisions of the EU pre-accession facility IPARD, when they are made available. All of these facilities are targeted to support the growth of the largest and more financially healthy enterprises, creating a growing and more clear-cut gap between subsistence and commercial farming, and between marginal/informal and formal/ medium-large (by Albanian standards) food processing companies.

Understanding the evolution of the agro-food sector over the last 20 years helps to identify some rather stable and predictable trends. First among these is that primary production will grow at a slower pace than in the 2000–2010 decade. Investments will be focused on intensification of production, such as better irrigation systems and simple greenhouse production. However, the still unresolved issues of the land market and the poor distribution system (in terms of skills, capital, and infrastructure) will affect the capacity of primary producers to further substitute imports, and to start exporting larger quantities of products in which Albania could be competitive (such as fruit and vegetables, MAPs, fish, and skins), at least in some markets and during some periods of the year. It is expected that in some sectors (fruit, vegetables, milk and dairy, fish, grapes and wine, and olives and olive oil), Albanian producers will manage to capture most or all the additional domestic demand, while their capacity to substitute imports will not be sufficient to reduce the value of imports, as demand for high quality and high value added products is and will remain mostly satisfied by imported products.

Another identifiable trend is that the strongest export oriented sectors, such as fish and fish products and MAPs, will have to face the increasingly strong depletion of natural resources. In the absence of effective policies and incentives for the sustainable use of such resources, the depletion of these reserves will continue, and enterprises will increasingly rely on aquaculture and on MAP cultivation and processing to maintain and increase their turnover.

At the same time, more sectors of the food industry will start a faster process of consolidation that will mostly involve milk and dairy, wine and fruit and vegetable processing industries. The introduction of the IPARD facility will be a game-changing factor, substantially speeding up this process, and creating a wider gap between larger companies and smaller ones, which will be gradually marginalized. The milling industry, the meat industry, and the olive oil industry are expected to maintain a structure similar to the present one, as the first two industries have already undergone this process of consolidation, and the olive oil industry has a peculiar production and distribution base that will allow the present fragmented structure to be kept.

The influence of supermarket chains will gradually change the consumption patterns of urban customers, pushing local producers and processors to introduce a wider range of products, and to become larger. From the consumer point of view, consumption will become less seasonal, and the range of products demanded will increase. However, in the next 10 years, the market share of supermarkets is not expected to grow dramatically, as traditional retail and green markets will continue to play a key role in supplying urban customers and green markets and personal consumption will continue to dominate rural demand. A side effect of this evolution will be the increase of losses in primary production of fruit and vegetables, as producers will be forced to widen the range of their products, but it will become much more difficult to balance the supply and demand of this wider range of products.

Overall demand will grow slowly, but in Albania, as in most countries, inequality is expected to grow too, with demand by most customers remaining stable or even being depressed by shrinking incomes, and demand by the small but richer segment of the population shifting toward more expensive products and services. This will also affect the business of restaurants, which flourished in the second part of the 2000–2010 decade, and will probably undergo a substantial process of selection, with a few trendy restaurants having good business, as well as taverns and fast food providing cheap meals, while a large number of restaurants, now targeted to the medium income segment of the population are expected to face a prolonged crisis.

In this framework, the supply chains facing the most serious challenges are: (i) the milk and dairy supply chain, as this sector has not yet managed to fully implement a much needed process of consolidation and improvement of standards; (ii) the primary production of cereals, as this continues to be non-competitive; and (iii) the fruit and vegetable processing industry, as it has not yet managed to find a sustainable path to growth, being focused on satisfying the basic needs of a small market, and finding itself not ready to cope with the new challenges and opportunities given by the growth of supermarket chains.

In the case of the milk and dairy industry, bolder action from the Albanian National Food Authority to curb informality in dairy processing units would be a major contribution to improving the situation. The Albanian economic environment is now strong enough to acquire more benefits (increased competitiveness of the formal dairy plants; better food safety; positive spillover on primary milk production, as a price-for-quality regime will be more consistently applied) than damage (less employment in the informal sector, which will be only partially balanced by increased employment in formal sector; loss of investments made in small and unsustainable dairy units; increase of average final price to consumers) from the formalization of the dairy sector. Less numerous, but stronger, milk processing units will also have the size and the financial means to obtain the maximum benefit from the EU-financed IPARD facility, which will provide grants for co-financing investments in the improvement of quality standards, but can be accessed only by enterprises that are already in the formal sector and have a sizable co-financing capacity. In parallel with the enforcement of rules (rules on food safety; labor and fiscal laws; rules on environment protection), actions should be also taken to reduce the production cost of raw milk, which is too high, considering the quality of milk (so that, in the case of investments in improving quality, the milk production cost would increase, too). The establishment of milk breeders' associative bodies for the common procurement of inputs is probably the single action that would contribute most to reducing production costs (FAO, 2008).

As for the fruit and vegetables processing industry, here the prospects are challenging. Informality and fragmentation are problems in this sector, and the growth of supermarkets will force also small formal enterprises to grow or to fall back into informality. The IPARD facility will also be available in this sector, but no more than five to ten enterprises will have the size and the means to be eligible for IPARD support. As a consequence, it is possible to expect that within 5–10 years, there will be no more than three or four national level competitors, all relatively small compared with regional competitors. An action that would help the growth of Albanian fruit and vegetable processing is support to the establishment of production contracts between producers and processors, a type of agreement whose importance and necessity is understood both by farmers and the processing industry, but which both sides are extremely reluctant to stipulate, knowing how difficult it is to establish mutual trust and to foresee contracts which would be at the same time flexible and applicable. The establishment of a well-functioning system of production contracts requires many years of experience and confidence building, and works only if there are strong representative bodies of farmers (producers' associations) and processing industries (sector associations) which can facilitate the standardization of contracts and conditions, and also provide assistance in the case of litigations or partial fulfillment of contractual provisions, cases that are extremely frequent in this field. The existing agribusiness organizations in Albania are not very effective, and so far have not managed to give any help in this sector.

In the primary production of cereals, Albania will never be really competitive, as its natural resources are more favorable to other crops. However, a more effective land market would contribute to creating the conditions for the establishment of larger and more efficient cereal growing farms.

In these three sectors, as in all the others, the role of the state and the impact of the EU process of integration is expected to speed up or slow down some processes that are already ongoing, but not to change the overall trends much. The experience of both past and ongoing state support schemes (mostly provided by MAFCP through the Rural Development Agency), and of the EU pre-accession schemes in other countries in Central Europe and in the Balkans, shows that these support schemes usually speed up already ongoing processes, and sometimes act as a trigger to anticipate the start of changes which had been already perceived as a need by all stakeholders in the supply chains, but the schemes do not generate change on their own.

The main common constraint that so far affected the development of all the sectors analyzed in this chapter is the lack of trust of farmers, processors, and other supply chain operators in the advantages of interaction between different supply chain operators, which leads most agribusiness entrepreneurs to focus on short-term profit and on keeping their businesses as simple and as integrated as possible (i.e. keeping direct ownership of production, processing, and the distribution network). This lack of confidence is particularly high when private entrepreneurs interact with the public sector.

Reckless exploitation of natural resources (fish, wood, MAPs), mistrust between farmers and processors (supply chains for fruit and vegetables, milk, grapes and wine, and olive oil), and weak supply chain governance (meat, milk, small ruminant breeding), have strongly conditioned the development of Albanian agribusiness, which, nonetheless, has managed to consolidate and grow in the long transition period. Currently, many of the basic investments required for the functioning of the agrofood system have been made, the 'easiest' efficiency gains have been achieved, and many market gaps have been filled, so Albanian entrepreneurs, regulators, and policymakers must realize that a new phase is beginning, characterized by slower, but more stable, growth, and by the need for a much closer interaction between agribusiness stakeholders in building up complex supply chains. Transition, at last, is largely over, and Albanian entrepreneurs and decision-makers can and should focus on longer term development perspectives.

References

Agrotec (2008) *Grape Production and Consumption Patterns*. Report for EC-CARDS (Community Assistance for Reconstruction, Development and Stabilisation) Project "Creation of a Vineyard Cadastre", Tirana, Albania. Agrotec, Rome.

Cela, R. and Hizmo, A. (2009) *Analysis of Renewable Energy and its Impact on Rural Development in Albania*. Report for Project "AgriPolicy" under the European Commission under the 7th Framework Programme. European Commission, Brussels. Available at: http://www.euroqualityfiles.net/AgriPolicy/Report%202.2/Agripolicy%20WP2%20D2%20Albania%20Final.pdf (accessed 10 September 2012).

Chan-Halbrendt, C., Imami, D., Zhang, Q., Zhllima, E. and Leonetti, L. (2010a) Latent class analysis of consumer preferences for wine in Tirana, Albania. Paper presented to: *20th IAMA [IFAMA] Annual World Forum and Symposium: Navigating the Global Food System in a New Era, 19–22 June, Boston, USA*. International Food and Agribusiness Management Association. Available at: https://www.ifama.org/events/conferences/2010/cmsdocs/134_paper.pdf (accessed 10 September 2012).

Chan-Halbrendt, C., Zhllima, E., Sisior, G., Imami, D. and Leonetti, L. (2010b) Consumer preference for olive oil in Tirana, Albania. Paper presented to: *20th IAMA [IFAMA] Annual World Forum and Symposium: Navigating the Global Food System in a New Era, 19–22 June, Boston, USA*. International Food and Agribusiness Management Association. Available at: https://www.ifama.org/events/conferences/2010/cmsdocs/135_paper3.pdf (accessed 5 September 2012).

DSA (2009a) *Market Survey and Value Chain Analysis of Apples in Albania*. Development Solution Associates, Tirana, Albania.

DSA (2009b) *The Olive and Olive Oil Value Chain in Albania*. Report from Project "Formulation of a Technical Assistance Project in Support of Albania's Olive Oil Sector". Development Solution Associates, Tirana, Albania.

DSA (2010) *Analyzing Consumer Preferences for Lamb Meat in Albania*. Report from Project "Improving the performance of livestock sector in Albania". Development Solution Associates, Tirana, Albania.

DSA (2011) *Agribusiness Sector Survey*. Report from Project "Italian–Albanian Program for the Development of Small and Medium Enterprises". Development Solution Associates, Tirana, Albania.

FAO (2008) *Terminal Report, Agriculture Production Support in Albania*. FAO (Food and Agriculture Organization) Country Office for Albania, Tirana Albania.

INSTAT (2000) *Households Budget Survey in Urban Areas*. Instituti i Statistikës (INSTAT), Tirana, Albania.

Kathe, W., Honnef, S. and Heym, A. (2003) *Medicinal and Aromatic Plants in Albania, Bosnia-Herzegovina, Bulgaria, Croatia and Romania*. BfN-Skripten 91. Study carried out by WWF Deutschland, Traffic Europe-Germany for Federal Agency for Nature Conservation (BfN), Bonn and Vilm, Germany. Available at: http://www.bfn.de/fileadmin/MDB/documents/skript91.pdf (accessed 12 September 2012).

Leonetti, L. and Kristo, I. (2005) *The Food Chain Structure of Small Ruminants Meat and Dairy Products in Albania*. Report prepared for UNDP (United Nations Development Programme) Project SMS, UNDP, Tirana, Albania.

Leonetti, L. and Kristo, I. (2010) *The Value Chain of Small Ruminants Meat Supplying the Tirana–Durres Urban Area*. Report prepared for UNDP (United Nations Development Programme) Project "Improving the Performance of the Livestock Sector in Albania". UNDP, Tirana, Albania.

Leonetti, L., Imami, D., Kongoli, R., Nani, A., Rapushi, P. and Xhafa, F. (2010) *Fruit and Vegetables Sector Study*. IPARD (Instrument for Pre-Accession Assistance for Rural Development) Report from Project "Capacity Building for Implementing the Rural Development Strategy", Tirana, Albania.

Mautner Markhof, M., Musabelliu, B. and Zusi, A. (2010) *Meat Sector Study*. Meat IPARD (Instrument for Pre-Accession Assistance for Rural Development), Report from Project "Capacity Building for Implementing the Rural Development Strategy", Tirana, Albania.

OIE (2008) *OIE Tool for the Evaluation of Performance of Veterinary Services (OIE PVS Tool)*, 3rd edn. OIE-World Organisation for Animal Health (OIE), Paris. Available at: http://web.oie.int/downld/SC/EN_OIE%20PVS%20Tool_2008.pdf (accessed 12 September 2012).

REC (2000) *Annex 1: Country report Albania within Strategic Environmental Analysis of Albania, Bosnia & Herzegovina, Kosovo and Macedonia*. Overall study prepared for SIDA by Scandiaconsult Natura and Regional Environmental Center for Central and Eastern Europe (REC). Country report Albania prepared by REC Country Office in Albania, Tirana, Albania. Available at: http://www.catsg.org/balkanlynx/06_conservation/pdfs/REC_2000_Country_report_Albania_Strategic_environmental_analysis.pdf (accessed 16 September 2012).

Schippmann, U., Leaman, D.J. and Cunningham, A.B. (2002) Impact of cultivation and gathering of medicinal plants on biodiversity: global trends and issues. In: *Biodiversity and the Ecosystem Approach in Agriculture, Forestry and Fisheries. Satellite Event on the Occasion of the Ninth Regular Session of the Commission on Genetic Resources for Food and Agriculture, Rome, 12–13 October 2002*. Inter-Departmental Working Group on Biological Diversity for Food and Agriculture, FAO (Food and Agriculture Organization), Rome. Available at: ftp://ftp.fao.org/docrep/fao/005/aa010e/AA010E00.pdf (accessed 12 September 2012).

Schröder, E., Uruçi, M. and Meço, M. (2010) *Milk Sector Study*. Milk IPARD (Instrument for Pre-Accession Assistance for Rural Development), Report from Project "Capacity Building for Implementing the Rural Development Strategy", Tirana, Albania.

UNDP (2005) *The Agribusiness Sector in Albania: Inward Investment Opportunities*. United Nations Development Program/European Union Trade Liberalization and Promotion Program. Report produced by independent consultants on behalf of the Albanian Foreign Investment Promotion Agency (ANIH), UNDP Albania and the European Union. UNDP, Tirana, Albania. Available at: http://www.trade.undp.org.al/u/d/trade/Albanian%20agribusiness%20sector.pdf (accessed 3 September 2012).

World Bank (2007) *Albania: Strategic Policies for a More Competitive Agriculture Sector*. Report No. AAA18-AL, World Bank Sustainable Development Sector Unit, Europe and Central Asia, Tirana, Albania. Available at: http://siteresources.worldbank.org/INTALBANIA/Resources/Strategic_Policies_for_more_competitive_Agriculture_sector.pdf (accessed 3 September 2012).

Zickel, R. and Iwaskiw, W.R. (eds) (1994) *Albania: A Country Study*. GPO [Government Printing Office] for the Library of Congress, Washington, DC. Available at: http://countrystudies.us/albania/index.htm (accessed 10 September 2012).

11 The Effects of EU Integration on a Transitioning Economy

Birgit Schaefer,[1] Roland Cela[2] and Shkelzen Marku[3]
[1]*Naccon GbR, Tuebingen, Germany;* [2]*Deutsche Gesellschaft für Internationale Zusammenarbeit (GIZ), Tirana, Albania;* [3]*Partnership for Development Foundation, Tirana, Agricultural University of Tirana and Mountain Area Development Agency, Tirana, Albania*

11.1 Agricultural and Rural Development Policy in the Context of the EU Integration Process

11.1.1 EU Common Agricultural Policy (CAP)—the concepts of the first and second pillars

The CAP of the EU (European Union) Member States (MS) celebrated its 50th anniversary in 2012. It is thus one of the oldest policies of the EU and is strongly rooted in the European integration project of having a reconciled, democratic, value-based, and enlarging community for increased prosperity and competitiveness. The strong rationale of the EU for pre-accession interventions in Central and Eastern Europe is to have an instrument enabling the realization of these objectives after the end of the Cold War and of dictatorial regimes such as those in Albania or Romania, as well as the end of the war in former Yugoslavia in the 1990s. Membership is subject to a gradual and carefully managed enlargement process. Accession to the EU will only take place when the candidate country meets the necessary legal, political, and institutional requirements, as well as the specific requirements in each sector, such as in agriculture and rural development. The different EU accession instruments that have been implemented since 1989 to link financial support to progress on political criteria and structural changes thus act as a supranational and interregional catalyst making use of the expertise of different administrations and agencies within the EU. Thus, the alignment of Albania's agriculture and rural development sector with the CAP requirements will not only gradually lead economically viable farms toward market orientation, but also provide food security, relative price stability, and a positive outlook for farmers' incomes in a rural economy.

Since its creation, the CAP has always been adapted to respond to the challenges of its time. In the 1960s, the CAP started operating with the policy focus of providing enough food at affordable prices for a Europe emerging from a decade of war-induced shortages. This included subsidizing production and supporting prices for farmers by buying up surpluses, which continued during the Cold War.

The first real attempt to reform farm policy and to increase the competitiveness of Europe's farmers was in 1992, and targeted at lowering institutional prices for key products and offseting the impact of these cuts on producer incomes by means of direct payments. To improve price competitiveness on both domestic and world markets, the Agenda 2000 reforms further reduced internal market prices and compensated farmers by increased direct payments under what became known as the first pillar of the CAP. Because production systems differ across the EU, parts of the direct payments in the beef and dairy sectors were paid to MS in the form of national financial allocations. From 2004 onward, MS, including the New Member States (NMS), were be able to distribute the aid according to their own priorities, subject to certain Community criteria to prevent distortions of competition.

With the reforms of the CAP in 2000, rural development became the second pillar of the CAP to support the broader rural economy. The multifunctionality of agriculture was recognized, and various structural and territorial measures were brought together in order to take into account consumers' and taxpayers' interests in food safety and environmental aspects. Significant reforms have been made in the past decade, notably in 2003, with the decoupling of subsidies from production in most sectors, and with the so-called 'modulation'. This is a mechanism to transfer funds from direct subsidy payments under Pillar 1 to rural development expenditures under Pillar 2. With the CAP health check in 2008, the Commission proposed reforms for the period 2009–2013 to further modernize, simplify and streamline the CAP and remove the remaining restrictions on farmers to help them respond to the growing demand for safe food.

Today, the two pillars work together in a complementary way toward the CAP objectives, with rural development responding to the needs for structural adjustment generated by reforms in the first pillar. Under Pillar 1, all previous production-linked payments are bundled into a Single Farm Payment (SFP). SFP is paid to farmers on the basis of payments received during a reference period, or the number of eligible hectares farmed during a reference period. To be eligible to receive SFP, farmers are obliged to meet animal welfare, plant health, hygiene and environmental minimum standards, and to keep their land in good agricultural and environmental condition (GAEC). This element is the 'cross-compliance' conditionality and has been presented as the public benefit delivered in return for SFP, out of which direct payments to farmers and market management measures are funded. Although the decoupling of subsidies was designed to encourage farmers to produce in response to market demand, a series of market management instruments have been retained, but limited. These include production quotas (with the abolition of milk quotas in 2015), non-mandatory set-aside (in which farmers must have left a proportion of their arable land fallow, or use it for non-agricultural purposes), and export refunds, as well as a gradual removal of the products from intervention storage, which is aid to public or private storage whereby the EU buys an agricultural product, such as, for example, beef, if prices in a MS or region fall under a certain value).

Rural development policy under Pillar 2 consists of a range of structural and environmental investment measures that are grouped into three thematic axes. Axis 1 covers measures designed to improve the competitiveness of the farming, food processing and forestry industries. Axis 2 covers environmental and land management schemes, while Axis 3 covers initiatives aimed at improving the quality of life and economic diversification in rural areas. A fourth element of the fund is reserved for Liaison Entre Actions pour le Développement de l'Économie Rurale (LEADER, or 'Linkage Between Actions for the Development of The Rural Economy') initiatives, whereby local action groups in rural areas can secure funding for local development projects. The policy operates on the basis of a multi-annual programming approach, in which MS design and co-finance Rural Development Programs (RDPs) tailored to their specific needs, but which must be approved by the European Commission (EC). There are 94 national and regional programs operating in the current period.

The CAP is financed by two funds: the European Agricultural Guarantee Fund (EAGF) finances Pillar 1 measures, while the European Agricultural Fund for Rural Development (EAFRD) finances the RDPs under Pillar 2. At around 53 billion euros a year, the CAP represents some 40% of the EU's long-term budget for 2007–2013, compared with nearly 71% in 1984, because other structural funds for economic and social cohesion objectives, such as for the EU's Regional Policy, have more than doubled over the same period. With the CAP post-2013 reforms, the figure is expected to fall further, to some 36%. The majority (over 70%) of CAP spending goes to direct payments for farmers, while some 20% of the CAP budget is spent on rural development measures. The rest is paid as export subsidies to food companies.

The CAP challenges across Europe for the upcoming programming period of 2014–2020 are linked to the long-term strategic objectives of the Europe 2020 Strategy of having smart, inclusive and sustainable growth. The CAP 2014–2020 legal proposal includes three main policy aspects (EC, 2010b). The first of these is contributing to the stability of the agricultural sector and rural communities in light of an enhanced volatility in the markets. This has occurred through the integration of global commodity markets with financial markets, and as a result of insecurities in agricultural production deriving from climate change. Fostering stability will also be achieved through a closer link between energy markets and agricultural markets. A second policy aspect is ensuring efficiency and equal distribution of support among MS, in particular under Pillar 1. The third aspect of the policy is environmental and rural preservation through fostering innovation, fighting climate change, and mobilizing the full potential of rural areas through a more effective application of the support instruments. This will be achieved through enhanced flexibility, administrative simplification, new delivery mechanisms, and further coordination with other EU funds (EC, 2011a; EPEC, 2011). The reform objectives of enhanced competitiveness, increased sustainability, and greater effectiveness reinforce the need for an outcome oriented approach. However, the different dynamics of structural changes in the agriculture sector of the NMS might be only partly reflected, as there are still tremendous disparities in the labor force, farming structure, and rural economy. Although all MS will be obliged to move gradually toward a uniform area payment under Pillar 1, at the same time farmers will have to deliver more public goods, and thus incur costs, in exchange for the subsidies.

11.1.2 The EU Instrument for Pre-accession Assistance—component V measures (IPARD 2007–2013)—latest developments

Since 2007, EU pre-accession funding in candidate and potential candidate countries has been governed by a single instrument, the Instrument for Pre-accession Assistance (IPA), with a financial envelope of 11.5 billion euros for the period 2007–2013. IPA interventions are grouped under five components: I. Transition Assistance and Institution Building; II. Cross-Border Co-operation; III. Regional Development; IV. Human Resources Development; and V. Rural Development. The instrument directly addresses the compliance of beneficiaries with the political and economic accession criteria, and the fulfilment of obligations as a future MS. Although it covers both the candidate and potential candidate that stand to benefit from the IPA, the last three components prepare for the implementation of EU cohesion and rural development policies, and are currently available to EU candidate countries only. Recognized candidate countries have reached the last stage of accession negotiation with the EU based on a country specific framework setting out the conditions and timing of the candidate's adoption, and the implementation, and enforcement of all the EU rules and institutional set-up (*acquis communautaire*) already in force. Potential candidates have signed Stabilization and Association Agreements (SAA) with the EU, which precede the lodging of membership applications. As of early 2012, Iceland, the former Yugoslav Republic of Macedonia, Montenegro, Serbia, and Turkey have candidate status. Albania, Bosnia and Herzegovina, and Kosovo (under United Nations Security Council Resolution 1244) are potential candidates. Croatia has successfully finalized its negotiations and has become an acceding country with EU membership on 1 July 2013.

The fifth IPA component—IPA Rural Development (IPARD)—encompasses the alignment of the CAP Pillar 2 *acquis communautaire*, in addition to the SAA through which the establishment of competition regimes are encouraged. IPARD is implemented by multi-annual RDPs, and in a fully decentralized manner guided by provisions laid down in bilateral agreements that are alike in MS. Measures available under the three priority axes of IPARD are based on the wide support open for MS, and selected based on the experience of the NMS through the Special Accession Program for Agriculture and Rural Development (SAPARD) during the period 2000–2006 (Schiessl, 2009).

IPARD Axis 1 measures aim to improve the market efficiency and implementation of Community Standards through assistance to farms and to the food industry to modernize, and to set up producer groups. Support under Axis 2 entails preparatory actions toward implementing agri-environmental measures to improve the environment and the countryside. It also encompasses actions to prepare for the LEADER approach to support local rural development strategies. Axis 3 aims to boost the rural economy through the improvement of rural infrastructure, the development of economic diversification activities, and the improvement of vocational training for farmers and rural entrepreneurs. The aid intensity of all income-generating investments under IPARD is, in general, up to 50% (the national contribution is 25% and the EU contribution is 75% of the total public share), as well as 50% co-financing from the beneficiary of the total eligible costs, which can decrease for young farmers and farmers in mountainous areas. In addition, the ministries of agriculture are eligible for technical assistance measures toward effectively implementing, monitoring, evaluating, and controlling the IPARD program.

Decentralized management is used for the implementation of IPARD programs, and the EC must confer the management of the RDP to the candidate country. The IPARD accreditation process takes several years, as it is demanding for the national institutions involved to fulfill the CAP *acquis communautaire* requirements once the country has been granted candidate status by the European Council. In the first step, the relevant Ministry of Agriculture has to develop and negotiate the multi-annual IPARD program based on the national strategy for rural development. This RDP must be adopted by the EC. The second step sets out the bilateral rules for co-operation concerning EC financial assistance for the candidate country in the IPA Framework Agreement (FA) and IPARD Sectoral Agreement (SA), which must be concluded between the EC and the national government. The third step is the establishment of an EU-conform management and control structure for the implementation of the IPARD program. This so-called IPARD Operating Structure (OS) consists of a Managing Authority (MA) and a Paying Agency (PA). Their accreditation contains the assessment of the readiness of the institutional structure, the quality and consistency of the written administrative procedures, as well as the qualification of the staff against the accreditation criteria laid down in particular in the SA. This is firstly done through a compliance assessment, and is followed by national accreditation (step 3) through the National Authorizing Officer (NAO) in the Ministry of Finance, in the case of successful results. For the third step, the NAO must have been accredited beforehand through the national Competent Accrediting Officer (CAO), which is a body responsible for issuing, monitoring and suspending or withdrawing accreditation of the NAO and the National Fund. After national accreditation of the IPARD OS, the auditors of the EC's Directorate-General for Agriculture and Rural Development (DG AGRI) undertake a pre-accreditation review. After a successful screening, the EC accredits the IPARD OS (step 4). The fifth step, which follows immediately after this conferral of management powers of EC funds to the IPARD OS, is the EC endorsement of the multi-annual Financing Agreement for IPARD. This is the starting point for the launch of the IPARD investment scheme (Schiessl, 2009).

Reform of the pre-accession instrument IPA for the new programming period 2014–2020 depends on the direction of the CAP policy and new IPA regulations based on the Europe 2020 growth strategy (EC, 2010b). Socioeconomic indicators show that, with the exception of Iceland, the enlargement countries are still well below the EU average and even below the level of the weakest MS, which calls for substantial investments to bring these countries closer to EU standards. This would allow them to take on board the obligations of membership, and to withstand the competitive pressures of the single market, as well as to withstand global challenges. Post-2013 pre-accession assistance is expected to have more flexibility by supporting all policy areas under the IPA components in all beneficiary countries, regardless of status, using a phased approach. This will make possible different scope and aid intensity on the basis of needs and technical and administrative capacities. Further reforms of the IPA concern an increased geographic flexibility to respond to interregional challenges, the use of innovative financial instruments to leverage more funds from other donors or the private sector, earlier use of the 'technical assistance' measure to foster capacity building, and the implementation of new measures, such as renewable energy or forestry, to better address the new challenges of 2020 (Dormal-Marino, 2011; EC, 2011a; Lazdinis, 2011). In its Communication of June 2011 'A Budget for Europe 2020,' the EC proposed to allocate an amount of 14.11 billion euros (current prices) to the new IPA for the period of 2014–2020.

11.2 Albanian Agriculture and Rural Development Policy in the Context of the EU Integration Process

11.2.1 Overview of the Albanian agriculture sector

Agriculture plays an important role in the economic development of Albania although declining trends have been reported. In 2010, the share of agriculture in the gross domestic product (GDP) was about 17%; 48% of the total labor force was engaged in agriculture and nearly 60% of the rural population gained their income from agriculture, which is high compared with the EU-27 (EU MS as of 1 January 2007) and neighboring countries (INSTAT, 2011). In the last 20 years, the main source of agricultural growth in Albania was total factor (land, labor, and capital) productivity growth. Macroeconomic stability and economic liberalization led to reallocation of resources from less profitable activities (cereals) to more profitable activities (horticulture and livestock), which allowed agriculture to generate growth with low rates of investment. However, there is evidence that the total factor productivity growth caused by post-transition reallocation is gradually coming to an end. More than 20 years after transition began in the early 1990s, the challenges for policymakers remain: how to transform agriculture from subsistence oriented production into a modern, resource efficient and competitive sector in the context of current and future EU requirements, while fostering alternative income opportunities for the rural population.

Agricultural land (about 1.1 million ha) covers about 39% of the total surface of the country, of which 699,000 ha, or 63%, is arable land and fruit trees. The rest is forest area (about 36%), and other land (about 25%). According to various studies, the broad land cover categories have changed relatively little during recent years. Over the last 15 years, the agricultural land area decreased by 1.4%; this is mainly attributed to the expansion of urban areas on former agricultural land (mostly in the surroundings of Tirana and other major cities), and there are no pronounced hot spots of change across the rest of the country. Albania has extensive underground water resources and a favorable climate, which provide a relative advantage for the production of certain agricultural products, and enable the early production of vegetables. Seven main rivers run from east to west of the country, and there are also about 650 small and medium reservoirs for irrigation, which have a total storage capacity of 560 million m^3 (Shelzen *et al.*, 2010).

Out of the 699,000 ha of arable land and orchard, about 561,000 ha (or 80%) are privatized and owned by rural families; about 137,000 ha (20%) are still state property, of which about 110,000 ha (15.7%) are managed by communes and municipalities, while about 27,000 ha (4.3%) are still public land. However, land management still faces significant problems. Some of the most important issues are: completion of the land registration process, and related legal and administrative regulation of land ownership rights and documentation; resolution of land property conflicts; development of the land market for agriculture use; improvement of the management and protection of privatized agricultural land; and improvement of land use profitability (Shelzen *et al.*, 2010).

In 2010, the gross agricultural output (GAO) amounted to 1.509 million euros, of which livestock contributed 53%, crop production 32%, and orchards 15%. Although only a limited number of farms are market oriented at this time, Albania's agriculture has some comparative advantages: based on the Albania to EU-27 yield ratio indicator for 2009, it becomes evident that the country has been efficient in the production of fruit (particularly tree nuts fruit) and vegetables. In terms of specific products, Albania has absolute advantages in the production of chestnuts, grapes, figs and cherries, cauliflowers and broccoli, and dry beans. Since 2002, a slow process of land consolidation and farm specialization has been taking place, in particular in the fruit, vegetable, and livestock sectors. Land renting has contributed to an increased efficiency of consolidated farms, in line with the evolution of market demand, and opportunities for raising productivity. For example, the number of larger cattle farms, with livestock of more than five and up to 50 head, has almost tripled since 2005. Also, the number of commercially sized flocks of sheep and goats has increased rapidly. Albania has a large and increasing trade deficit in agricultural and agri-food products, with a ratio of imports to exports of 9:1 in 2010. The share of this product group in imports has been quite stable in recent years (2005–2010) at around 17%, while its share in exports is fluctuating at about 6–8%. About one fifth of the total Albanian foreign trade deficit is due to the agricultural and agri-food sector. Between 2005 and 2009, the trade deficit in agri-food products increased by 58%; it worsened particularly in 2007 and in 2010 as a result of falling exports of some agricultural products, and rising imports. The trade deficit in agricultural products in recent years has been negatively influenced by the increase of import prices, which has been higher than the increase in prices for Albanian exports (INSTAT, 2011).

In 2010, the value of Albanian agri-food exports amounted to 66 million euros, of which about 76% went to EU countries (particularly Italy, Greece, Germany, and Austria), about 15% to the Central European Free Trade Agreement 2006 (CEFTA) countries, and 9% to the rest of the world. Neighboring Western Balkan economies are starting to become more important in Albania's trade in agricultural products, particularly due to free trade agreements (FTAs) signed by these countries. A significant increase (about 56%) in the volume of trade with these countries was observed in 2010, with exports increasing by about 60%, and imports by 55%. The main export products (about 50% of them) are tinned fish, medicinal herbs and spices (of which 80% of herbs and spices are exported to the USA), raw hides and skins, edible offal, and oilseeds. For the first time, in 2010 and 2011 exports to Lithuania, Hungary, and the Czech Republic increased at a remarkable rate. In 2010, the import of agricultural products amounted to 581 million euros; about 57% of these were imports from the EU (Italy and Greece, mainly), about 14% from CEFTA countries, and 29% from the rest of the world. The most significant imports of agricultural products in value terms were livestock and dairy products, cereals (wheat and maize) and fruits, alcoholic beverages, and tobacco and manufactured tobacco (INSTAT, 2011).

Despite some recent positive developments, Albania's exports of processed agricultural products remain low owing to the lack of technology, such as modern storage facilities, and facilities for the packaging and labeling of processed products. In addition, a substantial part of the country's farming sector remains predominantly subsistence oriented, and provides little income to rural families. As a result, less than half of field crop production, about two thirds of fruit trees, grapes, livestock products, and a low percentage of farm-processed products reach the market. Thus, the overall efficiency of small farmers in Albania is still low. According to a World Bank study (2007), the average farm is operating at only one third of its production potential. The four major sources of inefficiency are: (i) small size and fragmentation of land due to the equal land distribution policy in the 1990s; (ii) underutilization of modern production techniques; (iii) lack of proper irrigation systems; and (iv) limited access to extension services. Only a few farmers refer to the extension services for information and advice on seeds, crops, pest control, livestock breeding, and difficulty in access to finance, which hampers farm modernization and growth.

The Albanian agri-food sector needs significant increases in both public and private investment. However, future access to IPARD funds is linked to farm economic viability. The current level of investment at farm level is modest, mainly because of the low comparative profitability of farming (compared with emigration and non-agricultural activities), small and fragmented farm sizes and, sometimes, land ownership insecurity, poor infrastructures related to farming (irrigation and roads) and limited access to external financing (loans), mainly due to the lack of collateral. The main sources of investments for the majority of farmers are their own resources, which cover up to 61% of any investment. Commercial farmers have access to loans from commercial sources as well, with 28% of their investment needs covered by loans from credit institutions. Donors' projects represent an important investment source as well.

11.2.2 Strategic framework for agriculture and rural development

The Law on Agriculture and Rural Development, No. 9817 of 22 October 2007, provides the legal basis for agriculture and rural development, and for the transposition of the relevant CAP *acquis communautaire*, while a range of forthcoming bylaws are required for the implementation. This Law determines the main objectives and policy measures for agriculture and rural development, the provision of public services for agriculture, research and training, and the setting up of an information database. In addition, it provides the main elements required for the establishment, functioning, and roles of key institutions responsible for the design, implementation, monitoring, and evaluation of the agriculture and rural development policy and related support measures. According to the provisions of the Law, support to farmers is to be provided under the following objectives:

- *Improving the competitiveness of the agriculture and agri-processing industry.* Measures targeting human resources development (improvement of knowledge); land management (land improvement, irrigation, etc.); modernization of working capital and support to operational costs (buildings and equipment, provision of agriculture inputs, etc.); compliance with production standards related to the environment, food safety and animal welfare; agriculture and the food chain (better marketing infrastructure, producer associations).
- *Land management and improvement of the environment.* Measures related to different levels of direct payment, and projects that enhance the sustainable use of resources and improve the situation for less favorable production conditions, maintenance and preservation of the environment, biodiversity and landscape.
- *Improving the quality of life and encouraging diversified economic activities in rural areas.* Investment measures, as well as projects that enhance investments in the rural economy to improve quality of life, services and training, and preserve the rural heritage.
- *Development of local initiatives.* Measures for the development of local initiatives of farmers, people working in forest management, and other rural actors to preserve and develop the natural and cultural local heritage, improve the environment, and promote local brands, tourism and the use of renewable energy resources, as well as the creation of public–private partnerships, encouragement of entrepreneurship, etc.

The two main strategic documents that operationalize the agriculture and rural development objectives are the 'Agriculture and Food Sector Strategy 2007–2013' (AFSS) and the 'Rural Development Crosscutting Strategy for 2007–2013' (RDCS), prepared by the Albanian Ministry of Agriculture, Food and Consumer Protection (MAFCP) and approved by the Council of Ministers. The main objectives of the AFSS 2007–2013 are: (i) sustainable management of the agricultural land; (ii) improvement of employment, farm incomes and livelihood of rural families; (iii) improvement of the economic efficiency of the agriculture and food industry sectors, and increasing the productivity and quality of production; (iv) improvement of food safety and consumer protection; and (v) improvement of agricultural marketing. The key priority sectors defined for support are: fruits, olives and vineyards; livestock production; and the collection, storage and processing of fruit, grapes, vegetables and livestock products (milk and meat). The RDCS

2007–2013 takes account of the four priority axes of the EU CAP Pillar 2 for MS, and their associated support measures, rather than of particular measures required for Albanian agriculture and the rural economic sector to become more competitive during the pre-accession period. In line with the EU pre-accession process, the two national strategies (AFSS and RDCS) will need to be integrated into one coherent strategic document for agriculture and rural development, and complemented by the IPARD program to achieve an integrated programming approach. This would allow maximization of the impact of public expenditure in the agri-food sector and in rural development, as well as prepare Albanian farmers and the rural population to absorb upcoming funds from the EU to upgrade to Community standards. The 'National Strategy for Development and Integration 2007–2013' (NSDI), which contains short and middle term objectives for all sectors, and the Mid-term Budget Program, which identifies programs, objectives, products, and activities that are scheduled within the government fiscal plan, are two other important general strategic documents (Shelzen *at al.*, 2010).

The implementation of national agriculture policy measures is done based on Annual National Action Plans (ANAPs), and on annual budgets, approved by the Council of Ministers. The ANAPs define specific measures for implementation of the agricultural and rural development policy in the respective year. These measures mainly aim to: (i) target beneficiaries; (ii) present the conditions, criteria, and procedures for the implementation of the measures; (iii) give the financial plan for the implementation of general measures; (iv) present the institutional mechanisms; and (v) describe the procedures for the monitoring and evaluation of agriculture and rural development policy.

The key institutions responsible for the implementation of Law No. 9818 are the Inter-Ministerial Committee for Agriculture and Rural Development (IMCARD), the PA and the Rural Development Directorate (RDD) within the General Directorate for Agricultural Policy (GDAP) of the MAFCP. IMCARD is an overall coordinating and advisory body of the Council of Ministers which gives opinions on the most important decisions in the field of agriculture and rural development, and especially on: proposals on the Annual Plan for agriculture and rural development; monitoring and implementation of the Annual Plan; legal acts; and international agreements in the field of agriculture and rural development, etc. The RDD coordinates and chairs the drafting of the Annual Plan, and the guidelines for its application and instructions. Since the beginning of 2009, the Agriculture and Rural Development Agency (ARDA) has performed the functions of the PA. ARDA is in charge of organizing publicity and information campaigns, contracting, the execution of on-the-spot checks, verification of costs, ex-post control, payments, accounting, establishment of an integrated management and control system and database, establishment and implementation of market information systems, and elaboration of reports and analyses. The Regional Departments of MAFCP support ARDA with the publicity and information campaigns, admission and registration of applications, administrative checks and execution of ex-ante/ex-post controls, on-the-spot checks, etc. (Shelzen *at al.*, 2010).

In the interest of improving the competitiveness of Albanian agriculture, higher production should be the result of increased utilization of the agricultural land area, via both revitalization of land that has been left uncultivated and the cultivation of plants and breeding of animals that have higher productivity. This goal can be achieved through (direct) support (investments) and through technical assistance. The AFSS suggests support measures in the form of grants or credit, and interest rate subsidies (or credit guarantees) for investments in production technologies, new plantations or animal breeding, equipment, agricultural machinery and storage capacities, etc. Support is anticipated for both production and processing levels of the agri-food chain. However, such measures should also include improvement in food safety standards, which will depend on the degree of improvement of production and marketing technologies, the improvement of food safety legislation through approximation with the EU legislation, and proper implementation, including improvement in monitoring capacities in terms of both human and laboratory infrastructure.

The AFSS considers enhancement of the agricultural information (and statistics) system as a priority to enable improvements in marketing as well as agricultural land use.

Since 2000, in addition to the introduction of specific value chain based support measures, Albania has placed increasing emphasis on improving local governance, and the capacity for participatory local development planning and implementation. This idea was first supported by the government policy for administrative decentralization and increasing devolution of decision-making and resources to municipalities and communes, and later by the development of the regional development policy. At the national strategic level, territorial rural development, such as is also being implemented through the EU LEADER approach, was first introduced as a policy priority with RDCS 2007–2013. Given the lack of tradition in local development planning, the efforts have concentrated on piloting participatory territorial development projects, and learning from their experience in order to gradually develop local capacity and forms suitable to the socioeconomic and cultural context of Albania. Therefore, during the 2007–2013 period, the strategy prioritizes the support of initiatives for local development, funded within national or donor supported projects for territorial and regional development, within which initiatives for local development planning are incorporated.

The participatory approaches for local territorial planning were incorporated in the design of a number of projects implemented through national and donor support. Such EU approaches for local rural development were promoted by the establishment of a LEADER-type organization through training in and visits to LEADER local action groups (LAGs) in EU MS, supported by the Mountain Area Development Agency (MADA), the EU Prepare Network, the Hungarian Rural Parliament, the Italian Development Co-operation, or Deutsche Gesellschaft für Internationale Zusammenarbeit (the German Agency for International Cooperation, GIZ). An example of this approach is the development of local fora in the Mountain Area Development Program funded by the International Fund for Agricultural Development (IFAD). Although rural development objectives were not directly targeted by these projects, they were often implemented in rural areas, and thus supported the development of the culture and capacity for integrated rural development planning.

Apart from the budget policy, the harmonization of the trade policies in the field of agricultural products is also an important part of the Albania's integration process. These reforms started in the early 1990s, and have been part of the structural transformation and reform processes, with the aim of formulating and consolidating the overall policy framework for economic development to meet the integration challenges of the economy at regional, EU and global levels. The process of trade liberalization further developed when Albania became a member of the World Trade Organization (WTO) in 2000, and the SAA was put into force with the EU in 2009. These processes made possible a complete renewal of trade policy in general, including a fundamental restructuring of the trade policy in the field of agricultural products. As such, trade policies in agriculture are currently implemented in full compliance with: (i) the commitments of Albania to the WTO; (ii) SAA between Albania and the EU; (iii) the CEFTA 2006 FTA (in force since July 2007); and (iv) the FTA between Albania and Turkey (in force since May 2008). Also, Albania has signed an FTA with the EFTA (European Free Trade Association) countries (Switzerland, Lichtenstein, Norway, Iceland), which is in the ratification process in the Parliament of Albania. All of these commitments are reflected in the Government Programme 2005–2009, in the Sector Strategy for agriculture and rural development 2007–2013 (AFSS), and in all of other development policy initiatives of the MAFCP.

As a result of the trade liberalization process, Albania has developed, and now implements, a liberal regime in the trade of agricultural and food products, which is typified by several main characteristics. The first of these is a simple tariff system composed of five scales of tariffs (0%, 2%, 5%, 10%, and 15%) for the most favored trading nations, on an *ad valorem* basis. The second characteristic is one of not implementing 'tariff-quotas' for the most favored nations, followed by seasonal tariffs that are applied for a group of products (HS 07 and HS 08 in the Harmonized Commodity Description and Coding System, an international nomenclature designed to classify traded goods on a common basis that is developed and maintained by the World Customs Organization). Further characteristics are the implementation of a single value added tax (VAT) system of 20%, or the application of the tax to a group of agricultural products. In its trade regime for the most favored nation, Albania does not use incentive or supporting schemes for the import and export of agricultural products (i.e. it does not use state aid), and the country applies an import licensing system in full compliance with the General Agreement on Tariffs and Trade (GATT) of 1994 for import licensing, which is performed by the National Licensing Center.

11.2.3 Implementation of national agricultural and rural development support schemes

In order to meet the strategic objectives in agriculture and rural development set by the Albanian government in line with the SAA, since 2007 the MAFCP has implemented a yearly income support scheme with different support measures coupled to agricultural production activities. These are summarized in Table 11.1. The yearly scheme is administered by its PA (ARDA). In 2008, the MAFCP introduced an interest subsidy measure for agro-processors. Although constantly recommended by the EC, national rural development investment measures were not implemented in order to align with the *acquis communautaire* of the CAP. They are planned for 2012, only as co-financing, and part under an EU funded IPA 2011 Technical Assistance Project that will also implement a grant scheme component based on EU 'IPARD-like' rules.

Although the support measures have been diversified and the yearly budgets have increased during the period 2007–2010, the national scheme can be considered limited in terms of its effect compared with the economic and social importance of the agricultural and rural development sector in Albania. Looking at the figures for 2010, about 10,060 farmers and agro-processors applied, out of which 8043 were successful with a total volume of 7.176 million euros, which gives an average of 892 euros per applicant. In Table 11.1, the number of applications financed and their total volume are summarized for the years 2007 to 2010. Figures for 2011 were not available at the time, yet the budget for 2011 was similar to the one for 2010, but with new support measures. For example, a new measure was to encourage the planting of new nut trees. For 2012, about 9 million euros are planned for the national support scheme, and, in addition, 2.08 million euros (24 months in the period from 2012–2014) as a national contribution to the support from the EU in the frame of IPA 2011 grant scheme for IPARD-like measures. Thus, for the first time, the MAFCP will introduce investment measures for the agro-food sector, which follow the rules of the CAP Pillar 2 financial scheme, and bring the agriculture and rural development sector closer to the CAP *acquis communautaire*.

Several other general agriculture support measures exist as well. The first among these is related to the reduction of the amount of the obligatory social contribution for all persons who are self-employed in agriculture. The payment of the social contribution is obligatory in Albania for employed persons above 16 years old. The amount of the contribution for persons self-employed in the agriculture sector is a fixed amount, significantly lower than in other sectors, while receiving the same benefits from the social security system as those in other sectors. The total annual amount of government contribution to the social security system as compensation for reduced individual contributions from farmers is calculated on an annual basis based on the total number of insured persons in the agriculture sector. Exclusion from payment of the agriculture land tax is another support measure currently in use. The agriculture land tax is defined as a property tax, and it is collected and managed by the local government. In order to promote investment in the establishment of fruit trees and orchards, land area planted with fruit trees is excluded from payment of land tax for a period of 5 years from the date of tree planting.

Table 11.1. National support schemes for agriculture and rural development, 2007–2010. Source: MAFCP[a], Yearly Reports 2007–2011.

	2007		2008		2009		2010	
Measures	No. projects	000 euros	No. projects	000 euros	No. projects	000 euros	No. projects	000 euros
New planting of fruit trees, olive trees, vineyards	2109	2641	2580	4302	3537	4822	4870	5139
Drip irrigation systems	–	–	178	330	100	178	773	1093
Controlling the olive fly	–	–	3521	136	5	207	–	–
Certifying of bio-agricultural production	–	–	8	1	9	2	88	32
Production of autochthonous wine saplings	–	–	4	13	3	12	–	–
Establishing wells for irrigation	–	–	–	–	171	127	–	–
Providing new plastic sheeting for greenhouses	–	–	–	–	20	20	82	41
Improving heating technology in greenhouses	–	–	71	420	13	57	–	–
Mushroom production	–	–	–	–	2	7	–	–
Premium per dairy cow (farms >10 head)	–	–	843	889	1229	1218	108	352
Premium per milk sheep (farms >50 head)	–	–	384	194	1201	353	2083	498
Payment per beehive (producers >50 beehives)	–	–	–	–	831	213	6	2.1
Payment/ha for intensive breeding of snails	–	–	–	–	9	22	–	–
Premium/l extra-virgin olive oil production	–	–	3	20	7	70	3	15.1
Interest rate subsidy for loans for agri-food sector	–	–	11	25	4	9	30	3
Total	2109	3641	7603	6331	7141	7317	8043	7176

[a]MAFCP, Ministry of Agriculture, Food and Consumer Protection.

Measures related to the control of animal diseases and veterinary emergencies, aimed to reduce higher animal health risks and ensure consumer security, are also in use and are financed through an annual veterinary emergency budget. The control of high phytosanitary risks in agriculture, including the management of emergencies related to interventions for the control of most risky diseases and pests of crops is another support measure. For this measure, a government budget is defined every year within the overall budget of the MAFCP, to be used according to emerging needs during the year. This measure is accompanied by establishment of communal wholesale markets in the main agricultural production areas. This includes the construction of six wholesale markets during the period 2004–2009 in Vlora, Lushnje, Korçë, Gjirokaster, Shkoder, Kukes, and Berat. Finally, the establishment of new communal slaughterhouses during 2005–2009 aimed to support improvements in food safety and consumer protection related to animal slaughtering and meat standards (Shelzen *at al.*, 2010).

11.2.4 *Alignment with the CAP* acquis communautaire *and IPARD preparation process in Albania*

During the last 20 years, Albania has undergone large economic and political changes. The result of these changes is that the country is committed to establishing democratic principles and a market economy, toward which it has already made significant progress. Albania's ability to 'take on the obligations of membership' refers to its ability to adopt, implement, and enforce the set of policies, laws, rules, and institutional settings to implement the CAP. To facilitate the accession negotiations, the entire legal framework of the EU is divided into 35 chapters, each corresponding to one policy area. Chapter 11 of this legal framework covers all aspects of the agricultural and rural development sector. Various EU funded projects are now helping Albania to prepare for negotiation, as a candidate country, while also familiarizing the authorities and other relevant organizations, such as producers, processors, and professional organizations, with the methods used to implement EU support measures.

However, significant adjustments remain that are necessary for Albania to meet EU standards and adopt and implement the *acquis communautaire* in the field of agriculture. These include harmonizing legislation in the areas of food safety, veterinary services and control, phytosanitation, the traceability of the movement of animals and agricultural products, and the system for administration and control of payments, among others. Adjustments also need to be made to the laws governing management systems such as a PA and the Integrated Administration and Control System (IACS), and to the capacity to implement rural development actions, which is the goal of IPARD.

Albanian agricultural legislation is still only partly harmonized with the *acquis*, and the screening process can only start when receiving candidate status. Relatively good progress has been made in harmonizing food security and veterinary legislation with the endorsement of secondary legislation such the 'Food Law', 'Veterinary Law', 'Plant Protection Law', and others, whereby the implementation of veterinary and phytosanitary controls needs to be enforced. The National Food Authority (NFA) was established in 2010, as a key institution responsible for controlling the food safety situation in the country. Alignment of the legislation regarding Common Market Organizations (CMOs), and the alignment of CMOs over a range of agricultural products, including arable crops, sugar, animal products, and specialized crops still lag behind (Shelzen *et al.*, 2010).

Consequently, since 2009, the MAFCP has started to prepare for the implementation of IPARD measures with technical support from the EU funded IPA Project 'Capacity Building for Implementing the Rural Development Strategy.' This Project was co-funded by the German Federal Government and implemented by GIZ, and supported the MAFCP in achieving considerable progress in establishing the IPARD structures. The MAFCP and ARDA are in an advanced status of IPARD preparation, as was mentioned in the Yearly Progress Report 2011 by the EC. Yet, the EC concluded that the capacities for carrying out policy analysis, and designing rural development policies need to be further strengthened. The implementation of IPARD also requires that the MAFCP takes over a leading role in and ownership of policymaking in order to enable the best use of knowledge and technical support (EC, 2009, 2010a, 2011b).

The designated IPARD OS of MAFCP should be ready for the compliance assessment for national accreditation, as the major preparatory steps towards CAP Pillar 2 alignment have been completed. One major objective of the IPA Project was to train the staff of the IPARD OS in all relevant CAP *acquis communautaire* issues, as well as in IPARD implementation requirements. Another achievement is the establishment of the organizational structure and legal framework of the designated IPARD OS of the MAFCP (ARDA as the IPARD PA, and the Directorate for Rural Development Policy (DRDP) as the IPARD Managing Authority). In January 2012, the MA submitted the latest draft of the IPARD RDP

2011–2013 to DG AGRI for consultation. The draft program includes two measures under IPARD Axis 1: 'Modernization of agricultural holdings', and 'Improvement of processing and marketing of agricultural products' for investments under the priority sectors of milk, meat, fruits and vegetables. It also includes the measure 'Technical Assistance' under Axis 3, with the MAFCP as beneficiary. Based on the institutional structure and planned measures, ARDA and MA have developed a comprehensive draft accreditation package (administrative procedures, version January 2012), which should be ready for compliance assessment for national accreditation. The IPARD OS has also established horizontal coordination and communication lines with the other national IPARD actors, relevant technical bodies, and the EC. The MAFCP has developed a basic Agricultural Information System (AIS) for the needs of IPARD, and is aware of further requirements in the accession process. Furthermore, an ICT based management information system (MIS) for application processing, monitoring and control of rural development investment schemes fulfilling IPARD requirements is in place (Schaefer, 2012).

As the start of the accreditation process under IPA 2007–2013 is subject to the granting of candidate status, the Albanian authorities cannot submit the draft IPARD Rural Development Program 2011–2013 for approval by the EC, nor they can submit the draft accreditation package of the IPARD OS for national accreditation. Consequently, Albanian farmers and food processors do not have access to the investment scheme of an anticipated 30 million euros of IPARD public funds. The EC and MAFCP therefore initiated an IPARD-like pilot Technical Assistance (TA) Project, 'Support to Agriculture and Rural Development', under the IPA component I Albanian program for 2011. This is to implement a grant scheme for investments to modernize farms and food processing companies using more than 6 million euros of EU funds, according to IPARD compliant rules, and with maximum involvement of the designated, but not accredited, IPARD OS. With this Project, the established IPARD capacities within the MAFCP and ARDA can be sustained and further strengthened to manage much larger IPARD funds in the new programming period 2014–2020. Thus, Albania will be the first enlargement country that has EU funds available for IPARD beneficiaries without being recognized as a candidate country. Implementing this path-breaking pilot project for the new programming period 2014–2020 to manage EU funds through a national PA, the Albanian agricultural administration will contribute to the planned new approach of post-2013 funding through best practice experience.

11.3 Effects of EU Integration on the Agricultural and Rural Development Sectors of New EU Member States

With their accession to the EU, agricultural policies in the NMS changed significantly, as they are aligned with the CAP. These policy adjustments have significant implications for farmers and food consumers in Central Europe, for market balance and trade in agriculture, for budget expenditure and for macroeconomic conditions. The accession of former communist countries to the EU was generally a historical turning point for these countries toward democratic and market-oriented regimes. It played an important and highly positive role in the acceleration of economic development, and modernization in NMS. The agricultural sector was among the first and biggest financial beneficiaries of this process.

Like the majority of European countries, many NMS lost some of their agricultural area after EU accession owing to measures such as set-aside, early retirements in agriculture, and withdrawal from agricultural production and farm bankruptcy as a result of the increased competition. Latvia and Lithuania, however, increased their shares of total utilized agricultural area (UAA) from 2003 to 2007, while Cyprus, the Czech Republic and Poland maintained their agricultural areas. No trends can be found in changes in arable land: some countries increased their shares of such land, while others decreased their shares. As a result, after EU accession, the UAA of NMS decreased by 3% between 2003 and 2007, resulting in increase in unemployment in rural areas, while the average share of arable land increased by 1% through farm consolidation and concentration of production.

Agricultural employment in NMS continued to fall after EU accession. In Bulgaria and Lithuania, it decreased by close to 40%, and in Latvia, Slovakia, and Hungary by 20%. Poland is the only country where the number of agricultural employees increased somewhat after accession (Csaki and Jambor, 2009). The most important determinants of agricultural employment changes are labor saving technical progress, the macroeconomic environment, the farm structure, socioeconomic characteristics of the farmer, and agricultural support policies (Baum *et al.*, 2007).

EU accession has had modest but not uniform impacts on the production of major products, and overall agricultural output. Most NMS noticed a slight increase in cereal output, because the CAP created more incentives than existed before accession. Yet the overall production was influenced by extreme weather conditions in 2004 and 2007. After accession, the majority of NMS observed declines in the meat sector due to a growth in cereal prices and free trade, while production in the milk sector stagnated because of higher competition on the common market. Both of these effects have resulted in decreasing the number of animals. The share of the agriculture sector in the overall economy further declined in the NMS after EU accession, as the other sectors, in particular the service sector, grew faster. Labor and land productivity has increased measurably, indicated by agricultural output/ha and output/worker as a result of farm consolidation and technological progress (Baum *et al.*, 2007). Although the productivity gap between the NMS and the EU-15 countries (those in the EU before the accession of ten candidate countries on 1 May 2004) has decreased, the huge differences among NMS continue to exist (Csaki and Jambor, 2009). Unfavorable terms of trade in agriculture, especially in terms of non-tariff barriers for agricultural products, have not been changed after accession. There was a much faster increase in agricultural input prices than in agricultural output prices. In the Czech Republic, the agricultural price index in 2006 reached 40% of the 1990 level, while in Poland it was 69% of the 1995 level (Doucha *et al.*, 2007). CAP payments played an important role in stimulating agribusiness and spending on consumption in rural areas.

Agricultural output/ha increased significantly in the NMS. The average NMS agricultural output was 701 euros/ha in 2003 and 1013 euros/ha in 2007, indicating a 44% increase in 4 years. The highest value of agricultural output is observable in Slovenia (220% of the NMS average in 2007), while the lowest values are for Latvia (51% of the NMS average in 2007), over all years analyzed. These differences are mainly due to the development situation of the agriculture sectors of different countries at the moment of accession, and the type of CAP policy measures implemented after accession. However, the value of agricultural output/ha (gross value added, or GVA, at current prices per annual work unit, or AWU) grew in almost all NMS from 2003 to 2007, indicating that EU accession had a positive impact on GVA/worker. The majority of NMS countries experienced values of between just 1000–5000/AWU. Gross farm income of agricultural producers in most cases increased year by year. The highest increase was observable in Slovenia, where the full CAP of the EU-15 was introduced, while the lowest was in Slovakia, where only selected EU-15 CAP measures were implemented. Among NMS, Estonia, Lithuania, and Latvia experienced a higher increase in agricultural income. In 2007 alone, the income of people employed in agriculture in Poland rose by 13.7%, with an average rise in EU-27 countries of 5.4%. The increase in agricultural income led mainly to improvement in the living conditions of agricultural families, though to a smaller extent, it was also allocated to agricultural investments. The majority of investments were financed from EU funds, preferential loans (state aid), and commercial loans (Csaki and Jambor, 2009).

Market reforms and the opening-up the Central and Eastern Europe (CEE) economies in the 1990s have revealed the low competitiveness of the agricultural and food economy in these countries. Most of them became net importers of agri-food products, including Poland, the biggest country in this region. EU accession resulted in an increased trade deficit in agri-food and beverage products in all NMS except Poland, Hungary, and partly, Lithuania.

Cyprus, Romania and Slovenia almost tripled their negative balance from 2003 to 2007, which, for example, means a deficit of almost US$2.5 billion for Romania. The value of exports increased about 20% on the NMS level by 2007. All countries raised their export values by an average of 13–39% in 2000–2007, year by year, indicating different export performances. The greatest agricultural exporters in value were Poland, Hungary and the Czech Republic, which had started the accession period with better farm structures, and had also experienced a larger farm consolidation and modernization processes; Malta and Cyprus exported the least from 2000 to 2007. The adjustment of farmers to the enlarged integrated food markets is one of the most pressing demands of the post accession situation. The strong competition among the big chains in the retail trade has resulted in low product prices combined, sometimes, with not fully fair business practices (Csaki and Jambor, 2009).

After EU accession, the average size of small farms increased by 20% in all NMS but Poland, whereas that of large farms decreased by 10% on average, except in Latvia and Lithuania, and there was an increase of land use in many countries. In 2007, the average size of small farms exceeded 10 ha in the Czech Republic, Estonia and Latvia, while that of large farms was around 300–800 ha. Though a bimodal farm size structure remained after EU accession, the dualism in most cases became even stronger. Despite the reduction of the gap between the average size of the small and large farms, the share of large farms out of the total number of farms in land use increased in Bulgaria, Estonia, Hungary, Latvia, Lithuania, and Slovenia. This growth was around 10% in the Baltic countries, while in Romania there was a 9% drop.

The pre-accession programs (SAPARD; CARDS, Community Assistance for Reconstruction, Development and Stabilization; and PHARE, Poland and Hungary: Assistance for Restructuring their Economies) played an important role in the preparation of both the agricultural administration and potential beneficiaries of NMS for accession, and for the use of post-accession programs. They contributed to building institutions which were necessary for successful integration with the EU, and to an accelerated modernization of agriculture, food industry, and rural areas.

During 2000–2006, as well as during a transition period up to 2009 for the later acceding countries (Bulgaria and Romania), the agricultural and rural development sector of the 12 NMS, as well as that of Croatia (for the year 2006) benefited from the IPARD preceding pre-accession instrument SAPARD. In total, 2.6 billion euros was allocated for EU pre-accession aid in the CEE region for 2000–2006, of which 1040 million euros was allocated to SAPARD to introduce the CAP Pillar 2 measures. Although SAPARD resources allocated to candidate countries had been available since 2000, most countries started the expenditure within the SAPARD program much later, owing to the complexity in setting up the required legal, organizational, and administrative management and control structures (Bulgaria, Estonia, Latvia, Lithuania and Slovenia in 2001; Hungary, Poland, Romania and Slovakia in 2002; the Czech Republic in 2004). In Hungary and in Romania, after starting SAPARD in 2002, there was a break in expenditure in the year 2003. Bulgaria, which started the program in the earliest group of countries, had utilized only 36% of allocated money until 2005 as a result of difficulties in implementing some of the measures. Except for Latvia, the final allocated budgets have been spent (KPMG, 2010). In most countries the absorption was high in administratively simpler measures with higher investments, such as farm investments, food processing and marketing, as well as rural infrastructure. The average deadweight effect was about 10%, showing the high relevance of the program, although the strategic focus was generally inadequate as the intervention logic of the national SAPARD programs was not coherently planned (KPMG 2010).

The impacts of SAPARD were more significant in the food processing sector than in the agricultural production sector in the eight countries that were evaluated (without Bulgaria and Romania) in the ex post evaluation study commissioned by the EC (KPMG, 2010). SAPARD could reach 9.5% of the total enterprises in the processing sector, but only 0.8% of the total farmer population. The program had a positive impact on individual farmers in helping them to prepare for EU accession, comply with EU standard, and enter into new markets, especially in the EU-15. Some 45% of the farmers benefiting from the agri-investment measure, and 46% of the processing companies that received assistance from the processing investment measure, became fully compliant with EU standards (KPMG, 2010). The impact of SAPARD on the preparation of rural economies for accession was relatively low, as a whole, because the measures implemented in most countries were agriculture focused. At the individual level, the most positive effect was in the creation of new jobs (on average 1.4 new employees/project). Trade-offs between job creation and technological modernization, especially in the processing sector, were compensated for by market expansion. At the micro level, quality of life was improved through investments in the local infrastructure. However, due to the support intensity rate of below 80%, communities with a better financial situation benefited more from the support than did poorer municipalities (KPMG, 2010).

In summary, it can be stated that there is an overall positive impact upon the agricultural sector in NMS resulting in a consolidation of production, higher current prices, higher export and import quantities, and especially higher farmers' incomes, though different country performances prevail as a result of a range of reasons (Csaki and Jambor, 2009). At the same time, though, rural–urban gaps have widened and rural poverty has increased. There are a number of reasons why the performances of countries differed. The initial resources available for the sector were not uniform, and this has influenced agricultural productivity and income; such resources include the agro-ecological conditions, quality of land endowment, and agricultural employment. The performance level of the individual countries after accession also reflects the structure of farming: a more consolidated structure brought a higher level of asset endowment. In countries with a so-called dual farming structure, both ends of the farming are still suffering from the transition phenomena (Swinnen and Rozelle, 2006). The small farms are generally too small, and farmers are inexperienced and exhibit a lack of resources, while the large ones still have some heritage of the collective farming system, with embedded inefficiencies. National policy and the institutional framework also made a significant contribution. The assessment of national policies before and after accession, combined with the above mentioned factors, provides a better understanding of these differences. On the one hand, in some countries, such as Romania, Hungary and the Czech Republic, the level of agricultural support was close to or, as with Romania, even higher than EU levels in 2001 and 2002. Other countries, especially Poland and Slovakia, maintained a very low level of support. A higher level of support would have been beneficial for the post accession period if this support had focused mainly on the enhancement of competitiveness. On the other hand, countries such as Poland with a low level of producer support estimates (PSE), and minimum price and market support before accession, have benefited strongly from accession as they have provided visible incentives for farmers, leading to significant production and trade responses. Furthermore, restrictive land policies (e.g. in Hungary), and the lack of land and farm consolidation, have been limiting factors in the utilization of the advantages of the common market by significantly constraining the flow of outside capital to the agricultural sector (Csaki and Jambor, 2009).

The way in which countries used pre-accession EU provided facilities, such as SAPARD and other programs, also had an impact upon post accession performance. Countries focusing on the enhancement of competitiveness and improvement in production benefited more from these resources in respect of post accession sector performance (Swinnen and Rozelle, 2006). The way in which the political and institutional framework for CAP implementation was created also had significant influence in some countries. Delay in creating the required institutions, and the actual time of their commencement, has created some difficulties, especially in the first year of membership (Sallaku and Dashi, 2009).

After EU accession, the total number of farms decreased by 700,000 in NMS (Csaki and Jambor, 2009), while only a smaller share of farms could diversify their income structure.

In the EU-27, only about one out of every ten (9.9%) agricultural holdings reported other gainful activity directly related to the holding. The proportion is only slightly higher (13.5%) for commercial holdings than for family holdings (Eurostat, 2010). Economic diversification of farm households is still widespread in the EU, but is stagnant or even decreasing in the EU-15 (Fritzsch *et al.*, 2010; Ludvig *et al.*, 2010). Yet, individual MS were able to report positive individual farm diversification success stories, and an increased share of farms with more than one diversified activity, across all groups of farms, especially when linked to territorial business clusters. Differences, however, between the countries are striking, as a clear west–east and north–south divide is evident (Eurostat 2010; Ludvig *et al.*, 2010) suggesting that diversification of farm income is rather consolidated in the EU-15. Challenges for successful economic diversification are not only persistent in the NMS, but particularly in potential candidate and candidate countries in South East Europe (SEE), alongside the structural changes in the agricultural sector and rural economy. Looking at the average share of 35.2% of farmers (holders–managers) in the EU-27 countries in 2007, including other gainful activities not related to the agricultural holding, it is evident that regional employment opportunities are of major economic importance for the income portfolio of farm households in all types of rural areas (DG AGRI, 2009; Wehinger and Schäfer, 2011).

11.4 Lessons Learned and Policy Recommendations Related to the Challenges and Barriers to EU Accession

The exposure of Albanian farmers to an increased market orientation in the process of EU integration also encompasses their amplified exposure to more volatile agricultural markets, which makes long-term planning difficult. In light of missing comprehensive, policy-driven agricultural and rural development support schemes in Albania, combined with high input costs, lack of access to credit, and incomplete regional food-chains, the majority of farmers may not be able to undertake the necessary investments to reach and sustain the necessary level of modernization through funds such as those of IPARD. Thus, the structural distortion in the sector might even be aggravated.

The large number of subsistence and semi-subsistence farms is a special characteristic of the EU's NMS and enlargement countries in SEE, which are hardly comparable to any other segments of the EU-15 farming sector, and require special attention and cohesion policies. Until now, the CAP has not addressed the needs of these farms, which are more related to social than to agricultural policy. The challenge for Albanian policymakers, therefore, is to transform the underutilized potential in the subsistence-oriented agricultural production of the country into a sustainable, modern and competitive sector alongside value chains, while at the same time fostering alternative income opportunities for the rural population that is leaving (totally or partly) the sector for diversified rural employment. The utilization of the comparative advantages of the agricultural sector in Albania against the EU-27, in particular in the fruit and vegetable sectors, needs more access to both public and private investment, the development of value chains, improved market access, and a generally favorable economic environment.

In the future, the government should continue to avoid interventions in areas where the private sector is more effective, particularly given the resource constraints posed by a decline in the availability of general donor financing. Priority should be given to enhancing the policy environment and strengthening institutions, improving access to extension services, promoting and protecting land rights and markets, and improving food safety standards in order to further align to EU standards, and facilitate access to future IPARD funds. Thus, important constraints to competitiveness still need to be overcome to increase Albania's competitive advantage in agricultural production, against a backdrop of increasing regional competition, and the need to harmonize with EU legislation and standards. In order to meet these challenges, the sector must accelerate its transformation, and this will have important implications for rural areas.

Policy analysis capacities within the MAFCP to program agriculture and rural development measures need to be further strengthened and institutionalized in a policy advisory unit. The process should be based on a gradual harmonization of the national legal and technical frame, such as establishing a comprehensive AIS, farm register, minimum standards, and solving property rights. The MAFCP needs to lead the policy formulation process and maximize the impact of public expenditure in agriculture and rural development through integrated programming (Sallaku and Dashi, 2009). This is a precondition to preparing Albania to absorb EU IPA funds. IPARD measures have to be seen as complementary to national agriculture and rural development measures, but in light of the limited budget resources and a large agenda for the agricultural sector, strategic decision-making and appropriate sequencing of interventions is extremely important, as is the reallocation of inefficient expenditure.

Although the Albanian agricultural administration is at an advanced stage of IPARD preparations, the accreditation of the IPARD bodies is subject to EU candidate status. Thus, Albanian farmers and food processors cannot yet benefit from investment funds for modernizing and upgrading their holdings towards EU standards. The EU plans to amend this accreditation process in the new planning period 2014–2020. The determining factor for the accessibility of IPA structural funds will be the administrative capacities of the operating bodies, regardless of candidate status, in order to enhance agriculture and rural development through targeted investments before accession. Although some EU IPA funds will be made available for the agro-processing sector through a TA project in 2012 and 2013, these funds cannot solve the policy deficiencies of the MAFCP. Albania should continue to conduct policy reform to harmonize with EU legislation and standards, in a carefully phased approach. The new flexibility of IPA fund implementation starting from 2014 will facilitate this process if the gradual strengthening of the institutions required for progress on accession is done in a collaborative and smart way.

References

Baum, S., Stange, H. and Weingarten, P. (2006) Employment in agriculture in the enlarged European Union. *Geographische Rundschau* 59(2), 4–13.

Cela, R., Shelzen, M. and Imami, D. (2010) *Review of Agriculture and Agricultural Policy in Albania*. AgriPolicy, Enlargement Network for Agripolicy Analysis, Seventh Framework Programme, European Commission, Brussels.

Csaki, C. and Jambor, A. (2009) *The Diversity of Effects of EU Membership on Agriculture in New Member States*. FAO Policy Studies on Rural Transition No. 2009-4, Food and Agriculture Organization (FAO) Regional Office for Europe and Central Asia. Budapest. Available at: http://www.fao.org/fileadmin/user_upload/Europe/documents/Publications/Policy_Stdies/EUeffects_en.pdf (accessed 12 September 2012).

DG AGRI (2009) *Why Do We Need a Common Agricultural Policy? Discussion Paper by DG Agriculture and Rural Development*. Directorate-General of Agriculture and Rural Development, European Commission, Brussels. Available at: http://ec.europa.eu/agriculture/cap-post-2013/reports/why_en.pdf (accessed 23 September 2012).

Dormal-Marino, L. (2011) CAP, including rural development and IPARD post-2013. Presentation of the Deputy Director-General, DG AGRI to: *Fifth Annual Working Meeting of the Ministers of Agriculture from SEE, 11–12 November 2011*. European Commission Agriculture and Rural Development, Brussels. Available at: http://www.maia.gov.md/public/files/Evenimente/2011/17.11.2011/1_CAP___Ms._Dormal_Marino.pdf (accessed 12 September 2012).

Doucha, T., Divila, E. and Dravnicek, Z. (2007) *Farm Transformation and Restructuring in Czech Agriculture – Ten Years Later*. Working Paper, Research Institute of Agricultural Economics, Prague, Czech Republic.

EC (2009) *Communication from the Commission to the European Parliament and the Council – Enlargement Strategy and Main Challenges 2009–2010*. {COM(2009) 533}, Albania 2009 Progress Report SEC(2009) 1337/3. Analytical Report. European Commission, Brussels.

EC (2010a) *Commission Opinion on Albania's Application for Membership of the European Union.* {SEC(2010) 1335}. COM(2010) 680 final – Communication from the Commission to the European Parliament and the Council Commission. Opinion on Albania's application for Membership of the European Union {SEC(2010) 1335 final} COM(2010) 660. European Commission, Brussels.

EC (2010b) *Europe 2020: A European Strategy for Smart, Sustainable and Inclusive Growth*. Commission Communication COM(2010) 2020. European Commission, Brussels.

EC (2011a) *Proposal for a Regulation of the European Parliament and of the Council on the Instrument for Pre-accession Assistance (IPA II).* {SEC(2011) 1462 final} {SEC(2011) 1463 final}. COM(2011) 838 final of 07.12.2011. European Commission, Brussels.

EC (2011b) *Communication from the Commission to the European Parliament and the Council. Enlargement Strategy and Main Challenges 2011–2012.* SEC(2011) 1205 final (COM(2011) 666), Albania 2011 Progress Report. European Commission, Brussels.

EPEC (2011) *Evaluation to Support the Preparation of Pre-accession Financial Instruments Beyond 2013. Final Report, 13 June 2011.* Led by GHK and Technopolis, European Policy Evaluation Consortium, Brussels.

Eurostat (2010) *Europe in Figures – Eurostat Yearbook 2010*. European Commission, Brussels. http://ec.europa.eu/eurostat (accessed 17 September 2012).

Fritzsch, J., Wegener, S., Buchenrieder, G., Curtiss, J. and Gomez de Paloma, S. (2010) *Economic Prospect for Semi-subsistence Farm Households in EU New Member States*. European Commission, Joint Research Centre, Institute for Prospective Technological Studies, Brussels. Available at: http://ftp.jrc.es/EURdoc/JRC58621.pdf (accessed 12 September 2012).

INSTAT (2011) *Statistical Yearbook 2010*. Instituti i Statistikës (INSTAT), Tirana, Albania.

KPMG (2010) *Synthesis of SAPARD Ex Post Evaluations. Evaluation Report*. Prepared by KMPG Advisory Ltd for European Commission DG Agriculture and Rural Development, Brussels. Available at: http://ec.europa.eu/agriculture/eval/reports/sapard2010/fulltext_en.pdf (accessed 12 September 2012).

Lazdinis, M. (2011) CAP and agriculture and rural development of pre-accession countries. Presentation to: *DG AGRI to: G.4, Pre-accession Assistance to Agriculture and Rural Development at the PREPARE Gathering 2011, 29 June 2011*. Available at: http://www.preparenetwork.org/files/PREPARE%20GATHERING%202011/conference%202011/Prepare%20Gathering%202011%20Marius%20Lazdinis.pdf (accessed 12 September 2012).

Ludvig, K., Eberlin, R. and Percze, L. (2010) *A Concept for Rural Economy Diversification and Short Review of its Experience in Europe*. Food and Agriculture Organization (FAO) Regional Office for Europe and Central Asia, Budapest.

Sallaku, F. and Dashi, E. (2009) *Country Report – Albania. Support to Capacity Building for Agriculture and Rural Development*. Policy Analysis and Formulation to Support EU-Integration of the South-Eastern European Countries. Food and Agriculture Organization (FAO), Rome.

Schaefer, B. (2012) *Final Report of the IPA 2008 Project "Capacity Building for Implementing the Rural Development Strategy", Delegation Agreement No: 2009/212–470, Reporting Period 01 October 2009–31 January 2012*. Gesellschaft für International Zusammenarbeit (GIZ), Tirana, Albania.

Schiessl, W. (2009) Pre-accession support for the agricultural sector – IPARD. Presentation to: *Agriculture and Enlargement Meeting, Zagreb, Croatia, 12–13 June 2009*. Available at: http://ec.europa.eu/agriculture/events/zagreb2009/schiessl_en.pdf (accessed 12 September 2012).

Swinnen, J. and Rozelle, S. (2006) *From Marx to Mao to the Market: The Economics and Politics of Agricultural Transition*. Oxford University Press, New York.

Wehinger, T. and Schäfer, B. (2011) *Potential for Diversification of the Rural Sectors in Albania and Montenegro*. Report prepared by Naccon GbR for the European Commission, DG Agriculture and Rural Development, Brussels. Available at: http://ec.europa.eu/agriculture/analysis/external/pot-diversif-albania-montenego/full-text_en.pdf (accessed 12 September 2012).

World Bank (2007) *Albania: Strategic Policies for a More Competitive Agriculture Sector*. Report No. AAA18-AL, World Bank Sustainable Development Sector Unit, Europe and Central Asia, Tirana, Albania. Available at: http://siteresources.worldbank.org/INTALBANIA/Resources/Strategic_Policies_for_more_competitive_Agriculture_sector.pdf (accessed 3 September 2012).

Acronyms and Abbreviations

AFSS	Agriculture and Food Sector Strategy 2007–2013
AIS	Agricultural Information System
ANAP	Annual National Action Plans
ARDA	Agriculture and Rural Development Agency
AWU	Annual Work Unit
CAO	Competent Accrediting Officer
CAP	Common Agricultural Policy (of the EU)
CARDS	Community Assistance for Reconstruction, Development and Stabilization
CEE	Central and Eastern Europe
CEFTA	Central European Free Trade Agreement
CMOs	Common Market Organizations
DG AGRI	Directorate-General for Agriculture and Rural Development (EU)
DRDP	Directorate for Rural Development Policy
EAFRD	European Agricultural Fund for Rural Development
EAGF	European Agricultural Guarantee Fund
EC	European Community
EFTA	European Free Trade Association
EPEC	European Policy Evaluation Consortium
EU	European Union
FA	Framework Agreement
GAEC	Good Agricultural and Environmental Condition
GAO	Gross agricultural output
GATT	General Agreement on Tariffs and Trade
GDAP	General Directorate for Agricultural Policy
GDP	Gross domestic product
GIZ	Gesellschaft für International Zusammenarbeit (German International Cooperation)
GVA	Gross value added
HS	Harmonized Commodity Description and Coding System
IACS	Integrated Administration and Control System
ICT	Information and communications technology
IFAD	International Fund for Agricultural Development
IMCARD	Inter-Ministerial Committee for Agriculture and Rural Development
IPA	Instrument for Pre-accession
IPARD	Instrument for Pre-accession – Rural Development
IPARD OS	IPARD Operating Structure
LAG	Local action groups
LEADER	Liaison Entre Actions pour le Développement de l'Économie Rurale
MA	Managing Authority
MADA	Mountain Area Development Agency
MAFCP	Ministry of Agriculture, Food and Consumer Protection
MIS	Management Information System
MS	Member State
NAO	National Authorizing Officer
NFA	National Food Authority
NMS	New Member State
NSDI	National Strategy for Development and Integration 2007–2013
OS	Operating Structure
PA	Paying Agency
PHARE	Poland and Hungary: Assistance for Restructuring their Economies

PSE	Producer Support Estimates
RDCS	Rural Development Crosscutting Strategy for 2007–2013
RDD	Rural Development Directorate
RDP	Rural Development Program
SA	Sectoral Agreement
SAA	Stabilization and Association Agreements
SAPARD	Special Accession Program for Agriculture and Rural Development
SEE	South East Europe
SFP	Single Farm Payment
TA	Technical assistance
UAA	Utilized Agricultural Area
VAT	Value added tax
WTO	World Trade Organization

12 Albanian Research and Extension Services in Transition

Edvin Zhllima,[1] Drini Imami,[1] Tatjana Dishnica,[2] Carl Evensen[3] and Bahri Musabelliu[1]
[1]*Agricultural University of Tirana, Albania;* [2]*Ministry of Agriculture, Food and Consumer Protection, Tirana, Albania;* [3]*University of Hawai'i at Mānoa, USA*

12.1 Introduction

Albanian agricultural reform in the early 1990s resulted in a very fragmented agriculture sector. Extension service started to operate in 1993 with the support of the European Union (EU). Initially, due to limited funding, it was based centrally, but in 1998, regional advisory centers started to operate in the main agricultural production areas. Currently, the extension service is present in many parts of the country, although the supply and market for agricultural services in Albania is still not very well developed. Public extension services are not very efficient as a result of limited resources, while the number of farms (350,654) is large for such a small country (with a population of about 3 million) (MAFCP, 2011). Private extension services are still emerging and have been developing in some sub-sectors, especially in the most prosperous agriculture regions (Imami *et al.*, 2009).

In parallel with extension services, the agricultural research and education system in Albania has undergone drastic changes during the transition. The role of agriculture high schools has experienced a marked decrease in influence, and has changed from being one of the main outlets of agricultural education to having insignificant numbers of students. In contrast, higher (tertiary) education, such as that at the Agricultural University of Tirana (Albania's leading institution for public agricultural education and research) has been expanding, hosting a growing number of students. However, the quality of education and research remains a challenge, as there are both limited resources and a limited faculty, and quality of teaching directly affects the capacities of future extension services—today's students will form the basis for recruitment of research, extension and agriculture specialists in the future.

The objective of this chapter is to analyze the research, education and extension services in Albania in light of institutional and transitional economic changes. Section 12.2 analyzes the development of agricultural education, research and extension services, with a focus on the current gaps and challenges. Section 12.3 highlights developments in research and extension in other more developed countries that can serve as guiding models for the future development of the extension services in Albania. The last part of the chapter (Section 12.4) provides policy recommendations for improving Albanian agricultural research and extension services.

 Agricultural Markets in a Transitioning Economy (eds C. Chan-Halbrendt and J. Fantle-Lepczyk)

12.2 Development of Agricultural Education, Research and Extension Services in Albania

12.2.1 Education in agriculture

Agricultural educational institutions in Albania consist of vocational high schools and universities. During the communist regime, the state demand for educated people in agriculture increased because of the overall objective of the state to increase agricultural productivity. This increased demand was reflected in the number of vocational agricultural schools, which expanded from five in 1960 to 343 in 1990. At the beginning of the transition to a market economy there were 153,000 people (or about 5% of the total population in Albania) who had had secondary education in agriculture. About 6% of the agricultural (vocational) high school graduates proceeded with tertiary (university) education in agriculture, and the rest typically continued to work in agricultural cooperatives or on state farms (Veizi, 1997).

After the breakup of the communist regime, the agricultural secondary school system collapsed and there were drastic declines in enrolment. There was a higher overall enrollment decline in both vocational and general education in rural areas, where most agricultural vocational schools were located. Relative rural enrollment in high schools (both types) fell from 60% in 1990 to only 30% in 1998, and the decline was much sharper in rural vocational schools: from 49% in 1990 to only 2% in 1998 (Spiro, 2003). Many of the agricultural vocational schools were closed, or changed their profiles (e.g. transformed into general high schools).

Currently, the numbers of agricultural vocational high schools and of their graduates (Table 12.1) are insignificant compared with two decades ago, and also insignificant in terms of the needs of the sector. The agricultural vocational education system is composed of nine high schools. The fields of education are diversified. However, although all nine schools provide education in crop production, other fields of study are provided only by some of the schools, e.g. animal production (one school), veterinary science (two schools), agribusiness (five schools), silviculture, forestry and furniture production (one school), and agricultural machinery (two schools). The agricultural vocational schools are based in the rural regions of Berat, Elbasan, Korçë (Korça), Fier, Shkoder and Tirana.

There are several reasons for this low level of enrollment in agricultural vocational schools. There is a general trend of dropping out of high school (both general and vocational high schools) in rural areas, and students perceive a disadvantage related to attending vocational high school because it requires an additional year compared with attending a general high school. In addition, the slow development of the agriculture sector since land reform (it is still dominated by small fragmented farming) and lack of services in rural areas, make employment in this sector, and living in rural areas, less attractive.

In addition, the quality of education in such schools is not high, and the quality of students enrolled in agricultural vocational high schools is often thought to be still lower. Vocational education is poorly developed in Albania, despite recent reforms initiated with the support of donor organizations. This weakness is

Table 12.1. Number of graduates in agricultural high schools during 2005–2009. Source: MES[a] (data provided upon request).

Specialization	2005–2006	2006–2007	2007–2008	2008–2009
Agribusiness	117	172	98	130
Veterinary	38	47	42	30
Forestry	29	30	27	37
Farming (animal and crop production)	2	5	8	5
Total	186	254	175	197

[a]MES, Ministry of Education and Science.

due to constraints in human (teaching staff) resources, financial resources, and infrastructure, as well as the use of old curricula, which are not compatible with the demands of the labor market. Compared with the general education secondary schools, vocational education and training (VET) agricultural schools have fewer textbooks, and their training components are underdeveloped (Bezhani, 2008).

Given the failure of vocational education to match the needs of the emerging economic system since the transition, choosing between attending general and vocational education is obviously biased toward general education. Most students and their parents find the general education degree to be more flexible, which is important in an uncertain labor market (Danaj *et al.*, 2005; Jørgen and Larsen, 2008). Thus, general (non-agriculture vocational) higher education enables higher flexibility for future educational advancement and employment, and also offers better chances for employment and living in urban areas.

In order to tackle these problems of increasingly low enrollment, particularly in farming areas, the Albanian VET system is undergoing reform aimed at improving the quality of training, adjusting training to the needs of the labor market, and diversifying the overall instructional portfolio. In ten pilot schools supported by donor projects, there have been initiatives to introduce new curriculum structures, including in four agricultural schools, but all of these initiatives are still at a pilot scale. Support for the restructuring of VET has been provided through the assistance of international donors, who have recently provided significant funding to improve vocational education. The main donors have been the EU, Switzerland, Italy, and Austria (with specific vocational training on tourism and agriculture) (Government of Albania, 2008). The national strategy for improving VET schools is to increase program flexibility and accessibility to the youth of the country, and to provide work placements. However, according to Jørgen and Larsen (2008) the reform of VET schools is not following a clear pattern, which implies that better planning is needed for the reform.

Higher (tertiary) agriculture education is represented only by the Agricultural University of Tirana (established in 1951) and the Faculty of Agriculture of the University of Korçë (Table 12.2). Up until 1991, there were approximately 20,000 graduates from these universities, of whom 40% were agronomists and 30% were agroeconomists/accountants. In the early 1990s, a drop of 90% was experienced in the graduation rate, resulting mostly from a decline of enrollment (Spase and Veshi, 1998). At present, the Agricultural University of Tirana is the only higher (tertiary) education institution covering undergraduate and graduate studies, scientific research, training, and extension in all the main areas of agriculture and food, including agronomy, horticulture and plant protection, economy and agrarian policy, agro-environment and ecology, agro-food technology, animal husbandry and business, aquaculture and fisheries management, forestry engineering, and veterinary medicine. The Faculty of Agriculture of the University of Korçë offers only undergraduate courses in agronomy and agro-economics.

In contrast to the situation in the high schools, agricultural tertiary education is attracting a growing numbers of students. In 2000, only 3% of public university graduates were in the agricultural sciences versus 12% in 2008 (Table 12.2). In absolute figures, the number of graduates in agricultural disciplines has increased remarkably, by more than sevenfold for this period, as a result of increasing enrollments (Fig. 12.1). At present, the enrollment rate is quite high—more than 3000 students were enrolled in undergraduate studies in agriculture related disciplines in 2012. The gender bias in agricultural studies is a concern though—most of the students (almost two thirds) are male (MES, 2012).

12.2.2 Agricultural research and innovation

Developing and transitional countries typically face strong challenges and constraints in access to research resources and capacities in the agriculture sector (as well as in other sectors) given state budget constraints, a fragile private economy, and overall political, economic, and social instability. But transitional countries also have a great need to develop their research

Table 12.2. Number of graduates in higher education in agriculture, 2000–2008. Source: MoES[a] (data provided upon request).

Institution/No. graduates	2000	2005	2006	2007	2008
Agricultural University of Tirana	129	152	417	469	909
Faculty of Agriculture, University of Korçë	0	27	0	38	35
No. non-agriculture graduates	4606	5577	7213	7913	6745
No. agriculture graduates	129	179	417	507	944
Total	4735	5756	7630	8420	7689
Share of agriculture graduates	3%	3%	5%	6%	12%

[a]MES, Ministry of Education and Science.

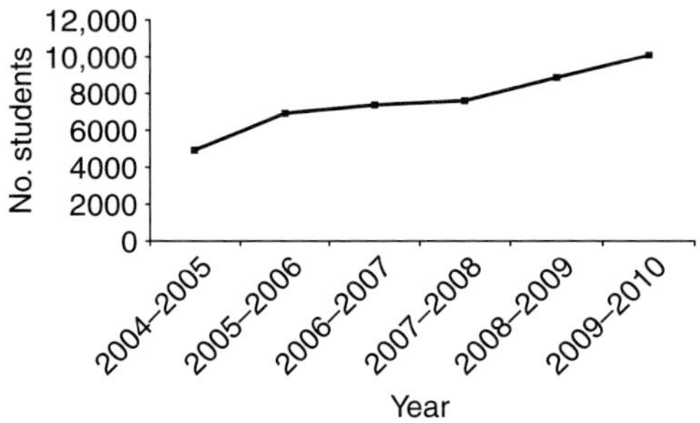

Fig. 12.1. University students enrolled in agriculture and veterinary sciences in the period 2004/05–2009/10. Source: Ministry of Education and Science (data provided upon request).

and innovation capacities to become more competitive in the global economy, and especially the Western Balkan countries (such as Albania) in the context of the EU integration process.

Albania has a relatively short history of agricultural research. Such research began in an organized fashion with the establishment of the Agriculture Scientific Research Institute in Lushnje, and the establishment of AUT (previously named the Higher Agriculture Institute) in the early 1950s. In 1970, there was an extended network of agricultural research institutes in various parts of the country (Dishnica, 2001). The research was coordinated, based on activities which were implemented in a top-down approach even at the secondary agricultural schools level. Spase and Veshi (1998) as well as Hotland (2001) observed that there was a gap between the research and field activities, and that research was limited to cooperation with institutions outside Albania.

The agricultural research institutes survived till the middle of 2000s. In 2006, national research reform nominally gave to the Agricultural University of Tirana the leading position for scientific research. However, based on the 18 agricultural research institutes, the Ministry of Agriculture established five Agricultural Technology Transfer Centers (ATTCs) charged which conducting applied research in various fields of agriculture (see Fig. 12.4). ATTCs are located in Fushe Kruja, Vlore, Shkoder, Lushnje and Korçë. Each of them addresses the needs and priorities identified by the regions in which they operate, as well as specialized applied research on selected agricultural activities: ATTC Lushnje specializes in vegetables, ATTC FusheKruja in soil analysis, plant fertilization, and animal feed, ATTC Vlore in fruit trees, especially olives and citrus, ATTC Shkoder in cereals and medicinal and aromatic plants, and ATTC Korçë in livestock, particularly small ruminants, fruits (especially apples and pears), and potatoes.

A bottom-up approach is used to establish the ATTC annual research agenda, which involves representatives of the farmers, extension agents, specialists, and other stakeholders. Typical problems experienced by farmers are identified through the extension network, and are then selected at Ministry level for inclusion in the research agenda of the ATTCs. The ATTCs transfer the research agenda into on-farm-research projects. The results of these projects are diffused into field demonstration, or through the cooperation of extension agents, in the form of leaflets and brochures. Most research projects are related to agricultural production and technology; less focus is given to farm management, markets, and environmental issues owing to lack of experience of ATTC staff (FAO, 2011). Studies in the agriculture sector mostly show very limited cooperation between the University and the Centers; few joint projects are taking place and few practical programs occur for students and professors (FAO, 2011).

One of the main challenges of research development is underfunding. Albania has particularly low funding for agri-food research compared with other Western Balkan (transitional) countries. The share of gross domestic product (GDP) allocated to research is lower in Albania than in the EU-27 countries (EU Member States as from 1 January 2007), as well as lower than that in the majority of the Western Balkan countries (except for Bosnia and Herzegovina) (Fig. 12.2). While the share of agricultural research expenditure in Albania is not known, the funding to ATTCs has not exceeded US$3 million annually (including salaries, operational expenses, and investments) for the past few years.

Over the last two decades, there has been a massive and continuous 'brain drain' experienced by Albanian research institutions, mainly toward neighboring and North American countries, accompanied by a 'brain waste', within Albania, when specialists left their professions for better paid jobs (UNESCO, 2007). Brain drain phenomena have also affected agriculture—about 60 experienced researchers have left the Agricultural University of Tirana to go to Western countries in the past two decades, and some of them have built well-established academic careers in Western universities. Many other researchers have abandoned research life for other more viable economic options, such as private business and politics, within Albania. As a result, there is a big gap in the age continuum of agricultural researchers (Borsi *et al.*, 2008; Leonetti *et al.*, 2011), with reduced numbers of middle-age researchers in research institutions. Scientific personnel are numerous in the higher education institutions, although limited research is being carried out in the public agricultural faculties (ATTCs) and the University of Tirana at the moment, due to limited financial resources and incentives. Private research is almost nonexistent. According to Borsi *et al.* (2008), the number of researchers in the agri-food sector in Albania who are affiliated with one of the research institutions is estimated at around 300 (Fig. 12.3), and almost four fifths of them hold a PhD degree.

12.2.3 Public versus private extension services in Albania

Before 1990, public agricultural extension did not exist as a separate organization but was linked to the research institutions and the agricultural units of production, such as the state farms and cooperatives (Qafoku, 1995; Daku, 1997). The linking of the extension service with research and farming communities was a practical framework, given the low number and

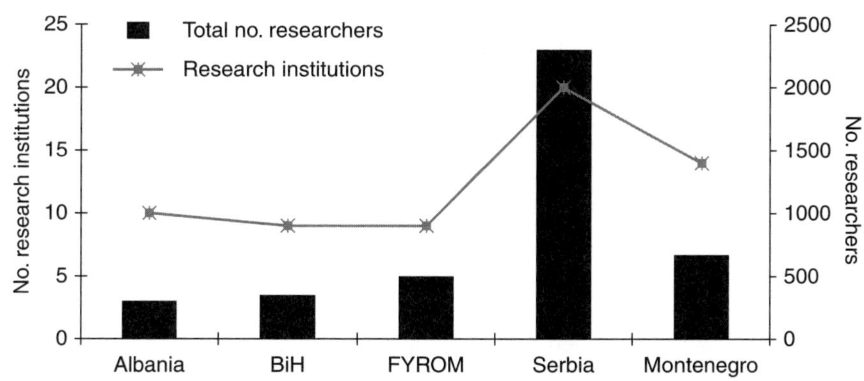

Fig. 12.2. Gross domestic expenditure on R&D as % GDP. Source: BAFN (2008); Jovanović and Despotović (2009). FYROM, Former Yugoslav Republic of Macedonia.

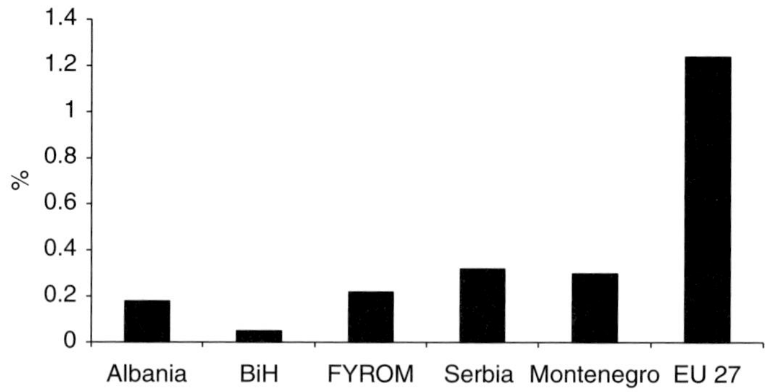

Fig. 12.3. Percentage of research institutions and numbers of researchers in the agri-food sector in Western Balkan countries and EU-27 countries. Source: Borsi *et al.* (2008); Jovanović and Despotović (2009). BiH, Bosnia and Herzegovina; FYROM, Former Yugoslav Republic of Macedonia.

large size of state farms and cooperatives. With the 1991 agriculture reform, though, another approach to extension services started to be implemented through donor projects (the EU-PHARE (Poland and Hungary: Assistance for Restructuring their Economies) program, Dutch Bilateral Aid, etc.) in 18 districts (EBRD and FAO, 2007; World Bank, 2007). The breakup of the state farms and cooperatives changed the position of the majority of agriculture specialists to that of simple farm operators. A few of the specialists continued to operate at the agricultural research centers of the Ministry of Agriculture and Food, and the newly created national extension service, from 1992 (Hotland, 2001). Donor intervention funded and facilitated the creation of coordinated extension service support in some regions of the country. Positive examples were seen, such as the semi-private Regional Agricultural Advisory Centers (RAACs) that provided advisory services to the commercially oriented farmers of the coastal plains (Hotland and Tarrelli, 2000).

Advisory services in Albania are organized at different levels and structures. There are 120 information centers, serving as core structures for information. There are two categories of the public extension service staff: subject matter specialists (e.g. plant production or protection, zootechnics, etc.) who provide specialized technical assistance; and Agricultural Information Center/field extension officers who transmit know-how on various issues to farmers. Agricultural Information Centers were established starting in early 2000. They currently cover the most important categories of agricultural production (e.g. livestock, fruits, vegetables, etc.). Each Agricultural Information Center employs an agricultural expert/extension officer, and they have the responsibility to provide farmers with information based on their individual requirements and needs (e.g. advice on plant protection measures).

Agricultural extension services are provided to farmers free of charge. Funding for this service is entirely provided by the Ministry of Agriculture, Food and Consumer Protection (MAFCP). Public extension services, in addition to providing direct technical assistance on production and management, also provide assistance to farmers in applying for subsidies offered by the MAFCP, in organizing events such as fairs, field demonstrations, and training, and in facilitating relations between different stakeholders. However, the approach is still very policy oriented rather than service oriented. As indicated by Daku (1997), the mission of state extension services is more to act as an agent for the implementation of the national agricultural strategies, and is not perceived as an assistance to farmers' needs for production and marketing.

According to the official figures, in 2011, some 19.8% of the total number of farmers received information from the extension specialists through several activities, such as field days and demonstrations, group meetings, and training exercises. Moreover, the extension specialists also support approximately 9000 farmers every year through the

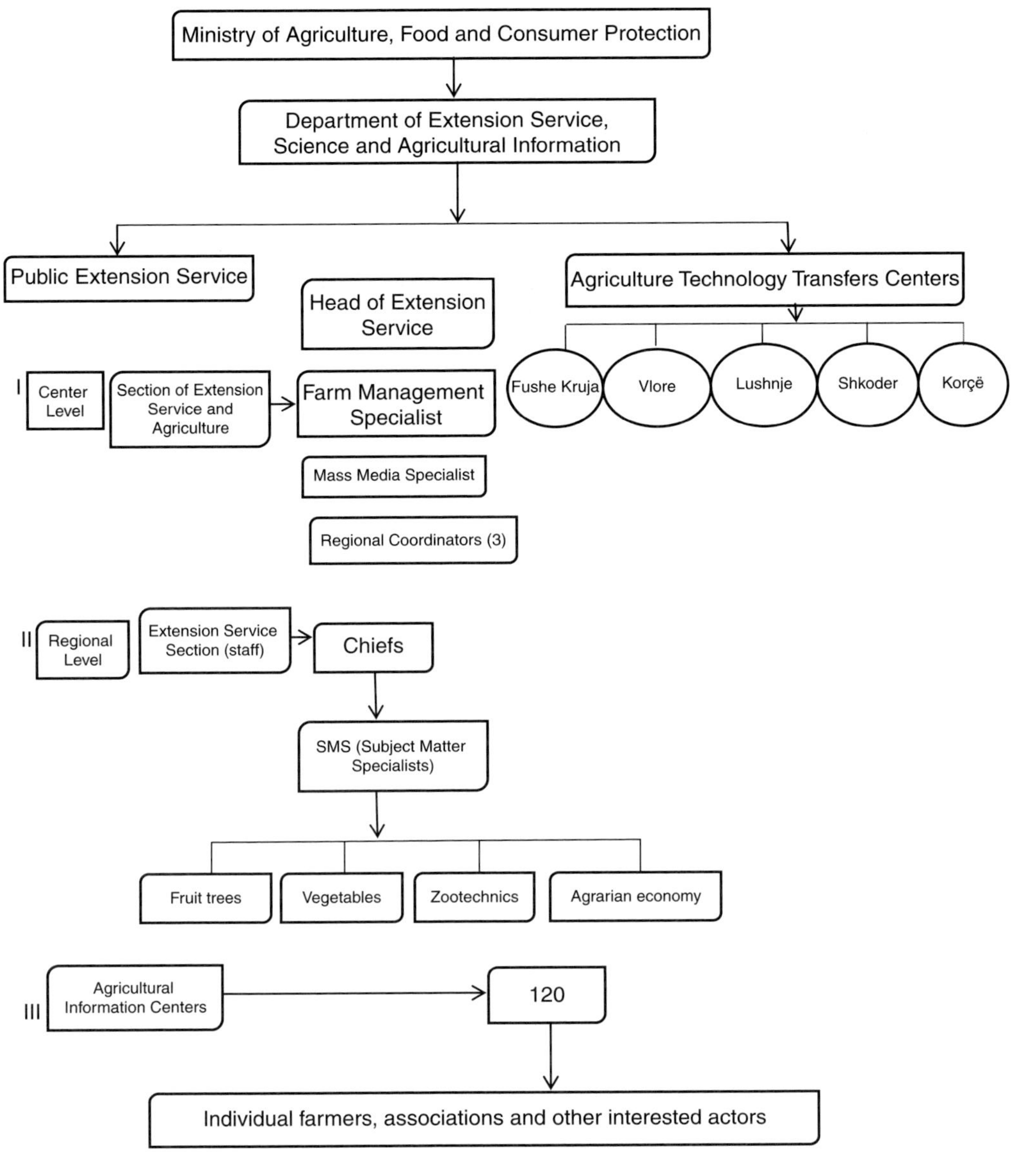

Fig. 12.4. The institutional structure of the State Agricultural Extension Service. Source: Ministry of Agriculture, Food and Consumer Protection (provided upon request).

national agricultural support scheme. Despite the potential and the role of the public extension service, its effectiveness is subject to both quantitative (e.g. number of staff) and qualitative (e.g. skills, know-how) limits. All sources (including the service providers themselves) confirm that the field activity of MAFCP extension specialists for the downstream value chain is quite limited, and that other sources of technical assistance and information are often more popular and available to farmers. According to a survey carried in 2005, only 34% of rural households in the country had direct access to extension (World Bank, 2007). In the less developed (typically mountainous) regions in Albania, there is less access to (both private and public) extension services. This low access to extension services is in contrast to

Western European or North American countries, where almost all farms have access to extension services. The very high number and the small size of farms in Albania means that there is a very low ratio of government extension specialists to farms. On average, each extension specialist in Albania covers more than 1000 farms, whereas in the USA, the ratio is about 1:245 (USDA, 2008, 2012).

Access to extension services is conditioned also by geographic factors. Typically, in mountainous areas, travel is more restricted by road conditions and constraints in means of transport. Therefore, despite the fact that mountainous regions, such as Kukes, Diber and Elbasan, have relatively higher numbers of extension staff (Table 12.3), in general, farmers in mountainous areas/regions have more limited access to such services. The Albanian Household Budget Survey of 2006 shows that the services are accessed much less by poorer farm households (28%) in the northern mountainous areas than they are by more prosperous farm households in coastal areas (41%) (World Bank, 2007).

Innovative reforms are being proposed, such as a virtual extension and research communication network (VERCON) platform using increased Internet availability in rural areas, and the establishment of connections between the information systems of MAFCP, the ATTCs, and other institutions linked with MAFCP (VERCON, 2011). As a result of the limited resources presently available, the public extension services play a relatively small role in providing support and technical assistance to off-farm sub-sectors (i.e. processors, wholesalers, retailers). This limited supply of knowledge is important owing to the need for more technological skills and the reduced education levels in rural areas. Figures reported from MAFCP show that surveys in the last decade find that approximately only one third of farmers have completed high school or university, 4% are not educated at all, and the remaining farmers have only a basic education (MAFCP, 2011).

In addition to MAFCP extension services, local governments (communes/municipalities) also employ agronomists and veterinarians. The agronomists of the municipalities mostly deal with land use, production practices, and tenure arrangements, while veterinarians focus on food safety, and the control of handling. Marketing, management, and some other specialized technical subjects are less covered by MAFCP extension services, leaving a gap.

Private operators and donor organizations have played a key role in filling gaps, and in some areas are considered the main source of technical assistance. Farm consolidation, and the diversification of product distribution channels, have raised the need for specific services at the production level, information at the market level, and accounting and safety standards services, as well as marketing and

Table 12.3. Extension staff distribution by region (qark) in Albania. Source: data provided by Extension Services Directorate, MAFPC[a].

Region	No. extension staff	No. farms	No. extension staff/10,000 farms
Berat	18	25,198	7.14
Diber	20	27,236	7.34
Durres	25	31,539	7.93
Elbasan	23	33,506	6.86
Fier	37	56,093	6.60
Gjirokaster	13	10,810	12.03
Korçë	27	30,908	8.74
Kukes	16	10,681	14.98
Lezhe	16	23,362	6.85
Shkoder	25	38,976	6.41
Tirana	31	32,870	9.43
Vlore	20	29,475	6.79
Total	271	350,654	7.73

[a]MAFCP, Ministry of Agriculture, Food and Consumer Protection.

other business development services. A study funded by SNV (Netherlands Development Organization) and implemented by DSA (Development Solutions Associates, Tirana) in 2010 found that the system of private extension services which support the sub-sectors of fruit, small ruminants and medicinal herbs is composed of various types of service providers (Leonetti *et al.*, 2009). In these three targeted value chains, the following categories of service providers were identified:

1. Professional or semi-professional agronomists, veterinarians, zootechnicians, and other providers of services at the local level (MAFCP extension service agronomists, private service providers working at the local level, and non-professional technicians specialized in pruning).
2. Input and equipment providers (suppliers of pesticides, seeds, seedlings; suppliers of agricultural machinery and repair services; suppliers of food processing equipment).
3. Regional and local development agency processors, who deliver services to small and medium enterprises (SME), including the preparation of business plans for loan applications, marketing studies, etc.).
4. National highly qualified specialists (greenhouse specialists, experts in international quality standards, international certifying bodies, post-harvest specialists, and engineers), which number more than 100 individual service providers. Most of these professionals are based in Tirana, but work throughout the country. The estimated number of highly qualified professionals given is exclusively based on peer review (based on the assessments of various professionals interviewed). Such specialists participate in or run nongovernmental organizations (NGOs) active in the agriculture sector, often financed by donor projects.
5. Donor-funded development projects. The main demand for qualified services comes from donor-funded development projects. The most qualified service suppliers agree that the demand from private enterprises is too small to justify full-time employment, so they tend to rely more on donor financing. While donors have played a crucial role in building up capacities, one unintended externality may be weak sustainability caused by donor dependence.
6. Business development specialists (chartered accountants, legal and fiscal advisors, land and farming asset surveyors). This category of experts rely mostly on private enterprise and farm payments, as such services (e.g. keeping accounting books) are required by law.
7. Marketing specialists (advertising agencies, marketing agencies, marketing consultants). Marketing specialists in the agriculture sector rely on a mix of donor and client (private enterprise) financing.

12.2.4 Current challenges in delivery of effective research and extension services

The current extension system in Albania has a long way to go before it produces the type of quality services needed by farmers. This could be due to the limited resources and limited professional training that is available for updating skills and knowledge. For example, some farmers interviewed complained that the agronomists/specialists working for the local and central public institutions do not offer them the services that they need in production, marketing, and financial advice; sometimes, they don't even visit them when requested. From their viewpoint, specialists working for public institutions claim that there are too few of them to provide their services to such a high number of farms—public extension services usually cover a group of (larger) targeted farms, and they may not offer their assistance to all farms. Furthermore, farmers interviewed said they need experts who live close to their villages, who can come and visit their farms when needed, and not just occasionally. Last, but not least, political appointments in public extension services are a major concern (Leonetti *et al.*, 2009).

Another fault within the system is the lack of incentives for extension agents to provide additional services. Public extension agents are remunerated at the same level whether they assist ten or 100 farmers, so there is little impetus for them to provide services to a greater number of farmers than they do already. Public extension employees also lack the capacity to provide technical assistance in some areas, such as marketing and management.

12.3 Comparisons of Research and Extension Programs in Other Countries

12.3.1 US Land-grant university system of integrated instruction, research and cooperative extension

In 2012, the USA celebrated the 150th anniversary of the Morrill Act, which created the historic partnership between the states and the federal government that formed the basis for the land-grant university system. This Act, signed by President Abraham Lincoln in 1862, forged an agreement between the national government and the states that established access to education by creating a nationwide system of Land-Grant Colleges and Universities. This partnership has been enhanced over the years by efforts to expand the scope and inclusiveness of the system to the 1890 (historically black) institutions dating from the 1890 Morrill Act, the tribal colleges dating from the 1994 Morrill Act, and Hispanic-serving institutions. This has been done by the creation of effective and often emulated extension programs. Since its inception, this land-grant system has graduated more than 20 million students, produced many scientific breakthroughs, greatly increased agricultural productivity, and significantly improved the lives of the American people (unpublished remarks by the Undersecretary of Agriculture at the Association of Public and Land Grant Universities Annual Meeting, Dallas, Texas, 15 November 2010; reported by Catherine Woteki).

A land-grant college or university is an institution that has been designated, either by its state legislature or by Congress, to receive the benefits of the Morrill Acts of 1862, 1890, and 1994. A system of research farms and agricultural experiment stations was established at these institutions through the Hatch Act of 1887 to ensure a steady stream of research dollars devoted to applied research conducted in the public interest. The Smith–Lever Act of 1914 created the Cooperative Extension Service to disseminate the science-based information derived from this research to individuals and communities nationwide. Most recently, the 29 native American tribal colleges established in 1994 were created to increase the reach of education and outreach to isolated and culturally diverse tribal reservations.

Due to reduced local funding, many extension programs in the USA have adapted to different modes of conducting their outreach activities. One such change has been the shift to the regionalization of extension delivery systems (Morse, 2011). In light of reduced budgets and staffing, many states have shifted from county based extension programs to regionally based systems in which a specialized extension faculty services multiple counties. This increases the outreach potential of the faculty, but reduces contact time with individual members of its clientele. As described by Morse, research is needed to evaluate the effectiveness of this extension delivery system. Another current adaptation of extension delivery involves the national Internet-based extension system. In this delivery method, 'communities of interest' are formed around various information topics, with contributions from researchers and extension educators from around the country. Extension clients access information through the community of interest, rather than through traditional, geographically based extension programs. Through these and other adaptations, the land-grant research and extension system in the USA continues to adapt to changing needs and conditions, and to remain relevant.

The model of integrated university instruction, research and extension, as expressed in the US land-grant systems, could be applied in Albania, which has all the requisite components, but simply lacks the coordination of these functions. This will be discussed further in Section 4. The relevance of regional or Internet-based extension delivery models (such as those being developed in the USA) should also be explored for Albanian extension services, as agricultural industries develop and information technology become increasing available in the country.

12.3.2 EU research and extension institutions

The EU research and extension service has been a reference point for the intervention of the majority of the donor projects in the agriculture sector in Albania. Similar to the situation in other Western Balkan countries, the reform process in Albania has merged the research and education systems according to EU trends. In the EU, the agri-food research sector, differently from Balkan countries, relies in a more systematic cooperation between research and extension. The research organizations and research groups have a trend toward the merging of entities (research institutions being included inside the universities), and the creation of research clusters (public institutes, universities and private companies) led by the creation of business 'incubators' and a further increasing of cooperation with the private sector. In research, there is a balance of power between the universities and the public research institutes, although there is a trend toward a growing importance of universities.

EU research is financed through a combination of national government and EU funding, and private sector funding. Public expenditure over the last few decades has been changing both in dimension and structure (Huffman and Just, 1998) with significant reductions in the rate of growth of public expenditure for agricultural research. In terms of structuring, traditional systematic funding has reduced significantly and shifted toward centrally controlled competitive grant programs. The average share of agriculture at GBAORD (Government budget appropriations or outlays for R&D) for 19 countries of the EU decreased from 9% to 6% between 1995 and 2006, indicating a relative decrease in support allocated directly to the agri-food sector (Chartier, 2007). Public research scientists are being encouraged to search for funds outside departments such as national governments, private companies and producer groups.

The CAP reform of 2003 influenced the orientation of research and extension toward multifunctionality of agriculture and rural development. Thus, research on agricultural competitiveness was accompanied by other more diversified research agendas, such as studies on the relationship of agriculture to biodiversity, sustainability, climate change, and energy (Chartier, 2007). In the past, extension services in the agriculture of EU countries have largely been publicly supported. In conjunction with the modernization of agriculture which commenced at the end of the 19th century and continued into the post-World War II period, other important factors leading to reform of agricultural extension services have been the CAP reforms (the reduction of policy support for production decoupling), the shrinking of public expenditure, trade liberalization processes (market protection), and the maintenance and strengthening of specific support for innovation, and for sustainable techniques. Since the 1980s, a pro-sustainable and multifunctional orientation of agriculture has increased the emphasis on sustainability and renewable energy.

It seems that the future of extension in EU countries will be faced with a number of challenges. Firstly, the role of input providers is expected to be biased by their interest in selling products. The influence of input sellers remains particularly strong. It is estimated that extension delivered by input providers may be three times as large as that delivered by public and freelance advisors, although statistics on this are not actually available (World Bank, 2007). The role of input providers seems particularly relevant when new technologies are proposed, e.g. biomass energy production has seen a very active private sector. The role is less and less marketing style, but is rather characterized by full design and project implementation (or at least implementation support), based on a strong technical expertise.

The second challenge is progress toward chain connections, or vertical integration, as this has long been perceived as a major need in the EU farming sector. This also implies the provision of extension services. Vertical integration is being supported by the promotion of contractual farming, particularly that based on input provision by the processing industries, which includes strong requirements in terms of input use and technology (Worley and McCluskey, 2000).

The third challenge entails the increase of the new requirements of the agriculture sector for the adoption of pro-environmental technology, and technological innovation. Hence, quality requirements, certification, and other requirements have attracted a lot of attention in recent years. Moreover, the process of decoupling and orienting the support policy to cross compliance (a combination of production and environmental criteria) has been a challenge for the farm advisory system, which has a duty to support funding efficacy (ADE, 2009). This requires that advisory staff should be up to date in their knowledge of innovations delivered by research activity, and of the market performance of global agricultural goods and services.

The fourth challenge is that highly educated and informed farmers actively seeking knowledge require more advanced services by advisors. These types of farmers are technology driven and demand different modes of communication with the extension service. Direct, face-to-face contact is still a relevant feature of extension, but modern technologies allow the expansion of information channels. The Internet and Internet forums are increasingly being used by farmers to purchase supplies and seek advice, as well as to share knowledge about specific input/production means. The above mentioned challenges are and will remain relevant for transition countries such as Albania as well.

12.4 Conclusions and Policy Recommendations for Improving Albanian Agricultural Research and Extension Systems

Agricultural research and extension services in Albania have undergone a radical change from the time of the state-organized economy to the present market economy. Public extension services were restructured during the 1990s, and donor organizations have played a key role in this context. At the same time, private extension services have emerged as contributors to the extension system. However, despite improvements in some private and public services, most farms in Albania are poorly served.

The effectiveness of the public research and extension system is limited by poor training and motivation of agricultural specialists who deliver extension information. Albania, as most Western Balkan Countries, possesses a weak agri-food research system, with fragmented research capacity characterized by obsolete programming and weak funding. Increased funding of agricultural research is required for salaries, staff training, equipment, facilities, and travel in order to strengthen relations with regional and EU research institutions. In addition, financial support of agricultural policies should be fostered by cooperation between research institutions and the private sector, research focused on actual market needs, and government support provided for business incubators and private–public partnerships.

There is an uneven distribution of extension coverage throughout Albania, with better availability of services mainly near cities and in more affluent areas. This situation promotes the continued stagnation of agriculture in underdeveloped areas, while exacerbating poor income distribution, particularly in the rural areas. In light of this lack of available services and integration with research, private operators and donor organizations have played a key role in filling the gaps in extension services and information, and in some areas are considered to be the main source of technical assistance. A good example of successful foreign assistance is the capacity building of agricultural research and instruction provided by the Albania/Hawaii Higher Education and Economic Development (AHEED) project (funded by the US Agency for International Development, USAID), with the Agricultural University of Tirana. This project, operating from 2008 to 2012, has improved graduate education programs in agricultural economics and agribusiness at the Agricultural University of Tirana through course development, promotion of research, publication by students and faculty, and increased linkages with Albanian government agencies.

The land consolidation and farm specialization processes that have characterized agriculture in Albania over the last decade could allow for greater competitiveness and efficiency. Consolidation and specialization has made technical assistance to farms both more indispensable and more affordable in the future. However, without dramatic changes in research and extension systems in Albania, the demand for extension services is expected to grow very slowly in the future, and in the short-term could even decrease owing to farmers' low expectations.

The specialists employed by donor projects do supplement the public sector's extension delivery, but for the sake of the greatest efficiency, their services should be carefully targeted and focused so as to complement or fill in gaps in extension coverage. Improved research to support extension needs should come from better cooperation and coordination between agricultural universities in Albania and MAFCP researchers. This is an area that requires government attention to ensure that research funding goes to the public universities to work on prioritized topics identified by the stakeholders.

Currently, MAFCP agricultural specialists are a weak link in the extension system because they are underqualified, underpaid, poorly assessed, and have limited accountability. There is a critical need to upgrade their qualifications, set performance criteria for them, and provide incentives for good performance, such as salary bonuses, or promotion opportunities. If it is not feasible to upgrade the current MAFCP specialists, it would be best over time to phase out these positions and to promote private sector extension services. This might be accomplished by implementing a system of subsidies, tax incentives, and specialized training with government certification for the private service providers. In either case, extension services cannot realistically be provided to all farms in Albania because of their very large numbers, and because many of them are primarily subsistence oriented. Public and/or private extension services should in the near-term focus on commercial farms, with priority assistance given to those meeting a specific annual income set by the government. In this way, limited government resources can be focused on improved agricultural production and economic development.

To help to fill in the gap in extension services to the remaining large number of small, subsistence oriented farms and rural households, programs to train and certify volunteer community educators could be created in Albania. These could include outreach programs, such as the training and support of master gardeners, master farmers, or paraveterinarians (Grieshop and Rupley, 1984; Catley *et al.*, 2004). These volunteers would receive special government training, certification, and regular refresher courses, along with honorific titles, local recognition, and perhaps other appropriate incentives. These 'train-the-trainer' programs could exponentially increase the coverage of extension services in Albania to reach many underserved households and communities. For the volunteers involved (some of whom could be retirees), the main incentives for participation would be increased access to current scientific information, local recognition, and the opportunity to help their neighbors and communities.

There is currently a serious disconnect between agricultural research and extension conducted by MAFCP and agricultural instruction and graduate education in the Agricultural University of Tirana, and the University of Korçë. MAFCP, which is the main agricultural research and extension agency in Albania, does not currently engage agricultural students and faculty to assist with research to improve agriculture in the country. Likewise, agricultural universities are not well grounded in current issues and needs in Albanian agriculture. To facilitate this connection, funding is needed for faculty and student projects (to be conducted in ATTCs and information centers) and faculty incentives to be developed, by such means as salary increases, bonuses, and tenure and promotion credit for advising students to undertake agricultural research and work with extension staff. Also, researchers in MAFCP should be engaged to assist university faculty to advise graduate students, through serving on student research committees. Strengthening the connection between universities and research and extension services would also help to attract the best and brightest students into research and extension careers.

Albanian research and extension should improve the institutional framework for interaction between all players concerned, including agricultural universities, agricultural research and extension centers, and private extension providers. A coordinated communication strategy (using mass media) has to be designed for extension to provide farmers with the information they need on inputs, markets, weather, etc. The ATTCs have to strengthen their networking capacity to identify priorities for research and technology transfer.

The extension service needs to identify appropriate methods for assessing farmers' needs for new technology in light of current trends, based on economic analysis of specific regions and crops. In doing this, the extension service has to increase its knowledge of the functioning of value chains (from production to marketing), and include farm economics and business management in all research and extension programs. Another step is to promote the best practices for encouraging risk-averse farmers to try new technologies/solutions. In order to achieve this goal, the extension service, in cooperation with ATTCs and universities, has to focus research and extension support on medium scale farms, which can better take advantage of new technologies (compared with subsistence farms) and do not have the resources of large farms for hiring private consultants. The problems of these farms need to be formulated broadly, taking into account their present practices, the farm's economics, and post-harvest issues. The service providers then have to build up their capacity to provide technical assistance, and to develop business plans; they also need to conduct regular external reviews to evaluate the impact of research and extension institutions and programs.

Government also has to give more importance—in terms of funding—to supporting research capacity by upgrading laboratory equipment, computers, agricultural machinery, and facilities. The ATTCs and extension have to improve priority setting and funding for applied research and extension via better coordination among research and extension centers, agricultural universities, and other government and private agricultural agencies.

While these recommendations are primarily focused on the opportunities and needs for improved agricultural research and extension in Albania, they have broad relevance for other Balkan countries, as well as for developing economies around the world. Many models of agricultural research and extension are in use worldwide, but all successful programs have the following features: (i) research must be relevant to current and emerging needs on farms; (ii) information transfer is two-way, with new technology and information delivered from research institutions to farms, and feedback on its successful application and new needs returned to researchers; (iii) agricultural university programs must be engaged with other government agencies and agricultural industries to keep the instruction relevant, and to train future research and extension professionals; and (iv) extension outreach programs must remain nimble and open to the use of new technologies to provide information in forms desired by farmers. The policy recommendations presented in this section have been developed with these broad principles in mind.

References

ADE (2009) *Evaluation of the Implementation of the Farm Advisory System*. ADE, in collaboration with ADAS, Agrotec and Evaluators.EU. A.D.E. s.a., Leuven-la-Neuve, Belgium. Available at: http://ec.europa.eu/agriculture/eval/reports/fas/report_des_en.pdf (accessed 13 September 2012).

BAFN (2008) *Agri-food Research in the Western Balkan Countries: Current Situation and Future Needs*. Synthesis Report from the Project "Support the Opening of the European Research Area by Developing a Sustainable Network in Agricultural and Food Sector in the Western Balkans". Balkan Agro Food Network, European Sixth Framework Programme. Available at: http://www.euroqualityfiles.net/bafn.eu/BAFN%20Sythesis%20Report.pdf (accessed 13 September 2012).

Bezhani, M. (2008) *Stocktaking of Potential Training Providers: North Eastern Region of Albania*. Report for the Swiss-Danish Formulation Mission, Tirana, Albania.

Borsi, B., Udvardi, A. and Vanitscek, M. (eds) (2008) *Agri-food Research in the Western Balkan Countries: Agrifood Research in Albania, National Mapping Report, Balkan Agro Food Network (BAFN)*. Published as part of project "Setting Up of an Agricultural and Food Research Network in the Western Balkan Countries", supported by the European Commission through Contract no. 026361 under the Sixth Framework Programme for Research, Technological Development and Demonstration Activities (2002 to 2006), and its specific programme "Integrating and Strengthening the European Research Area – Specific Measures in Support of International Co-operation". Available at: http://

www.euroqualityfiles.net/bafn.eu/Mapping%20reports/BAFN_Albanian_Mapping_Report.pdf (accessed 16 September 2012).

Catley, A., Leyland, T., Mariner, J.C., Akabwai, D.M.O., Admassu, B., Asfaw, W., Bekele, G. and Hassan, H. (2004) Para-veterinarian professionals and the development of quality, self-sustaining, community-based services. *Revue Scientifique et Technique* (OIE) 23(1), 225–252. Available at: http://web.oie.int/boutique/extrait/18catleymr.pdf (accessed 13 September 2012).

Chartier, O. (2007) *Agri-Food Research in Europe: Final Report. Overview of the Agri-food Research Landscape in Europe*. EU AGRI MAPPING Project, FP6-506087. European Commission, Brussels. Available at: http://www.euroqualityfiles.net/agrifoodresearch/reports/EUAM%20final%20report.doc (accessed 13 September 2012).

Daku, L. (1997) Investing in agricultural extension: the case of Albania. Research project submitted to the Faculty of the Virginia Polytechnic Institute and State University in partial fulfillment of the requirements for the degree of Master of Science in Agricultural and Applied Economics. Virginia Polytechnic Institute and State University, Virginia. Available at: http://scholar.lib.vt.edu/theses/available/etd-02898-154928/unrestricted/ETD.PDF (accessed 13 September 2012).

Danaj, E., Zhllima, E., Guxho, A., Lika, M. and Festy, P. (2005) *Becoming an Adult: Challenges and Potentials of Youth in Albania*. Instituti i Statistikës (INSTAT), Tirana, Albania. Available at: http://www.instat.gov.al/graphics/doc/downloads/Studimet/anglisht/adult-challenges.pdf (accessed 14 September 2012).

Dishnica, T. (2001) *Agriculture Research in Albania: A Monograph*. Julvin, Tirana, Albania.

EBRD and FAO (2007) *A Systematic Analysis of the Agribusiness Sector in Transition Economies. The Albanian Olive Oil Sector*. Commissioned by the European Bank for Reconstruction and Development (London) and the Investment Centre of the Food and Agriculture Organization of the United Nations (FAO), Rome. Available at: http://www.eastagri.org/files/ALB_Oliveoil_VC_final.pdf (accessed 13 September 2012).

FAO (2011) *Assessment of the Human Capacity Development Needs for, and Gaps in, the Agricultural Advisory Services in Western Balkans, Final Report*. FAO Technical Cooperation Programme Facility Project "Technical Support for Human Resources Development of Agricultural Advisory Services in Albania, Bosnia and Herzegovina, Bulgaria, FYR Macedonia, Montenegro, Serbia and UNMIK Kosovo". Available at: http://www.fao.org/docrep/015/i2119e/i2119e00.pdf (accessed 13 September 2012).

Government of Albania (2008) *External Assistance Orientation Document*. Council of Ministers, Tirana, Albania. Available at: http://www.dsdc.gov.al/dsdc/pub/external_assistance_orientation_document_10_1.pdf (accessed 13 September 2012).

Grieshop, J.I. and Rupley, V. (1984) How Do You Spell Relief? Master Gardening! *Journal of Extension* 22(4) 4FEA3. Available at: http://www.joe.org/joe/1984july/a3.php (accessed 13 September 2012).

Hotland, G. (2001) *The Role of the Fier Agricultural Programme in the Development of an Albanian AKIS: Continuity and Change in the Albanian Agricultural Knowledge and Information System During the Transition in the 1990s*. Working Paper, Tirana, Albania. Produced for the EU-PHARE (Poland and Hungary: Assistance for Restructuring their Economies) Program; available from the author.

Hotland, G. and Tarelli, I. (2000) The Albanian AKIS in transition: institutional development and organisational changes in the Albanian extension service in the 1990s. Paper presented at: *International Conference on Eastern European Agricultural Extension, Gogollo, Hungary, 13–18 May 2000*.

Huffman, E.W. and Just, R.E. (1998) *The Organization of Agricultural Research in Western and Developed Countries*. Staff Paper No. 303, Iowa State University. Available at: http://ageconsearch.umn.edu/bitstream/18262/1/isu303.pdf (accessed 13 September 2012).

Imami, D., Leonetti, L., Stefanllari, A. and Zhllima, E. (2009) Public and private agriculture extension services in Albania – current situation and future perspectives. In: Labar, K., Petrick, M. and Buchenrieder, G. (eds) *Conference on Modern Agriculture in Central and Eastern Europe (MACE) 2010. "Challenges of Education and Innovation". Proceedings of the Fourth Green Week Scientific Conference 2010, 13th–14th January 2010, Berlin, Germany. Studies on the Agricultural and Food Sector in Central and Eastern Europe Volume 56, IAMO (Leibniz-Institut für Agrarentwicklung in Mittel- und Osteuropa), Halle (Saale), Germany*, pp. 33–47. Available at: http://www.iamo.de/dok/sr_vol56.pdf (accessed 13 September 2012).

Jørgen, T.S. and Larsen, O. (2008) *VET for the Agribusiness Sector in the Northeast of Albania*. Report produced on behalf of SDC [Swiss Agency for Development and Cooperation] for the Swiss-Danish Formulation Mission: Formulation of a Swiss–Danish Programme Component for Development of Vocational Education and Training (VET) in Albania.

Jovanović, M. and Despotović, A. (2009) *The Agrifood Research Environment in Montenegro*. Available at: wbc-inco.net/attach/1AGRIFOODRESEARCHMNE09.pdf (accessed 13 September 2012).
Leonetti, L., Imami, D., Zhllima, E. and Stefanllari, A. (2009) *Survey and Analysis of Agriculture Service Providers*. Final report prepared for SNV (Netherlands Development Organisation).
Leonetti, L., Zhllima, E. and Imami, D. (2011) *Survey on Veterinarian Extension Services*. Final report for the EU-financed PAZA-Project: Improving Consumer Protection Against Zoonotic Diseases. Development Solutions Associates, Tirana, Albania.
MAFCP (2011) *Agricultural Statistical Yearbook for the Year 2010*. Ministry of Agriculture, Food and Consumer Protection, Tirana, Albania.
MES (2012) *Midterm and Longterm Strategy on the Development of Albanian Preuniversity Educational System (Shorted Version)*. Ministry of Education and Science, Tirana, Albania. Available at: http://www.see-educoop.net/education_in/pdf/book-alb-enl-t02.pdf (accessed 16 September 2012).
Morse, G.W. (2011) Regionalization with or without specialization: a call for a national research agenda. *Journal of Extension* 49(2), 2COM1. Available at: http://www.joe.org/joe/2011april/comm1.php (accessed 14 September 2012).
Qafoku, N. (1995) Problems related to food security and agricultural research in Albania. *Cahiers Options Méditerranéenes* 1(5), 81–84. Available at: http://ressources.ciheam.org/om/pdf/c01-5/93400051.pdf (accessed 14 September 2012).
Spase, I. and Veshi, H. (1998) Pre-University Agricultural Education. Institute for Pedagogic Studies, Tirana.
Spiro, M (2003) Reforming the study programs and curricula at the University of Tirana from the perspective of the provisions of the Bologna Declaration. A case study (January 2003). Presented to: *UNESCO-CEPES/European University Association (EUA) Seminar on the External Dimension of the Bologna Process, Southeastern European Higher Education and the European Higher Education Area in a Global World 6–8 March 2003, Bucharest*. Available at: http://www.see-educoop.net/education_in/pdf/unesco-cepes-8-alb-oth-enl-t02-t05.pdf (accessed 14 September 2012).
UNESCO (2007) *Science, Technology and Economic Development in South Eastern Europe*. Science Policy Series 1, UNESCO Office in Venice, Italy. Available at: http://unesdoc.unesco.org/images/0014/001414/141495e.pdf (accessed 14 September 2012).
USDA (2008) *Salary Analyses of Cooperative Extension Service Positions*. US Department of Agriculture, Washington, DC.
USDA (2012) State Fact Sheets. Department of Agriculture Economic Research Service, Washington, DC.
Veizi, P. (1997) Re-qualification of teachers and extension workers as a crucial element in the restructuring of agricultural education. In: Halla, I. (ed.) *Agricultural School in Agricultural Development*. Golem, Tirana, Albania.
VERCON (2011) *Communication Strategy for the Albanian Agricultural Extension-Reserach [Research] Network*. [A project proposal. Virtual Extension, Research and Communication Network]. Available at: http://www.gazdahalo.hu/uploads/media/comstrat_demo.pdf (accessed 14 September 2012).
World Bank (2007) *Albania: Strategic Policies for a More Competitive Agriculture Sector*. Report No. AAA18-AL, World Bank Sustainable Development Sector Unit, Europe and Central Asia, Tirana, Albania. Available at: http://siteresources.worldbank.org/INTALBANIA/Resources/Strategic_Policies_for_more_competitive_Agriculture_sector.pdf (accessed 3 September 2012).
Worley, T. and McCluskey, J.J. (2000) Production contracts as a means of vertical coordination with application to the wheat industry. *Journal of Food Distribution Research* 31(1), 215–224.

Index

AFSS *see* Agricultural and Food Sector Strategy (AFSS)
Aggregate measurement of support (AMS) 160, 161
Agribusiness
  apple 100–102
  dairy industry 95
  domestic production 94
  fruit trees 99
  grapes, wine and raki 102–104
  green markets 100
  imported milk quantity and domestic production 96
  investments and production, milk industry 96, 97
  meat 98–99
  milk collection system 94–96
  milk production, 2000–2009 95
  olive oil 96–98
  vegetables 99
  wholesalers/importers 99–100
Agricultural and Food Sector Strategy (AFSS) 82, 90, 225–228
Agricultural extension
  farmers 249
  graduates, schools 242
  public university graduates 243, 244
  public *vs.* private *see* Public *vs.* private extension services
  research and innovation 243–246
  VET system 243
  vocational schools 242
Agricultural labor productivity (ALP) 25, 26
Agricultural production
  economic and political system 75
  field crop 78–80
  fruit and vegetable 80
  livestock 76–78
  'mixed farming system' 76
  privatization process 76
  sectors 75
  value of fruit production 76
Agricultural trends and projections
  2011 census 215
  crop production *see* Crop production
  evolution, agro-food sector *see* Agro-food sector evolution
  finance 216, 217, 218
  fisheries *see* Fish and seafood industry
  land privatization 177, 178
  livestock production
    broilers and egg production 192–193
    cattle production 193–195
    leather and hides 203–204
    meet consumption 198–201
    milk and dairy consumption 201–203
    sheep and goat production 195–198
  medicinal and aromatic plants 211–214
  planned economy supply chains 177, 178
  supply chain 215, 216
  wholesale market 215
  wood 210–211
Agricultural University of Tirana (AUT) 43, 64, 71, 244
Agriculture industry
  constraints 60
  description 52–53
  economy, Albania's
    contraction 53
    GDP per capita annual growth 53, 54

Agriculture industry (*continued*)
gross national income 53
import oriented 54
performance indicators 53–54
place 54–56
service and industrial sectors 53
transition, 1992 53
GDP 52
growth and development 52
limitation 52
milestones
CEE countries 59
FTA 59
joining WTO 59
land privatization 58–59
large cooperatives and state farms 58
per capita privatization 59
pre-transition activities 58
SAA 59
trading partners and agreement 60
WTO membership 59
ownership 60
privatization 60
production and processing
bread and flour 57
corps production 56
cultivated area and techniques 56
impressive corps yields 56
livestock 56
MAFCP, 2009 56
marketing 56
milk and meat 57
registered companies 57
tax burden 56–57
sector characteristics 57–58
SWOT 68–71
value chains analysis *see* Value chain analysis
Agri-food products, purchasing behavior and consumer preferences
fruit 141–143
meat 138–141
milk 143–144
olive oil 132–135
wine 135–138
Agro-food sector evolution
commercial sized operators 177
consumption life styles 174, 175
demand 173–174
demand structure 175
domestic products 174
economic downturn, 1997 175
food processing industries 173
hamper sector development 173
harmonized laws, food safety 176
integration, EU 176
internal market 174
IPARD 176
market and lifestyles 175
migration flows 174
MPS 174
political stability 175
processing activities 174–175
producers capacity 176–177
purchasing power development 175
recovery, primary production 174
revolutionary changes and recurrent reforms 173
socioeconomic changes, 1991–2005 173
stable growth 173
supermarket chain 176
surveys 176
uncoordinated activity, farms 174
AHEED project *see* Albania/Hawaii Higher Education and Economic Development (AHEED) project
Albania/Hawaii Higher Education and Economic Development (AHEED) project 64, 252
Albanian agricultural sector
agri-food products and exports 225
economic stability and liberalization 224
GAO 224–225
GDP 224
investment 225–226
land cover categories 224
land renting 225
loans, commercial formers 225
small farmers and crop production 225
transition 224
underground water resources 224
Albanian Meat Processors Association (AMPA) 85
Albanian Trade and Transport Facilitation Body (ALBAPRO) 166
ALBAPRO *see* Albanian Trade and Transport Facilitation Body (ALBAPRO)
Alliance for Vocational Agricultural Training in the Albanian Region (AVATAR) project 41
ALP *see* Agricultural labor productivity (ALP)
AMPA *see* Albanian Meat Processors Association (AMPA)
AMS *see* Aggregate measurement of support (AMS)
ANAPs *see* Annual National Action Plans (ANAPs)
Annual National Action Plans (ANAPs) 227
Aquaculture
before 1991 206
development, 2005–2010 207
establishment 206–207
exports and imports 208–209
investments 208
marine 208
production 207
product quality 207
AUT *see* Agricultural University of Tirana (AUT)
AVATAR project *see* Alliance for Vocational Agricultural Training in the Albanian Region (AVATAR) project

Broilers and egg production 192–193

CAP *see* Common Agricultural Policy (CAP)
CAP reform, 2003 7, 251
Cattle production
breeding 193
FAOSTAT 194
meat and milk, 1988–2010 193, 194
per capita consumption 193–194
stagnation 194–195
Central and Eastern Europe (CEE) transition
accession
intensifies pressures, reform *see* European Union (EU) accession
Western Balkans *see* Western Balkans
adaption, single market *see* Single market adaption, Bulgaria and Romania
agricultural performance 6
beef output rises, Poland 7
Bulgaria and Romania 6
CAP reform, 2003 7
competition 5–6
declaiming cattle and beef production 7
democratic and free market economics 1
direct payments 16
and FADN 5
farm income 1
Hungarian cattle breeders 7
impacts, global recession 2008 16
livestock sectors 7
market oriented reforms 16
NMS framers 16
Polish exports shift, cattle-beef 8, 10
Polish live cattle exports 7, 9
quality standards and delayed payments 6
rapeseed output 6
reforms and restructuring agriculture 1
supply and demand side shocks
bipolar farm structure 3
communism effects 2
consumer income and purchasing power 2
early 1990s 3
fruit production and export market 2–3
land consolidation 3
market-based economies 2
prices liberalization, end of 1990s 2
privatization and land redistribution 3
production and consumption 2
Cereals
fragmentation, production 177–178
GDP 179
imports 179–180
and industrial corps 180
marketing 178
milling industry 180
'pyramids' schemes 179
scarce competitiveness 179
traditional consumption pattern 179
trends 178–179
wheat and maize 180
Common Agricultural Policy (CAP)
acquis communautaire and IPARD preparation process 231–232
Albanian agriculture sector 224–226
Cold War 220
description 220
effects, integration 232–236
efficiency and equal distribution 222
farmers 220–221
framework 226–229
funding 221–222
GAEC 221
implementation, national support schemes 229–231
integration 222
interventions 220
in 1960s 220
Consumer preferences
agricultural privatization 127
agri-food products 128
application 144
capita income and expenditure 128
cereals and fruit supply 130, 131
consumer basket food supply 129
consumption trends 128
FAOSTAT 128
farmers 144
GDP 127
grain production 129
MAFCP 128
meat per capita supply 130
policymakers and donor programs activation 144
and purchasing behavior *see* Agri-food products, purchasing behavior and consumer preferences
quality and safety systems 144
retailers and intermediaries 144
supply and consumption 128–129
vegetable supply 130, 132
Crop production
cereals *see* Cereals
citrus, stone and pome fruits
conflicts, farmers 182
Greek producers and exports 183
international trade deficit 183
investments 182
reinforcement 183
storage facilities and marketing 183
support schemes 183
transition 182

Crop production (*continued*)
trend and development patterns 182
wholesale market 183
fruits and vegetable processing industry
competitive with imported products 184
development 183–184
environment 184
impact, large companies 184–185
investment 184
supermarket chains 184
grapes *see* Grapes production
industrial crops 180
oilves and olive oil 189–192
vegetables *see* Vegetables, potatoes and beans production
wine *see* Wine production

Distributions and markets, Albania
agribusiness sub-sectors *see* Agribusiness
agricultural products 93
economy 106
FAO 105
import-export business model 106
large-scale farms 104, 105
legal and administrative rules 107
macro and micro policies 93
Oxfam collaboration 105
retail establishment and producer's sales skills 93
selling problems, farmers 105
small sized farms 104–105
social organization 106

Education and extension services, Albania
agricultural development *see* Agricultural extension
AHEED project 252
EU research and extension institutions 250–252
farm economics and business management 254
land consolidation and farm specialization processes 252
policy recommendations 254
state-organized economy 252
US land-grant university system 250
volunteers 253
EU integration, transitioning economy
CAP *see* Common Agricultural Policy (CAP)
IPA component 222–224
legal proposal 222
recommendations 236–237
reforms, rural development 221
rural economy 220
SEP 221
structural adjustment 221

European Union (EU) accession
acquis communitaire 4
agricultural policy reform 4
CAP 4
CEE prices 3
costs 4
direct payments, CEE producers 4, 5
early 2004 4–5
early 1990s 3
farmer income 3–4
price and income support programs 4
price gaps 4
Western Balkans *see* Western Balkans
yielding, CEE countries 4

FADN *see* Farm accountancy data network (FADN)
FAO *see* Food and Agriculture Organization (FAO)
FAOSTAT *see* Food and Agriculture Organization Statistics database (FAOSTAT)
Farm accountancy data network (FADN) 5
FBE *see* Food and Beverage Enterprises (FBE)
Fish and seafood industry
canning factories 206
consumption 206–207
demand 207, 208
export and import flows 207–208
freshwater fish 206
international trade 207
marketing 207
trade deficit 207–208
Food and Agriculture Organization (FAO) 105, 109–110
Food and Agriculture Organization Statistics database (FAOSTAT) 128–130, 194
Food and Beverage Enterprises (FBE) 110, 119
Food processing industry
advantages 84
AFSS 90
agricultural and agro-processing sector 90
Albania exports herbs and spices 89
Albanian agribusiness sector 84
Albanian agricultural sector 81
consolidation processes 81
Deutsche Investitions- und Entwicklungsgesellschaft mbH (DEG) 82–83
employees 82
fresh agricultural products 80–81
fruit and vegetables sector 90–91
import-export business model 91

meat and milk 81
policymakers and experts 83
prioritizing governmental support 90
privatization 81
quality management systems 82
raw materials 82
sub-sectors
alcoholic beverages production 88
canned fruit, vegetables and oil products 86–87
dairy products 87–88
description 84
meat and fish products 84–86
substantial investments 91
SWOT analysis 83
Forestry and non-timber forestry products
medicinal and aromatic plants 211–214
wood 209–210
Free Trade Agreement (FTA) 162–165
Fruit, purchasing behavior
consumer apple preference model 142
consumer groups 142
consumption, apple 141
domestic and imported apples 143
marketing and policy viewpoints 143
per capita trend 141
FTA *see* Free trade agreement (FTA)
supporting private sectors 168
transports 166
weakness 166–167
national economy 165
trade liberalization *vs.* facilitation
border crossing costs 165
business surveys 165
CEFTA and FTA 165
cooperation, public and private sector 166
corruption and reduce transaction costs 166
customs taxation 165
economic transition 165
International Monetary Fund (IMF) 165
objectives, ALBAPRO 166
Shock therapy 165
Grapes production
after 2005 185
cultivated area and data 185–186
domestic market 186
early production 185
growth 186
and wine *see* Wine production
Gross agricultural output (GAO) 25, 224–225
Gross domestic product (GDP) 24, 52, 75, 127, 179, 224

GAEC *see* Good agricultural and environmental conditions (GAEC)
GAO *see* Gross agricultural output (GAO)
GDP *see* Gross domestic product (GDP)
Geographic distribution, agricultural trade
distance 153
exports and imports, trading partners
in 2010 152
Brazil and Russia 152
labor intensive products 153
transport and communication cost 153
policy 151–152
size 153–154
traditional and new partners 151
Good agricultural and environmental conditions (GAEC) 221
Government strategies and trade balance
export competitiveness *vs.* import substitution 168–169
export promotion *vs.* enhancement of export competitiveness
agricultural and food products 166
AIPA 168
expansion, incentive programs 166
improving policy and administrative conditions 168
METE design 167–168
strengths 166, 167

HACCP *see* Hazard analysis and critical control points (HACCP)
Hazard analysis and critical control points (HACCP) 83, 85, 109, 110, 118, 122

IBRD *see* International Bank for Reconstruction and Development (IBRD)
IDA *see* International Development Association (IDA)
IFC *see* International Finance Corporation (IFC)
Instrument for pre-accession assistance (IPA) component V measures
beneficiaries 222
funding 222
geographic flexibility 224
IPARD 222–223
reforms 223–224
SSA 222
International Bank for Reconstruction and Development (IBRD) 41
International Development Association (IDA) 41
International Fertilizer Development Centre 41
International Finance Corporation (IFC) 41
International Organization for Standardization (ISO) 109, 110, 122
IPA *see* Instrument for pre-accession assistance (IPA)

IPARD *see* IPA rural development (IPARD)
IPA rural development (IPARD)
  description 176
  framework and sectoral agreement 223
  investment scheme 223
  investments, food processing 123
  and measures 222–223
  operating structure 223
  programs 223
ISO *see* International Organization for Standardization (ISO)

Land privatization
  agribusiness sectors 177
  agriculture sector efficiency
    ALP 25, 26
    CEEC 26
    GAO 25
    GDP 24
    private enterprises 24
  fragmentation and abandonment 26–27
  gender inequality exists 28–30
  insecurity, investments and market 27–28
  security 27
Land reform and agriculture, Albania
  in 1991
    legal instrument 24
    MAFCP 24
    national committee 23
    'On Constitutional Principles' 23
    political and economic regime 23
    restitution and compensation process 23
  after World War II
    economic cooperation partnership 22
    farms and cooperatives 21
    government, farmers' motivation 22–23
    land expropriation and redistribution 20
    private ownership 22
    Soviet regime ideology 21
  CEEC 19
  customary rights 31
  'equally per capita' principle 30
  farmers and landholders 18
  Ottoman legacy 19–21
  privatization *see* Land privatization
  property rights 30
  redistribution, farmers 30
  rental market activities 31
  security 30
  socioeconomic studies 31
Leather and hides 203–204

MAFCP *see* Ministry of Agriculture, Food and Consumer Protection (MAFCP)
MAPs *see* Medicinal and aromatic plants (MAPs)
Meat consumption and industry
  annual report 198–199
  avian flu crisis 199
  beef 198
  broilers 200
  cold cuts markets 201
  consumers 198
  description 198
  economic downturn 201–202
  FAOSTAT data 199
  imports 199–200
  patterns and lifestyles 198
  pig breeding facilities 200
  pork consumers 199
  poultry stock 199
  purchasing behavior *see* Purchasing behavior, meat
  sheep/goat 199
  structure and processing 200–201
  supply chain 200
  survey, LSMS and HBS 199
  trade balance deficit 199, 200
Medicinal and aromatic plants (MAPs)
  collection and distribution 211
  competitors 211
  consumption 214
  control 212
  cost and scarcity 212–213
  export 211
  exporters 213
  growth and developments 214
  harvesting 174, 212
  joint venture establishment 212
  markets 212
  supply and value chains 211
MIGA *see* Multi lateral Investment Guarantee Agency (MIGA)
Milk
  and dairy consumption 201–203
  purchasing behavior
    cattle production 143
    conjoint choice experiment study 143
    dairy activities 143
    per capita trend 143
    preference model 143, 144
    UHT 143
    urban districts 143, 144
Ministry of Agriculture, Food and Consumer Protection (MAFCP) 24, 42–43, 56, 76, 225, 246–248
Multilateral Investment Guarantee Agency (MIGA) 41

National agriculture and rural development support schemes
  in 2007–2010 230–231
  land tax 229

measures 229
self-employment 229
SSA and MAFCP 229
National Food Authority (NFA) 112–113, 116, 117
New Member States (NMS) framers 16
NFA *see* National Food Authority (NFA)
NGOs *see* Non-governmental organizations (NGOs)
Non-governmental organizations (NGOs) 37, 45

Olives and olive oil
development support scheme 189–190
domestic demand 189, 191
farmers risk 190
government support 190
imports 191
industry 190
price cycle and international market 189
price, raw olives 190
production and weather conditions 189
purchasing behavior *see* Purchasing behavior, olive oil
reorientation business 190
supply chain 192
trade balance 191
trends, production base 191
Organized extension systems
in 1945–1990
Albanian government/government funding 39
artificial insemination 40
research institutes 40
State Agricultural Enterprises 40
World War II 38–39
Albania
AUT 43
AVATAR project 41
cooperative farms 41
farmer organizations 43–44
farms 37
IDA and IBRD 41
IFC and MIGA 41
IFDA 41
MAFCP 42–43
market reforms 40
SARA project 41
communications, farmers 48
Eastern Europe, extension
Bulgaria 44–45
Hungary 45–46
institutions 44
PHARE and CEEC 44
Poland 46
Romania 46
SAPARD and IPA 44
economy and democratic institutions 36
information and technology 35
land-grant system 36
market economies 47
NGOs and European and American agencies 37
organized extension systems 35
pre-war
farmer organizations 38
Ottoman Empire, 1912 37
Poland 38
societies and farmer institutes 36
US Department of Agriculture 49
Ottoman legacy 19–20

PHARE *see* Program of Community aid to the countries of Central and Eastern Europe (PHARE)
Planned economy supply chains 177
Pre-accession assistance for rural development (IPARD) 176
Private veterinarian practitioners (PVP) 115
Program of Community aid to the countries of Central and Eastern Europe (PHARE) 44, 45
Public *vs.* private extension services
Agricultural Information Centers 246
donor organizations 248
extension staff distribution 248
fruit, small ruminants and medicinal herbs 249
geographic factors 248
MAFCP 246–248
quantitative and qualitative limits 247
research and farming communities 245
VERCON 248
Purchasing behavior
meat
in central/southwest Albania 140
consumer groups 140
consumer lamb meat preference model 139–140
estimated consumption 138
lamb and goat kid 138
origin identification 141
per capita trend, lamb/mutton and goat 138
prices and weight constraints 140
production 138
public veterinarian and food safety regulation system 141
olive oil
conjoint choice experiment 132
consumer preference model 134–135
domestic production and income patterns 132
per capita trend 132, 133

Purchasing behavior (*continued*)
  wine
    consumer preference model 136–137
    consumption 135
    market information 137
    per capita trend 135
    product attribute-price strategies 137–138
PVP *see* Private veterinarian practitioners (PVP)

Quality assurance and food safety, Albania
  agricultural policies 123
  awareness and implementation survey 120–122
  certification organizations 118
  Codex Alimentarius 109
  consumer protection 115–116
  consumers' demand 109
  cost-benefit analyses 110
  domestic agri-food processors 113
  FAO 109–110
  FBE 110, 119
  food-borne diseases 114
  fruit and vegetable processing sector 117
  HACCP and ISO 109, 110
  incentives and deterrents 124–125
  institutions and public services 116–117
  IPARD 123
  laboratories works 118–119
  legal framework 111–113
  legal provisions 110
  market segmentation 114
  meat and dairy products chains 113
  milk industry 122
  NFA 114
  planned economy 113
  private consultants and consulting enterprises 118
  PVP 115
  quality assessment (QA) 113
  SME training 122–123
  SVS 115
  training needs assessment (TNA) 113
  veterinary stamp 119
  World Bank 111

RDCS *see* Rural Development Crosscutting Strategy (RDCS)
Rural Development Crosscutting Strategy (RDCS) 225

SAA *see* Stabilization and association agreement (SAA)
SAPARD Program *see* Special Accession Program for Agriculture and Rural Development (SAPARD)
SARA project *see* Support for Agriculture Restructuring in Albania (SARA) project
SEP *see* Single farm payment (SEP)
Sheep and goat production
  changes, economic and social environment 195
  commercial breeding 197–198
  investments, dairy processing plants 198
  meat 195
  migration 195
  milk 195
  price and demand shifts 197
  stock, milk production and yields 196
Single farm payment (SEP) 221
Single market adaption, Bulgaria and Romania
  agricultural trade deficit 10, 11
  corruption and fraud 11–12
  pork imports 10, 11
  quality standards 12
  rural development funds 12–13
  SAPARD Program 12
  small producers 12
Small and medium-sized businesses (SME) 113, 122–123
Special Accession Program for Agriculture and Rural Development (SAPARD) 12, 44
SSA *see* Stabilization and association agreement (SAA)
Stabilization and association agreement (SAA) 59, 222
State veterinary service (SVS) 115–117
Strategic framework, agricultural and rural development
  AFSS and RDCS 225–227
  ANAPs 227
  CAP *acquis communautaire* 225
  competitiveness 227
  coordination 227
  food safety standards 227
  GATT 229
  IFAD 228
  integration 227
  liberalization process 228–229
  local development planning and implementation 228
  objectives 225
  policy and environmental measures 228
  policy measures 225
  pre-accession process 227
  trade policy and harmonization 228
  value added tax 229
Strengths, weaknesses, opportunities, threats (SWOT) 83
Support for Agriculture Restructuring in Albania (SARA) project 41
SVS *see* State veterinary service (SVS)
SWOT *see* Strengths, weaknesses, opportunities, threats (SWOT)

Total quality management (TQM) 90
TQM *see* Total quality management (TQM)
Trade
  agricultural dynamics
    imports 150
    low export base 151
    national trade flows, 1996–2010 150
    performance 150
  deficit 148
  description, national trade 148
  export and import ratio 148
  geographic distribution *see* Geographic distribution, agricultural trade
  government strategies *see* Government strategies and trade balance
  import substitutions 170
  liberal policy 148
  MAPs and fish processing industry 169
  market economy, early 1990s 148
  national trade dynamics
    data, 1996–2010 149
    economic self-sufficiency 150–151
    export and imports 149–150
  patterns *see* Trade patterns, Albania's agriculture
  policy *see* Trade policy
Trade patterns, Albania's agriculture
  exports and imports
    agricultural products and top three trading partners, 2010 154, 155
    cereals 155
    labor-land ratio, trading partners, 2008 155, 156
    meat and fruits 155
  factor endowment theory
    data, exports labor intensive products 156
    description 155–156
    imports 156
    labor-land ratio, trading partners, 2008 155, 156
  intra-industry trade
    agricultural products 158
    data 157
    Grubel and Lloyd index 157
    measurement 157, 158
    product differentiation 158–159
  labor productivity and Albania's agricultural trade
    advantages 156
    factor endowment theory 157
    fresh fish and MAPs exports 156
    importing 156–157
    Ricardian model 156
    wages 157
    yield ratio and product 157
Trade policy
  domestic support measures
    AMS 160, 161
    farmers 161–162
    financial assistance 160
    government 161
    product specific and non-product specific support 160, 161
    schemes 162
    stabilize market and price 160
  excises tax 162
  free trade agreement 159
  import prohibitions and licensing 162
  non-discriminating and preferential trade regime 159
  non discriminating principle 159
  preferential trade regime and outcomes
    Albania's customs tariffs, 2009 163–164
    enter into force, CEFTA-2006 164
    FTA 162–163
    imports, Turkey 164
    trade deficit countries 164–165
    trade flows 164
  tariff protection 159–160
  VAT 162

UHT *see* Ultra high temperature (UHT)
Ultra high temperature (UHT) 143, 144
UNDP *see* United Nations Development Programme (UNDP)
United Nations Development Programme (UNDP) 81

Value added tax (VAT) 88, 162
Value chain analysis
  add value, product/service 61
  agribusiness 65
  competitiveness 62
  definition 61
  distributors 65
  dynamic nature, agriculture 62
  environments 62, 64
  farm management training 64
  findings 67–68
  framework 61
  fruits and vegetable industries 62, 63
  grape and wine production 65–66
  harvesting capacity 63–64
  identification, constrains 62
  intellectual property protection 66
  IPARD program 64
  marketing 65
  meat and milk industry 65
  olives and oil industry 65
  operation and interactions, individuals 60–61

Value chain analysis (*continued*)
- poor irrigation infrastructure 66
- processors 64
- producers 62–63
- review, similarities and differences 66
- service based components 61–62
- small ruminants subsector 65
- supports 61, 64
- technology improvements 64

VAT *see* Value added tax (VAT)

Vegetables, potatoes and beans production
- domestic market 181
- features 181
- growth 180
- high and low-tech greenhouses 182
- industrial processing 180–181
- and international fruit 181–182
- marketing 180
- trends 181

VERCON *see* Virtual extension and research communication network (VERCON)

VET system *see* Vocational education and training (VET) system

Virtual extension and research communication network (VERCON) 248

Vocational education and training (VET) system 243

Western Balkans accession
- cattle production 13–14
- farmer training 15
- fragmented farm structure 14
- Haphazard, ad hoc policymaking 15–16
- inadequate collection and statistics reporting 15–16
- lack of market information 14
- lack of quality and food safety standards 14–15
- poor infrastructure 14
- potential candidates 13
- reforms 13
- small producers 13
- strengths, food processing sector 13
- weak land markets 14
- weak tradition farm associations and cooperatives 15

Wine production
- bottled red wine 186
- changes and development, 2003 onwards 186–187
- changes, consumption pattern 189
- competition 186
- demands 189
- domestic producers and foreign exports 189
- end of 2007 186
- factories and production 186
- government measures 188
- imported wines 188
- international trade balance 188
- licenses 187–188
- low price 189
- low quality 188
- positive market trends 187
- processing 186
- purchasing behavior *see* Purchasing behavior, wine
- supply chain, 1997–2010 188

Wood
- classification, forest 210
- decade, 1991–2001 209–210
- and forestry industry product outcomes 210
- and hydropower 210
- illegal logging 209
- international trade 211
- lumber industry 209
- modernization efforts 210–211
- production areas 209
- source 209